Carolina International Symposium on Neutrino Physics 2008

AA002635

Journal of Physics: Conference Series Volume 173

Columbia, South Carolina, USA
15-17 May 2008

Editors:

Frank Avignone
Richard Creswick
Kuniharu Kubodera
Milind Purohit

ISBN: 978-1-61738-310-6

Printed from e-media with permission by:

Curran Associates, Inc.
57 Morehouse Lane
Red Hook, NY 12571
www.proceedings.com

Some format issues inherent in the e-media version may also appear in this print version.

Printed by Curran Associates, Inc. (2010)

Institute of Physics
Dirac House, Temple Back
Bristol BS1 6BE UK

Phone: 44 1 17 929 7481
Fax: 44 1 17 920 0979

techtracking@iop.org

TABLE OF CONTENTS

PREFACE

The Carolina International Symposium on Neutrino Physics, 2008 (CISNP'08) was organized and held at the University of South Carolina by the Department of Physics in May 2008, to celebrate the 75th birthdays of Professors Frank Avignone (South Carolina) and Ettore Fiorini (Milan) and to commemorate the 75th birthday of the late Peter Rosen (DOE). Although much of the work done by these luminaries has been in non-accelerator areas such as double beta-decay, the meeting covered many topics in neutrino physics as well, including neutrino oscillations, supernova explosions, neutrino nucleosynthesis, axions, dark matter, dark energy, and cosmology. Talks included presentations of recent theoretical progress, experimental results, detector technology advances and a few reminiscences. This is the second such symposium held at Carolina, the first was held in 2000. We were fortunate to have attracted many top speakers who gave scintillating presentations, most of which have been put in writing and are presented in this volume.

Many thanks go to various people involved in this conference, including of course Drs Avignone, Fiorini and Rosen whose efforts over the years provided us with the opportunity, and all the speakers, many of whom took time out of their very busy schedules to come to Columbia and give talks and then to write them up. Thanks also to our Department Chairman, Professor Chaden Djalali, and to our support staff which included Mr Robert Sproul, Ms Mary Papp, Ms Beth Powell and Mr R Simmons. Finally, we must thank our funding agencies which are the South Carolina EPSCoR/IDeA Program, The Oak Ridge Associated Universities, and the University of South Carolina.

The Editorial Team:
Frank Avignone (USC)
Richard Creswick (USC)
Kuniharu Kubodera (USC)
Milind Purohit (USC, Chief Editor)

CISNP Scientific Advisory Committee:
Wick Haxton (Seattle)
Barry Holstein (Amherst)
Kuniharu Kubodera (USC)

CISNP Organizing Committee:
Richard Creswick (USC)
Chaden Djalali (USC)
Vladimir Gudkov (USC)
Kuniharu Kubodera (USC, Chair)
Milind Purohit (USC)

All papers published in this volume of *Journal of Physics: Conference Series* have been peer reviewed through processes administered by the proceedings Editors. Reviews were conducted by expert referees to the professional and scientific standards expected of a proceedings journal published by IOP Publishing.

Carolina International Symposium on Neutrino Physics IOP Publishing
Journal of Physics: Conference Series **173** (2009) 012001 doi:10.1088/1742-6596/173/1/012001

Some Practical Applications of Dark Matter Research

Presented at the CISNP Conference (Columbia 2008)
in Honor of Frank Avignone, E. Fiorini, and Peter Rosen

L. Stodolsky
Max-Planck-Institut für Physik (Werner-Heisenberg-Institut)
Föhringer Ring 6, 80805 München, Germany

E-mail: les@mppmu.mpg.de

Abstract. Two practical spin-offs from the development of cryogenic dark matter detectors are presented. One in materials research, the other in biology.

1. Introduction

I know it's hard to believe and I certainly would never have believed it myself when the field started back in the 80's, but there can be what could be called a "practical" aspect to research on direct detection of dark matter.

This is because with a very small energy left behind in a dark matter interaction, we must develop very sensitive detectors. In particular there has been an intensive development of cryogenic detectors which in virtue of their operation at very low temperature show a strong repsonse to very small energies[1]. This can have unexpected consequences. I'd like to present two we have been involved with, one related to materials science and one in biology.

2. Cracks-the scare

The first begins with the early runs of the CRESST Cryogenic Dark Matter Detector in Gran Sasso in 1999 [2]. Much careful effort went into the design and construction of a low background setup, aimed at achieving only a few events per day. When it finally ran in Gran Sasso we found– to our horror– rates in the 1000's per hour instead.

There followed several months of feverish search with all kinds of hypotheses, some plausible and others less so ...electronics?...suspensions? ...somebody touch the crystal with bare hands?... correlation with traffic in the Gran Sasso tunnel?...

The worst nightmare would of course have been a radioactive contamination. But on this there was one thing that saved us from complete desperation. Even if there is a fearsome radioactive background you don't know anything about, there is one thing you *do* know: it must be Poisson distributed in time. However, the mysterious events were not

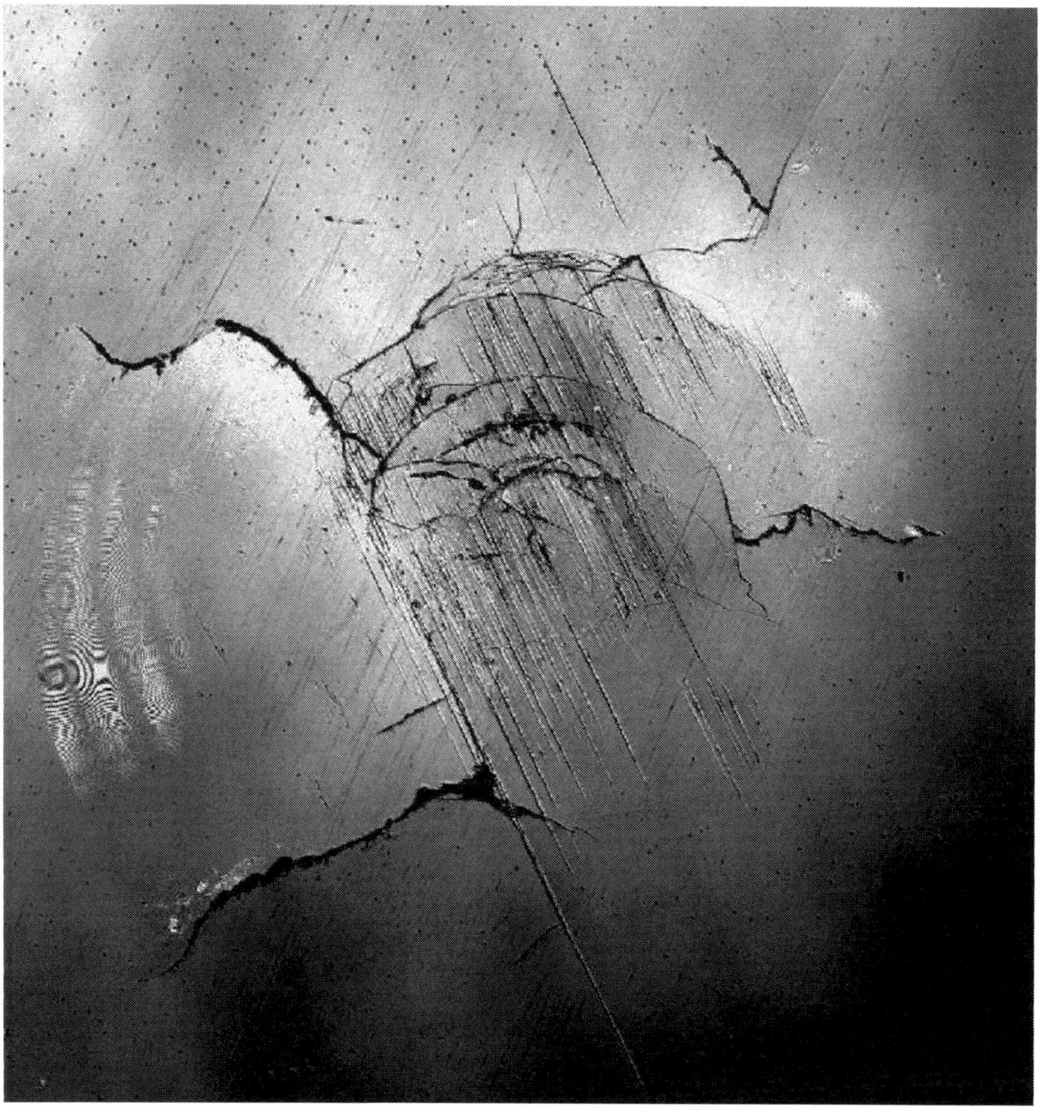

Figure 1. Sapphire crystal surface at contact point with sapphire ball

Poissonian. They seemed to rather come in 'bursts', and this was confirmed by statistical anaylsis. So it wasn't a radioactive contamination. But what the devil was it?

Fig. 1 shows the culprit. The detectors were sapphire (Al_2O_3) crystals held tightly in place by small (~mm) sapphire balls. One sees some kind of fracture at the point of contact– "cracks" we called them. Apparently the "tight holding" which is used in cryogenic work to avoid problems with "microphonics" was *too* tight; enough to crack the very hard material sapphire. As soon as the sapphire balls were replaced by plastic stubs, which are evidently somewhat softer, the rate went down to to the expected level.

We all breathed a sigh of relief and that seemed the end of the story.

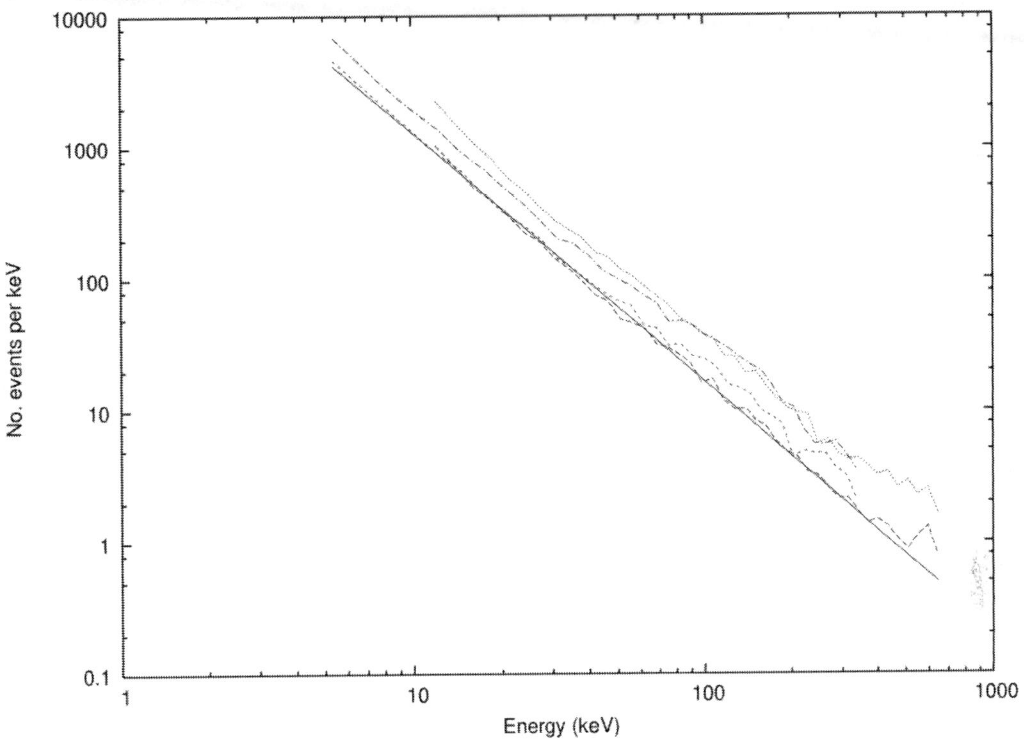

Figure 2. Energy spectra from 'crack' events with a power law fit to one of the curves, yielding $E^{-1.9}$ (solid line).

3. Cracks–the analysis

Several years later, however, the realization sudenly hit that we must have the worlds' greatest data sample on 'cracks'. Since the rate went down from thousands per hour to a few per day, the pulses detected in that period must have been essentially all "cracks". And these were taken with good energy and time resolution, under low background conditions and with many, many, thousands of events. This splendid collection of well observed "cracks" ought to be of interest to somebody.

Indeed, several years later, in Finland, we found people knowledgeable about the subject and we began to look at the data [3].

The first thing we did was to plot the energy spectrum. This is shown in Fig. 2. "Looks just like earthquakes"was the first observation. Indeed the spectra seem to follow a power law, $dN/dE \sim E^{-\beta}$ as do the magnitudes (Richter scale) of earthquakes.

Interpretation here is simpler than in seismology. In seismology it is non-trivial to find the energy spectrum for quakes since working back from the Richter scale (a kind of amplitude) to the whole energy of the event involves various assumptions and calculations. But the cryodetector is a type of calorimeter, one is just measuring the total energy directly. (On the other hand, it must be said that earthquake data has a much larger range, covering five or six orders of magnitude compared to the one or two here.)

But if we follow the standard lore of how to translate the Richter scale to an energy scale, we get a power $\beta \approx 1.7$ for earthquakes. Over different CRESST data sets the power was $\beta \sim 1.7$– 2.0, suggestively close to the earthquake number. I'm not aware of any simple, basic, explanation for this power, nor for any of other ones we found in the analysis.

With this wealth of data various other interesting statistical aspects of the data can be studied. One is the "waiting time", a statistic suitable for intermittent phenomena and often used in this kind of work. To every event we associate the time until the next event, and then plot the distribution of these "waiting" times.

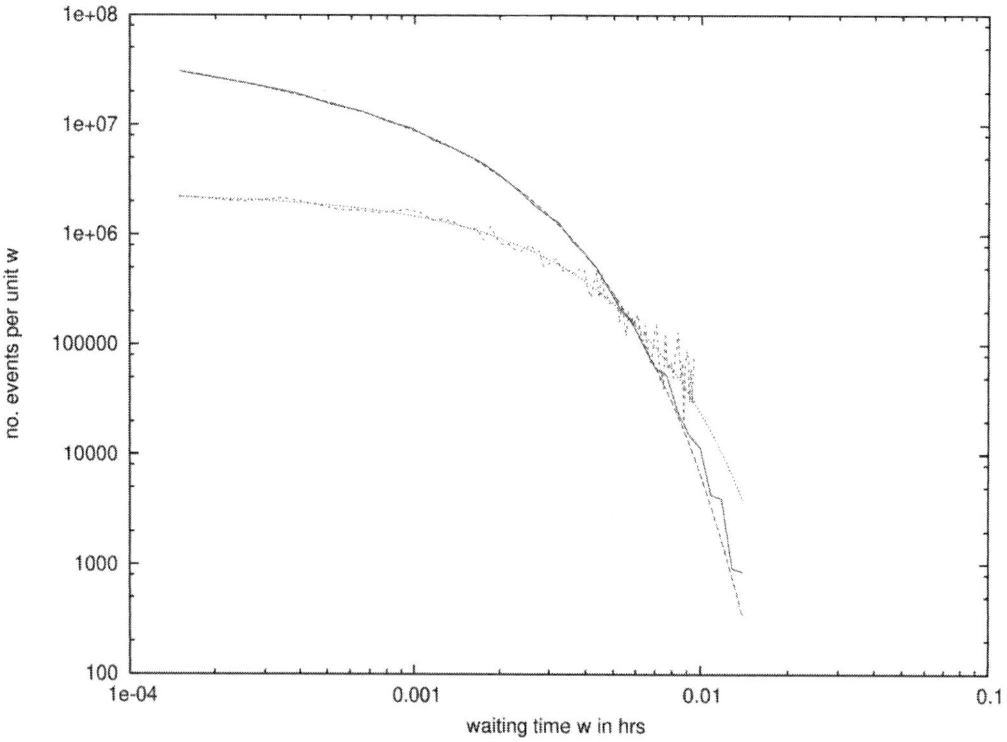

Figure 3. Distributions of waiting times for 'cracks' (upper curve) and photons from an external calibration source (lower curve). The lines are fits to (lower) a simple Poissonian e^{-w/w_o} and to (upper) $w^{-\alpha}e^{-w/w_o}$ with $\alpha = 0.3$.

For an ideal Poissonian source the waiting time distribution should be e^{-w/w_o} where w_o is the average waiting time or the inverse of the event rate. Fortunately we had such data available, since CRESST detectors are periodically calibrated with an external γ source–necessarily Poissonian. This is shown in Fig 3 by the lower curve, and it has indeed the expected Poissonian form. On the other hand, the same plot for the cracks (upper curve) is well above a Poissonian at small waiting times, and in fact is well fit by a Poissonian times a power law, namely

$$w^{-\alpha}e^{-w/w_o} \tag{1}$$

with $\alpha \approx 0.3$. Interestingly enough, an analysis of earthquake waiting times came up with the same fit with the same power [4]. Our power is not very well determined and this could be a coincidence, but it certainly is intriguing.

Another point concerns not earthquakes but material studies. As can be seen from Fig 2 the energy threshold in this data was in the keV's. This corresponds to breaking only some hundreds of bonds in sapphire. It turns out this is many orders of magnitude more sensitive than previous work in the subject, where it's more like 10^7 bonds [3]. Possibly, with a dedicated setup, one could get down to the single bond level. This would be an exciting possibility and we have some thoughts about what such an apparatus might look like [5].

Briefly, we can advertise the following points from this study and for the cryodetector :

• A new technology for studying microfracture. With unparalled sensitivity. Perhaps to the few atom level with a dedicated setup.

• The method provides a direct, absolute measurement of the *total* energy, as opposed to previous work either in seismology or materials study.

• There are striking similarities with earthquakes. Despite the stupdenous difference in energy scale, and big material differences, there appear to be close and even quantitative similarities. Something universal must be at work. This is a challenge to theory. Is there, for example, a relation between the exponents α and β?

4. The cryodetector in mass spectrometry

My second story begins even earlier, in 1991. It rests upon a very deep physical insight, namely:

$$20\ \text{keV} = 20\ \text{keV} \tag{2}$$

The history of this profound observation is the following. Mass spectroscopy with macromolecules is a valuable and frequently used tool in molecular biology. In such fields as genomics and proteomics time-of-flight studies are performed with very big molecules and their fragments. The biologist will have a, say, 20 keV accelerator in the basement. As opposed to the particle or nuclear physicist who at most will deal with heavy nuclei, the biologists is concerned with macromolecules in the many kD range. (One D= Dalton= 1 H atom.) With such enormous masses, and given that $E = \frac{1}{2}Mv^2$, a chunk with 20 keV will not be moving very fast at all.

Now in practically all familiar detectors, Geiger counters, scintillation counters,.. the initiating event is the ejection of an electron, leading to ionization, scintillation... But as we learn in elementary atomic physics, the cross section for hitting an electron depends on *velocity* and is maximum when $v(projectile) \approx v(atomic\ electron)$. Thus with big, slow, molecules, as we get into the hundreds of kD or more, detection becomes more and more difficult and inefficient.

Mulling upon this problem now and then, one day I suddenly came upon the deep insight Eq 2. The cryodetector, being a kind of calorimeter, doesn't care if the molecule is slow. It works on heat, *energy*, not velocity. For a cryodetector a huge, slow, 20 keV protein is the same thing as a 20 keV electron !?

...Theoretically. As many of our senior, seasoned participants at the conference will know, most good, simple-sounding ideas have a catch. Could this really work? But after discussion with my brother Marv, who is a microbiologist and who found the idea interesting, it was actually tried out by Damien Twerenbold and collaborators [6]. It actually worked!

This in turn has lead to the production of commercial devices. A result with one of these is shown in Fig 4, where the results with a cryodetector are compared with that for a conventional detector in looking for rare proteins associated with the liver disease HELLP [7]. A couple of nice peaks (masses in boldface), lost in the noise with the conventional detector, show up nicely with the cryodetector.

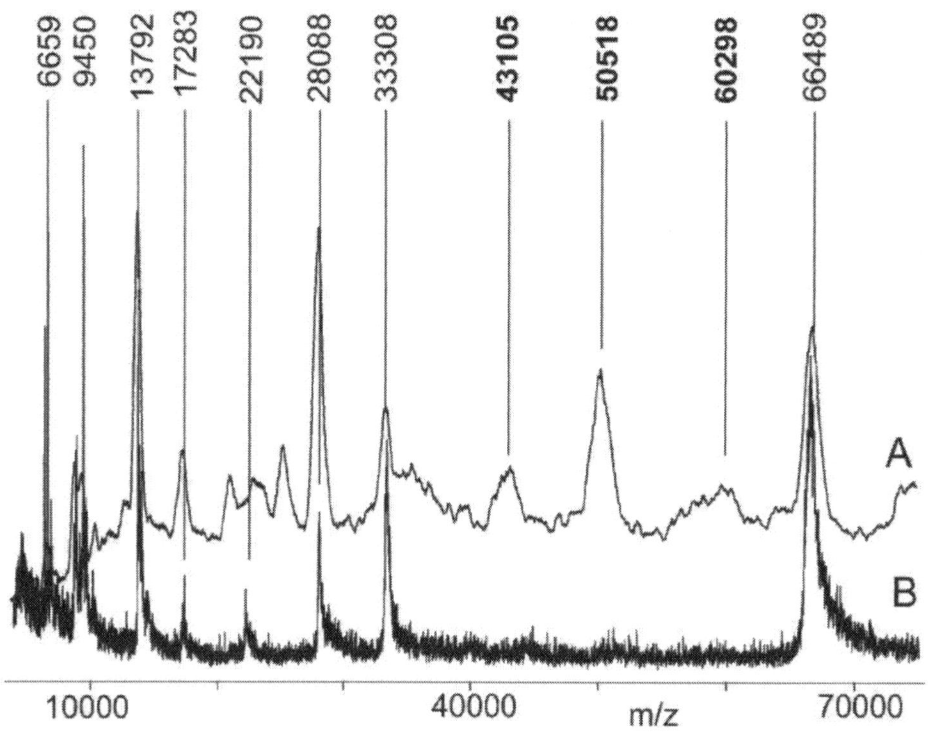

Figure 4. Time-of-flight spectra comparing the cryodetector (upper curve) and a conventional detector (lower courve) in the study of proteins associated with the liver disease HELLP. Numbers attached to the peaks are the masses. Fig 2 of ref [7].

Our group at the MPI, has furthermore developed improved detectors optimized for good timing for improved time-of-flight accuracy. This also naturally leads to background reduction and tests showed very high sensitivty, reaching attomoles [8].

For the application of the cryodetector in mass spectrometry we can thus note

• It *really* appears to be true that "20 keV ≈ 20 keV" for the cryodetector

• In principle the technique has no mass limitation (one of our students once tried to launch a whole virus–results unclear)

• The technique has a very high sensitivy, useful in rare protein studies and diagnostics

• Good timing can give high mass resolution; one could possibly see the modification of a single base

5. Conclusions

These two examples of 'cracks' and mass spectrometry are not the last we'll see from this still relatively young technology and many are being actively discussed and tried out [9]. It will be interesting to see what's still coming.

[1] For a popular introduction to cryodetectors see L. Stodolsky, Neutrino and Dark Matter Detection at Low Temperature *Physics Today*, August, 1991.

[2] For recent CRESST results on dark matter see G. Angloher et al., arXiv:0809.1829 [astro-ph].

[3] J. Astrom et al., Phys. Lett.**A356** 262 (2006); arXiv:physics/0504151.

[4] A. Corral, Phys Rev. **E 68**, 035102 (2003); A. Corral, Physica (Amsterdam) **A 340**, 590 (2004).

[5] "Brittle fracture down to femto-Joules – and below", J. Astrom, P. C. F. Di Stefano, F. Probst, L. Stodolsky, and J. Timonen, arXiv:0708.4315.
[6] D. Twerenbold et al, Appl. Phys. Lett. 68, 3503 (1996).
[7] C. Koy et al., Proteomics **5**, 3079 (2005).
[8] P. Christ et al., European Journal of Mass Spectrometry 10, 469-476 (2004).
[9] Recent developments on cryodetectors, including diverse applications, are presented at the LTD meetings. See LTD-11 in Tokyo (2005) and LTD-12 in Paris (2007).

SNO and the new SNOLAB

A B McDonald

Queen's University, Kingston, Ontario, Canada K7L 3N6

art@snolab.ca

Abstract. The SNO experiment has completed data taking for the last of three phases in which an array of ultra-low radioactivity ^{3}He-filled proportional counters were deployed in the heavy water medium to observe neutrons from the neutral current reaction on deuterium. Data analysis is nearly completed and will provide a new independent measure of the neutral current reaction for ^{8}B solar neutrinos with completely different systematic uncertainties from previous measurements. The new SNOLAB international underground science facility, expanding the experimental area 2 km underground, is nearly complete and will be ready for experiments this year. Brief descriptions of several experiments planned for this facility will be provided, including dark matter measurements by PICASSO, DEAP-1, Mini-CLEAN, DEAP/CLEAN and Super-CDMS; double beta decay measurements by SNO+ and EXO-gas and solar, geological, reactor and supernova neutrinos by SNO+ and HALO.

1. Introduction

Thank you for the opportunity to speak at this conference honoring Frank Avignone and Ettore Fiorini, who have made and continue to make major contributions to our field and Peter Rosen who was a pioneer of the field and is missed by all of us. These are long-time friends of mine and it is a pleasure to speak in their honor today.

2. The Sudbury Neutrino Observatory (SNO)

The Sudbury Neutrino Observatory [1] is a water Cerenkov detector 2 km underground that used 1000 tonnes of heavy water on loan from Canada's reserves together with an array of about 9500 photomultipliers (PMT's) (see figure 1) to observe solar ^{8}B neutrinos through several reactions with different sensitivities.

$$\upsilon_e + d \rightarrow p + p + e^- \quad (CC)$$

$$\upsilon_x + d \rightarrow p + n + \upsilon_x \quad (NC)$$

$$\upsilon_x + e^- \rightarrow e^- + \upsilon_x \quad (ES)$$

where the charged current (CC) reaction is sensitive specifically to electron neutrinos and provides an energetic neutrino that produces Cerenkov light in the heavy water, the neutral current (NC) reaction is sensitive equally to all active neutrino types above a threshold of 2.2 MeV and the elastic scattering (ES) reaction is sensitive to all active neutrino types with about 6 times larger sensitivity for electron neutrinos. The CC reaction produces electrons with an energy nearly equal to the incoming neutrino less the 1.4

MeV Q value and has a small directional sensitivity. The ES reaction is strongly directional, producing electrons travelling in nearly the same direction as the incident neutrino. The NC reaction is detected by the observation of the free neutron that is quickly thermalized in the heavy water.

In the first phase of SNO with pure heavy water, these neutrons were observed through the light from Compton scattering of the 6.25 MeV gamma ray produced when the neutron is captured by deuterium. In the second phase, about 2 tonnes of NaCl was added to the heavy water so that the neutrons from the NC reaction were mainly captured on Cl, producing several gammas with total energy of about 8.6 MeV. The total capture probability was enhanced and the pattern of light produced by these gammas was much more isotropic than the Cerenkov light cone produced by the electrons from the CC reaction, enabling the two reactions to be distinguished on a statistical basis in this phase. In the third phase, an array of about 400 meters of ^{3}He-filled proportional counters was placed in pure heavy water in the central acrylic vessel of the SNO detector [NCD paper]. These so-called Neutral Current Detectors (NCD's) provided a measure of the neutrons from the NC reaction that was independent from the light observed by the PMT's, providing a very separate but simultaneous measurement of the CC and NC reactions.

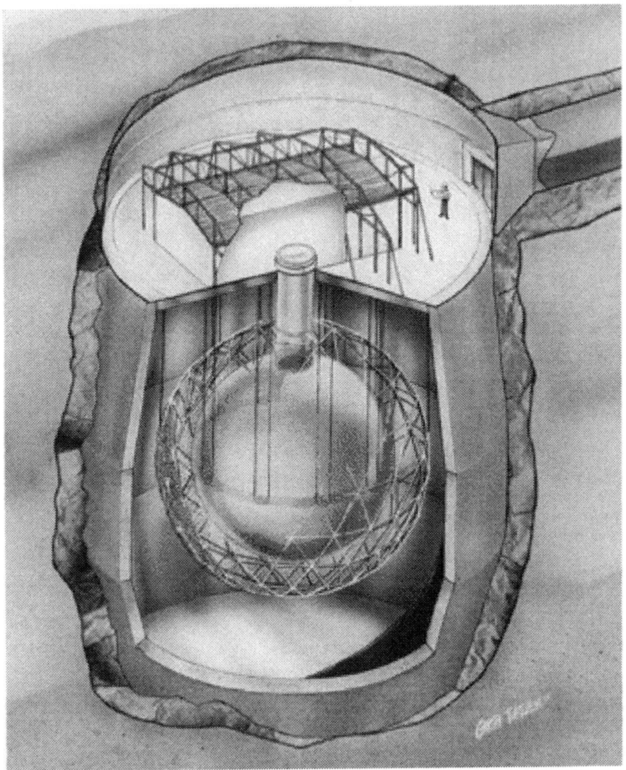

Figure 1: Schematic picture of the SNO detector, showing about 9500 PMT's looking inward towards a transparent acylic vessel holding 1000 tonnes of heavy water. The space outside the acrylic vessel is filled with about 8000 tonnes of ultra-pure light water.

The CC and NC reactions provide a determination of the solar electron neutrino flux and the total flux of active neutrinos, respectively. Therefore their ratio can determine the fraction of solar electron neutrinos reaching the earth. The results from the first phase of SNO gave a clear indication of solar neutrino flavor change for ^{8}B neutrinos as a hypothesis test of non-oscillation at the 5.5 sigma level. [2]. The second phase with salt provided an independent measure of the CC and NC reactions, [3] reinforcing the evidence for flavor change and providing a measure of the original ^{8}B electron neutrino flux in agreement with solar model calculations [4,5].

In order to accomplish the measurements, extreme care was taken with the selection of detector materials and in the cleanliness of the laboratory during construction. Gamma rays greater than 2.2 MeV are capable of photo-disintegrating deuterium, creating a free neutron that mimics the NC reaction. By maintaining the laboratory as a Class 2000 clean room throughout construction, it was possible to restrict mine dust to a total of less than about a gram distributed on the detector parts. Water purification and monitoring systems were developed to measure the residual U and Th radioactivity in the light and heavy water with sensitivities of better than 10^{-14} gm U (or Th) per gm of water.

The fluxes determined from the salt phase data (in units of 10^6 cm^{-2} sec^{-1}) are:

$$\phi_{CC} = 1.68 \ {}^{+0.06}_{-0.06}(\text{stat.})^{+0.08}_{-0.09}(\text{syst.})$$
$$\phi_{NC} = 4.94 \ {}^{+0.21}_{-0.21}(\text{stat.})^{+0.38}_{-0.34}(\text{syst.})$$
$$\phi_{ES} = 2.35 \ {}^{+0.22}_{-0.22}(\text{stat.})^{+0.15}_{-0.15}(\text{syst.})$$

A comparison of the CC and NC fluxes shows that about two-thirds of the electron neutrinos have changed their flavor to other active neutrino types, violating a hypothesis test for no flavor change at greater than 7 σ. The observed total flux of active neutrinos (NC) is in excellent agreement with the flux of ^{8}B neutrinos obtained from solar models [4,5]:

$$\Phi_e = 5.82 \pm 1.3[4]; 5.31 \pm 0.6[5]$$

Oscillation purely to sterile neutrinos is strongly disfavored. By comparison of the active total flux of ^{8}B solar neutrinos observed by the NC reaction with the ^{8}B flux calculated with solar models, restrictions are placed on sub-dominant oscillations to sterile neutrinos.

A two neutrino oscillation analysis [6] including the MSW effect in the Sun [7] for this data plus previous solar data, including Phase 1 of SNO and terrestrial reactor neutrino data from KamLAND [8] provides the following best-fit parameters:

$$\Delta m^2 = 8.0^{+0.6}_{-0.4} \times 10^{-5}\, eV^2$$
$$\theta = 33.9^{+2.4}_{-2.2}\ \text{deg}.$$

This data shows agreement only for the Large Mixing Angle solution for solar neutrinos and confirms the 1-2 mass hierarchy with mass 2 larger than mass 1. The result for the mixing angle is primarily defined by the CC/NC ratio from SNO and shows that the mixing angle is non-maximal by more than 5 standard deviations. A further measurement has recently been reported [9] by KamLAND that provides a significant improvement in the mass uncertainty.

The third and final phase of SNO included data obtained from November 2004 to November 2006. The NCD's provided a signal for the NC reaction on deuterium from neutron capture on ^{3}He producing a triton and a proton with a combined energy of 764 keV. A blind analysis of the data is being carried out by adding a number of extra pulses from neutrons produced by atmospheric muons and removing a fraction of events. Figure 2 shows a spectrum of neutrons from the NCD's showing the neutron peak and the underlying alpha particle background corresponding to only about 0.5 counts per hour in the region of interest. This illustrates the ultra-low radioactivity of the NCD detector array [10]. A few weeks after the presentation of the present paper, the results of this analysis were submitted for publication [11] showing independent results in good agreement with previous phases of SNO and providing improved best-fit parameters for neutrino oscillation.

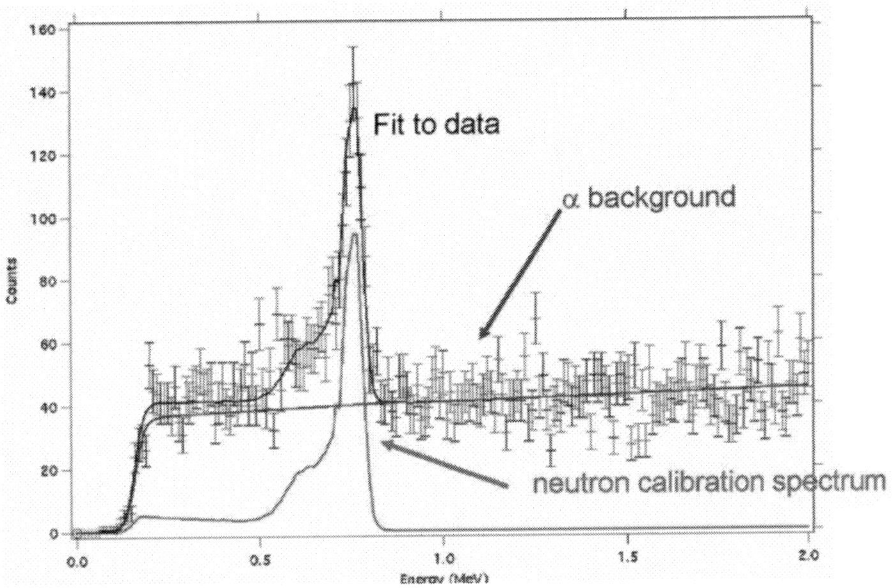

Figure 2: Blinded neutrino data from the ^3He-filled proportional counters in Phase 3 of SNO. The neutron calibration spectrum shows the response from a neutron source producing protons and tritons by capture on ^3He. The alpha background is from residual activity in the detectors.

3. The new SNOLAB International Underground Science Facility

The new SNOLAB facility has been created by excavating three new experimental areas near the original SNO cavity 2 km underground at Vale-INCO's Creighton mine near Sudbury Ontario, Canada. The excavation is very nearly complete and the new areas will be ready for experimental occupancy towards the end of 2008. The newly excavated volume is almost two times the original volume for the SNO experiment and provides space for an additional 5 or more experiments. The entire laboratory will be maintained as a Class 2000 or better clean room, as has been done for the SNO area.

Many letters of interest or intent have been received from experiments wishing to be sited in SNOLAB. These requests have been studied by an International Experiment Advisory committee and recommendations have been made to the Director for siting of an initial set of experiments. The following describes a number of experiments that are already in operation in the underground location or are planned for the future, covering measurements that will benefit significantly from the 2 km depth (about 6000 meters of water equivalent, presently the deepest operational international laboratory) and the cleanliness of the laboratory. These experiments involve direct searches for dark matter with spin-dependent and spin-independent interactions (PICASSO, DEAP-1, Mini-Clean, DEAP/CLEAN, Super-CDMS), searches for neutrino-less double beta decay (SNO+, EXO-gas), searches for Supernovae (HALO, SNO+), solar, geo and reactor neutrinos (SNO+).

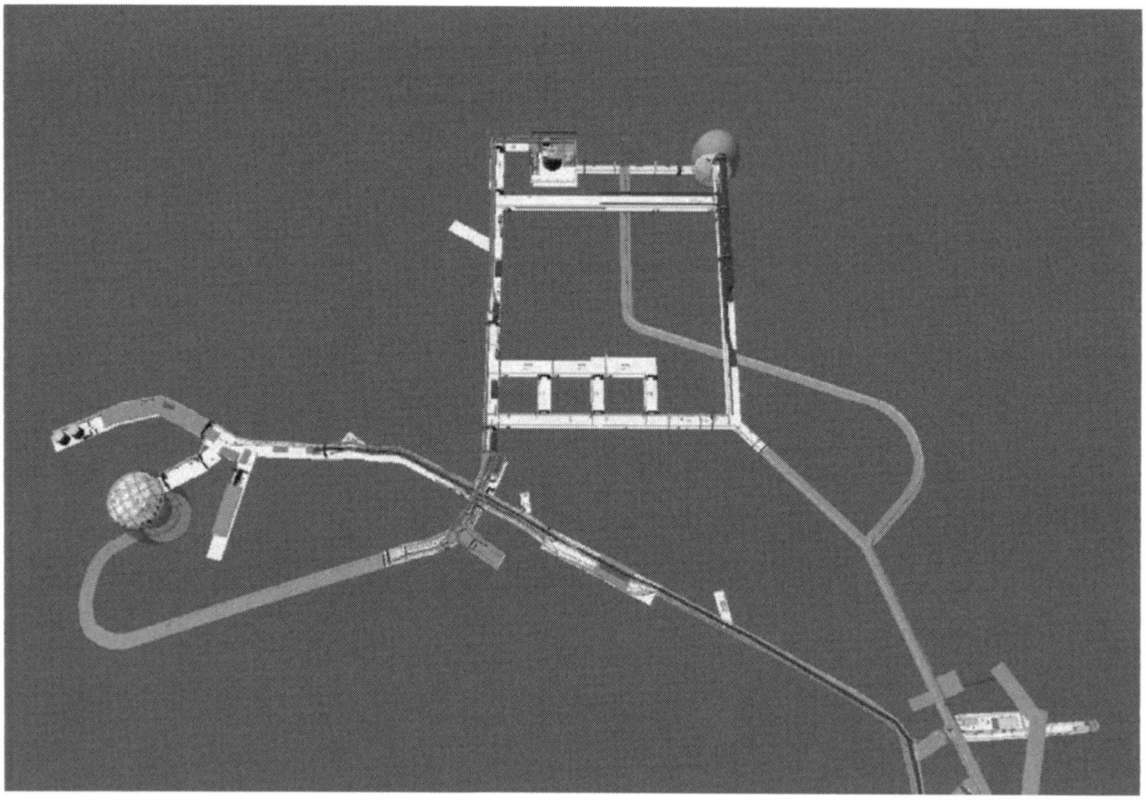

Figure 3: The new SNOLAB facility provides three new experimental areas 2 km underground near the existing SNO cavity (bottom left). These areas in the center of the figure can house 5 or more new experiments in a total newly excavated volume about twice the original volume in the SNO cavity. The entire underground laboratory will be maintained as a Class 2000 clean room facility.

3.1. Dark Matter Experiments

The PICASSO experiment [12] uses super-heated droplets containing fluorine compounds in a gel to search for dark matter via the spin-dependent interaction. They are presently operating about 2 kg of active material in an array of detectors in an auxiliary area near the SNO detector. This array is providing results at sensitivity levels at or better than current limits for spin-dependent detection. Future plans involve scaling up by factors of 100 or more and siting in one of the new underground experimental areas.

The DEAP-1, Mini-CLEAN-360 and DEAP/CLEAN-3600 detectors are being designed by the DEAP/CLEAN international collaboration to use pulse shape discrimination in liquid noble gases (Ar, Ne) to search for dark matter via the spin-independent interaction. The DEAP-1 detector is currently in operation underground in the former SNO control room with 7 kg of liquid argon viewed by 2 photomultiplier tubes (PMT's) in a shielded enclosure. The large difference between the timing for light output by nuclear recoils and ionizing events is being used to discriminate between these types of events. This discrimination will be used to deal with the beta decays from ^{39}Ar that provide a background of about 1 Bq/kg in atmospheric argon [13]. The DEAP-1 detector has been operated on surface to demonstrate discrimination of gammas at the level of 6×10^{-8}. The lower background levels underground will enable this sensitivity measurement to be extended beyond 10^{-9} with a triggered gamma source. It also provides an excellent opportunity to study processes for lowering radioactivity on inner surfaces. The

Mini-CLEAN 360 kg detector is being prepared for siting in 2009 with the ability to define a 100 kg fiducial volume as well as pulse-shape discrimination. It will have a potential sensitivity with liquid argon about 10 times beyond current limits [14] for spin-independent dark matter detection and the capability to be operated with liquid Ne as an alternate detection medium. The DEAP/CLEAN 3600 kg detector is being designed for siting in 2010, also with the ability to perform pulse-shape discrimination and define a 1000 kg fiducial volume to remove effects of residual radioactive contamination on inner surfaces. The projected sensitivity is 100 times better than current limits obtained with the spin-independent interaction. Figure 5 shows the preliminary layout for these two detectors in the new Cube Hall in SNOLAB. It is intended to construct the external water shielding for the two experiments at the same time to minimize interference.

The CDMS experiment [14] currently has the lowest limits for dark matter detection via the spin-independent interaction through measurements at the Soudan mine. The experiment uses simultaneous measurement of charge and heat (through bolometry at very low temperature) in Ge to discriminate between nuclear recoils from dark matter particles and ionizing background events. The Super-CDMS collaboration has sought and received approval from the SNOLAB Experiment Advisory Committee for siting of up to 25 kg of Ge at SNOLAB in order to take advantage of the lower background available at the greater depth. This will enable significant improvements in the present experimental sensitivity.

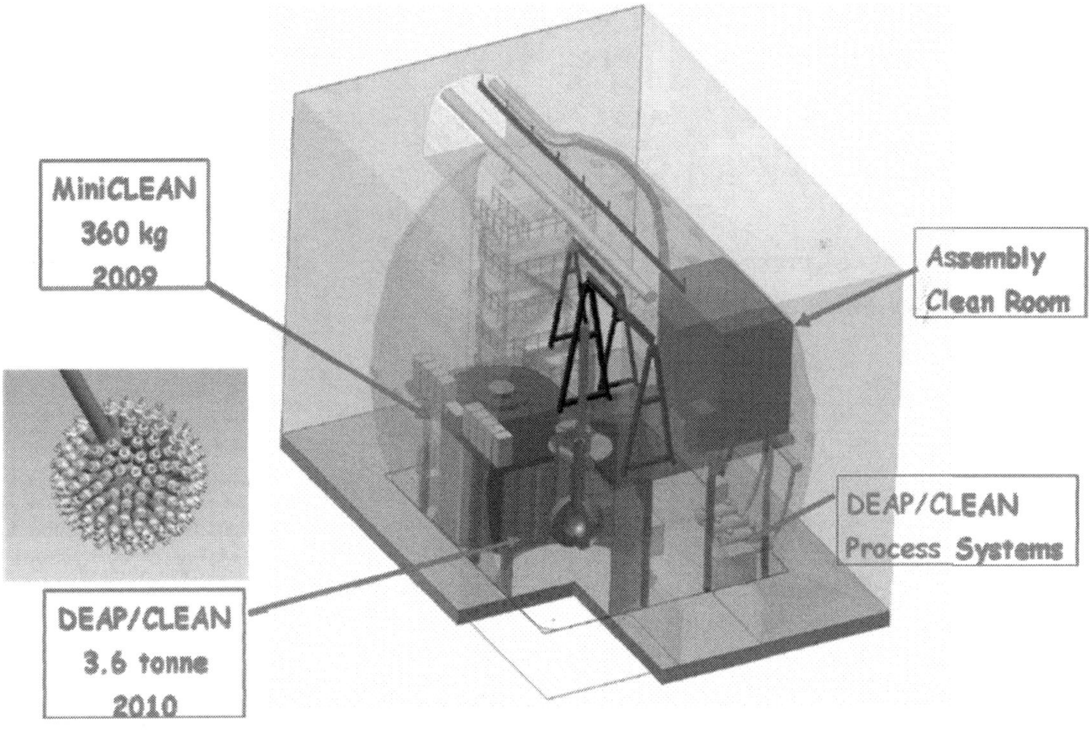

Figure 4 Preliminary layout of the Mini-CLEAN-360 and DEAP/CLEAN-3600 detectors in the new SNOLAB Cube Hall (about 20m by 20 m by 22 m high).

3.2. Neutrino-less Double Beta Decay Experiments

The principal initial objective of the SNO+ experiment [15] will be a study of the neutrino-less double beta decay of ^{150}Nd. The SNO+ experiment is being designed to use the existing SNO experiment equipment with the heavy water replaced with 1000 tonnes of an organic liquid scintillator, Linear Alkyl Benzene (LAB), with a metallo-organic compound containing 1 tonne of Nd dissolved in the LAB. The LAB is lighter than the water surrounding the central acrylic vessel holding it and so the existing ropes holding up the vessel must be augmented with additional low-radioactivity ropes to hold it down when filled. A purification system for LAB is being designed to meet the stringent background requirements for the experiment. However, the neutrino-less double beta decay summed energy for the emitted electrons is 3.3 MeV and so the region of interest is above the background radioactivity from the Uranium chain, including ^{222}Rn, so the main concern is radioactivity from the ^{232}Th decay chain. Levels of radioactivity already demonstrated by the KamLAND and BOREXINO collaborations for liquid scintillator would be acceptable for the proposed measurement.

The high value of decay energy provides an additional advantage compared to other proposed experiments because the decay rate for a given effective neutrino mass has a strong dependence on this energy. With one tonne of natural Nd dissolved in the LAB, there would be 56 Kg of ^{150}Nd isotope present and this is the initial configuration planned for SNO+. However, an international collaboration has been formed to pursue the potential production of ^{150}Nd at a former U laser isotope enrichment facility in France. It could be possible to produce hundreds of kg of isotope from the converted facility. Figure 5 shows a simulation of the spectrum for neutrino-less double beta decay for an effective mass of 0.15 eV for one year of counting for 500 kg of ^{150}Nd. For 1 tonne of natural Nd it would be possible to set a 5 sigma upper limit of 0.1 eV for the effective neutrino mass after 3 years of counting and for 500 kg of ^{150}Nd it would be possible to set a 5 sigma upper limit of 0.04 eV after 3 years.

In addition to the development of a 200 kg liquid ^{136}Xe detector at WIPP, the EXO collaboration is investigating the neutrino-less double beta decay of ^{136}Xe in a gaseous xenon ionization chamber. Work at SNOLAB is currently exploring gaseous xenon detectors which offer tracking information as well as the prospect of a very high degree of background rejection through the identification of the daughter Ba ion. A concept is being investigated wherein the Ba ions are drifted out of the decay region to a location where single atoms can be identified by laser fluorescence. This approach will provide a specific identification of the Xe decay process.

3.3. Neutrinos from the Sun, the Earth, Nuclear Reactors or Supernovae

If very high radiopurity can be obtained for a fill of pure LAB, the SNO+ experiment will have the capability of observing low energy solar neutrinos, particularly pep and CNO neutrinos because the background from muon-induced ^{11}C beta decay will not be a significant background because of the depth of SNOLAB [15]. This measurement is not compatible with the presence of ^{150}Nd as the two-neutrino double beta decay would interfere, so it would be pursued after the measurements with Nd. The pep neutrinos are of particular interest because their energy of 1.4 MeV is in a transition region between vacuum oscillation for pp neutrinos and significant matter-enhanced oscillations for ^{8}B neutrinos. The initial pep flux is also defined by solar models [4,5] to better than 2% so they can provide a valuable test of the MSW effect or other sub-dominant physics possibilities [16] such as non-standard interactions, sterile neutrinos or mass-varying neutrinos.

The SNO+ detector will also provide an interesting measurement of electron anti-neutrinos from the U and Th chain decays in the Earth. The mid-continental location provides a measurement with well defined and different characteristics than other reactor measurements such as Kamland. Measurements of neutrinos from the major reactors in Ontario will provide a test of the KamLAND reactor anti-neutrino oscillation measurements with an effective baseline about 50% longer.

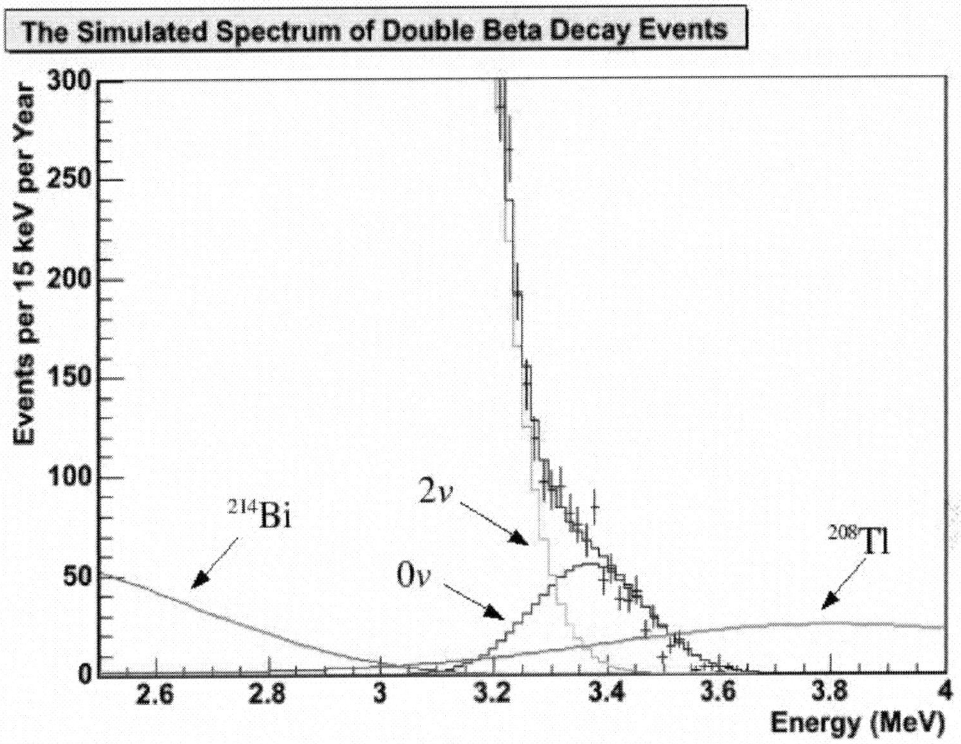

Figure 5 Simulation of the SNO+ detector response for 500 kg of ^{150}Nd, assuming an effective neutrino mass of 0.15 eV, matrix elements from ref [17] and backgrounds as achieved at the Kamland experiment [8].

The SNO+ detector will be sensitive to neutrino bursts from Supernovae at distances out to the Large Magellenic Cloud. The SNO experiment was a participant the Supernova Early Warning System (SNEWS) and it is planned for SNO+ to re-establish that participation.

The Helium and Lead Neutrino Observatory (HALO) is a supernova detector to be created from 80 tons of lead, together with the 400 m of ^{3}He proportional counters and associated electronics used in the final phase of the SNO experiment. It is estimated that this detector could observe about 50 events from a supernova at the center of our galaxy with the sensitivity dominated by electron neutrinos, unlike other supernova detectors. Implementation of the detector is planned within the next year.

References

[1] SNO Collaboration, Boger J et al. 2000 *Nucl. Instrum. Meth.* **A449**, 172
[2] SNO Collaboration, 2001 *Phys. Rev. Lett.* **87**, 071301
SNO Collaboration, 2002 *Phys. Rev. Lett.* **89**, 011301
SNO Collaboration, 2002 *Phys. Rev. Lett.* **89**, 011302
[3] SNO Collaboration, Aharmim B et al 2005 *Phys. Rev.* **C72**, 055502
[4] Bahcall J N and Pinsonneault M H 2004 *Phys. Rev. Lett.* **93** 121301
Bahcall J N and Pena-Garay 2003 hep-ph/0305159
[5] Turck-Chieze S et al 2004 *Phys. Rev. Lett.* **93**, 211102

[6] Maki Z, Nakagawa N and Sakata S 1962 *Prog. Theor. Phys* **28,** 870
Pontecorvo B 1957 *J. Expt.Theoret. Phys.* **33**, 549; 1958, *J. Expt. Theoret. Phys.* **34,** 247
Gribov V and Pontecorvo B 1969 *Phys. Lett.* **B28,** 493

[7] Mikheyev S P and Smirnov A Y 1986 in Massive Neutrinos in Astrophysics and in Particle Physics, Proceedings of the Moriond Workshop, edited by O. Fackler and J. Tran Thanh Van, Editions Frontieres, Gif-sur-Yvette, 335
Mikheyev S P and Smirnov A Y 1985 *Sov. J. Nucl. Phys.* **42,** 1441
Wolfenstein L 1978 *Phys. Rev.* **D 17,** 2369

[8] KamLAND Collaboration Araki T et al 2004 hep-ex/0406035

[9] KamLAND Collaboration Abe S et al 2008 hep-ex 0801.4589v2

[10] Amsbaugh J F et al 2007 *Nucl. Instr. andMeth.* **A579**, 1054

[11] SNO Collaboration, Aharmim B et al 2008 nucl-ex/0806.0989

[12] Barnabé Heider M et al 2005 hep-ex/0508098; hep/ex/0502028

[13] Boulay M and Hime A 2004 astro-ph/0411358

[14] CDMS Collaboration Ahmed Z hep-ex/080222

[15] Chen M C 2005 *Nuclear Physics B (Proc. Suppl.)* **145** 65–68

[16] Friedland A et al 2004 hep-ph/0402266
Miranda O et al 2004 hep-ph/0406280
Barger V et al 2004 hep-ph/0502196
De Holanda P and Smirnov A 2003 hep-ph/0307266

[17] Eliott S R and Vogel P, 2002 *Ann.Rev.Nucl.Part.Sci.* **52** 115-151

Carolina International Symposium on Neutrino Physics IOP Publishing
Journal of Physics: Conference Series **173** (2009) 012003 doi:10.1088/1742-6596/173/1/012003

Testing Relativity with Neutrinos

Brett Altschul

Department of Physics and Astronomy, University of South Carolina, Columbia, SC 29208, USA

E-mail: `baltschu@physics.sc.edu`

Abstract. Recent evidence appears to confirm that the ultra-high-energy primary cosmic ray spectrum consists mostly of protons. The fact that these protons can traverse large distances to reach Earth allows us to place bounds on Lorentz symmmetry violations. The fact that the protons do not β-decay constrains the maximum achievable velocities of neutrinos at the 10^{-10} level.

The observation [1] of the Greisen-Zatsepin-Kuz'min (GZK) cutoff [2, 3] and the discovery that ultra-high-energy cosmic rays are associated with nearby active galactic nuclei [4] resolved a major puzzle in physics. A number of exotic physical scenarios had been suggested to explain the apparent absence of the GZK cutoff in earlier observations. Such exotic physics would now seem to be unnecessary; however, it is interesting to turn things around and ask what constraints can be placed on exotic theories based on our improved understanding of cosmic ray physics.

One exotic idea which has attracted significant interest in the last decade (and which was put forward to explain the apparent anomalies in the cosmic ray spectrum [5]) is Lorentz violation. If a violation of this very basic symmetry were discovered, it would be of tremendous importance. It would be an incredibly important clue as to the nature of Planck scale physics. Many candidate theories of quantum gravity suggest the possibility of Lorentz symmetry breaking in certain regimes. For example, Lorentz violation could arise spontaneously in string theory [6, 7] or elsewhere [8]. There could also be Lorentz-violating physics in loop quantum gravity [9, 10] and non-commutative geometry [11, 12] theories, or Lorentz violation through spacetime-varying couplings [13], or anomalous breaking of Lorentz and CPT symmetries [14] in certain spacetimes. Ultimately, the correctness of Lorentz symmetry must be verified experimentally. To date, there have been many high-precision experimental tests of Lorentz invariance.

Lorentz violation can be described by an effective field theory, the standard model extension (SME), which contains all possible local Lorentz-violating operators constructed from standard model fields [15, 16]. The minimal SME, which is the standard theory used to parameterize Lorentz tests, is a version of the SME containing only gauge invariant and power-counting renormalizable operators. Many of the minimal SME coupling constants have been constrained extremely tightly, but others have not. The neutrino sector is among the least constrained. (For details, see [17]. Up-to-date information on minimal SME bounds can be found in [18].)

It turns out that present cosmic ray data can be used to constrain Lorentz violations for neutrinos. It may seem surprising that bounds on any neutrino coefficients are possible, since neutrino physics ordinarily has little to do with primary cosmic rays. However, Lorentz violation could allow otherwise forbidden proton decays. We therefore wish to emphasize the rather interesting observation that measurements that are entirely concerned with one species of particles (here, protons) can be used to place strong constraints on exotic physics involving different particles (neutrinos or other species).

The experimental confirmation that the GZK cutoff does indeed exist at $\sim 6 \times 10^{10}$ GeV is good evidence that ultra-high-energy primary cosmic rays are mostly protons. That protons should dominate the spectrum was expected originally, but was called into question by the observation of a few cosmic ray events well above the cutoff. However, we can now feel fairly confident (although not absolutely certain) that we understand the protonic cosmic ray spectrum up to the highest energies. We shall assume that most of the highest energy primary cosmic rays are indeed protons. However, further confirmation of the protonic nature of the rays would be helpful.

The minimal SME Lagrange density for each spin-$\frac{1}{2}$ fermion species includes terms of the form

$$\mathcal{L} = \bar{\psi}[i(\gamma^{\mu} + c^{\nu\mu}\gamma_{\nu} - d^{\nu\mu}\gamma_{\nu}\gamma_{5})\partial_{\mu} - m]\psi. \tag{1}$$

There is a different set of c and d coefficients for each species. There are many additional terms in the most general minimal SME $\mathcal{L}$, but c and d are the most relevant at high energies; the effects of other terms are suppressed in relative importance by powers of m/E. For neutrinos, which are almost entirely left-handed, only the combination $c_L = c + d$ is relevant. There are also analogous coefficients for scalar and vector particles. Since Lorentz violation is known to be a small effect, all calculations can be done to leading order in the violation coefficients.

When Lorentz symmetry is broken by c or d, novel effects may appear. For example, it may become possible for sufficiently energetic protons to decay. Obviously, this is a unique signature of Lorentz violation. In a Lorentz-symmetric theory, if the decay is forbidden when the proton is at rest, it is forbidden for all proton momenta; and even if the proton is not absolutely stable at rest, Lorentz invariance dictates that it will survive longer when it is highly boosted, because of the time dilation effect.

Of interest are decays of the proton into two or more particles. At high energies, the Lorentz violation changes the energy-momentum relation for each particle involved to

$$E_w(\vec{p}) = \sqrt{m_w^2 + [1 + 2\delta_w(\hat{p})]\vec{p}^{\,2}}, \tag{2}$$

where w labels the species, and the maximum achievable velocity (MAV) for a species in the direction $\hat{p}$ is $1 + \delta_w(\hat{p})$. Other forms of Lorentz violation may modify the energy-momentum relation at lower energies, but any such effects are suppressed in relative importance by powers of m_w/p. The parameter δ_w can be spin dependent, and for spin-$\frac{1}{2}$ fermions, it is equal to [19]

$$\delta_w(\hat{p}) = -c_{00}^{w} - c_{(0j)}^{w}\hat{p}_j - c_{jk}^{w}\hat{p}_j\hat{p}_k + sd_{00}^{w} + sd_{(0j)}^{w}\hat{p}_j + sd_{jk}^{w}\hat{p}_j\hat{p}_k, \tag{3}$$

where s is the helicity, $c_{(0j)} = c_{0j} + c_{j0}$ [likewise for $d_{(0j)}$], and the superscripts indicate that the coefficients are those for the species w. Owing to its appearance in (2), $\delta_w(\hat{p})$ is the most important parameter describing Lorentz violation for a given species of particles at relativistic energies.

Proton decay above a certain threshold could occur if there were a mismatch between the δ_p for the proton and the corresponding parameters for the daughter particles. It is not clear what the preferred kinematics for this reaction are, so we shall consider the kinematic situation which leads to the most conservative bounds on the Lorentz violation parameters. In the relevant configuration, one of the daughter particles will carry away practically all of the proton's initial momentum $\vec{p}$, leaving the remaining one (or more) essentially at rest. The moving particles are essentially collinear, so only the $\delta(\hat{p})$ values for a single direction enter. Energy-momentum conservation requires that for a two-body decay,

$$(1 + \delta_p)p = (1 + \delta_1)p + m_2, \tag{4}$$

where 1 and 2 label the daughter particles. We have dropped terms that are suppressed relative to the particle masses by powers of m/p. Moreover, because only one direction is involved, we have omitted the dependence of δ_w on $\hat{p}$.

Apparently, the process can occur at a momentum p if $\delta_p - \delta_1 > m_2/p$. A positive δ_p indicates that the initial proton with momentum p has more energy than it would have in the Lorentz-invariant theory. If this momentum is transferred to a particle with a smaller δ_1, there may be enough energy left over to create a particle of type 2. If $\delta_p < \delta_1$, however, the process will never be allowed, so the observed absence of these kinds of decays can only give one-sided bounds on the various $\delta_p - \delta_w$. However, one-sided bounds on the $\delta(\hat{p})$ for different directions $\hat{p}$ can be used to place two-sided bounds on some of the individual c and d coefficients.

The generalization to more than two decay products is straightforward. At threshold, one particle still carries away all the momentum, and the process can occur if $\delta_p - \delta_1 > (\sum m_w)/p$, where $\sum m_w$ is the sum of the masses of all the other daughter particles. Usually this sum will be dominated by the mass of the heaviest product other than the one carrying away the momentum. It follows that the strength of the bounds we can place on the SME coefficients relevant to a particular species is determined by the masses of the other species that are produced along with it. The bounds for a particular species are better if the other particles involved are lighter.

Of particular interest is the three-body β-decay $p^+ \to n + e^+ + \nu_e$. The fact that protons of energy E traveling in the direction $\hat{p}$ do not decay in this fashion places a bound

$$\delta_p(\hat{p}) - \delta_{\nu_e}(\hat{p}) < \frac{m_n}{E}. \tag{5}$$

For the most energetic observed cosmic rays, $m_n/E < 10^{-11}$. It is interesting that the constraint (5) is only $\mathcal{O}(m/E)$, whereas most previous bounds on δ_w coming from high-energy observations were $\mathcal{O}(m^2/E^2)$; this new energy dependence arises from the involvement of a nonrelativistic particle in the decay reaction.

Since the decay $p^+ \to n + e^+ + \nu_e$ is a weak interaction process, there might be a concern that it will not be rapid enough to occur during the primary proton's flight to Earth. However, a proton traveling 10 Mpc to Earth with an energy of 10^{11} GeV experiences a proper time span greater than 10^4 s. This is much longer than the typical lifetime of weakly decaying hadrons, longer in particular than the neutron lifetime of approximately 880 s. The matrix element for the neutron β-decay process ought to be similar to the matrix element for the proton decay considered here; moreover, an increase in the parent proton energy above threshold by one percent gives the proton decay greater available phase space than the decay of a neutron at rest. This indicates that the weak nature of the decay should be no impediment to its ready occurrence during the time it takes for a cosmic ray proton to reach Earth.

A related process is proton vacuum Cerenkov radiation, $p^+ \to p^+ + \gamma$. This is allowed if the velocity of the initial proton exceeds the phase velocity of light in the same direction. If the process were allowed above some threshold, a proton with a higher energy would radiate away the excess energy extremely quickly [20]. The Cerenkov process actually falls into the general category of processes considered above, with a proton going to two particles. However, since the photon is massless, the approximations needed to obtain equation (4) are not valid. The mass m_2 vanishes, so the $\mathcal{O}(m_p^2/p)$ terms that were neglected in (4) would need to be included. If protons with energies up to E are observed not to emit vacuum Cerenkov radiation, this translates into a bound on the quantity $\delta_p - \delta_\gamma$ that is $\mathcal{O}(m_p^2/E^2)$. The quantity $\delta_\gamma(\hat{p})$ is analogous to the fermionic $\delta(\hat{p})$, and it governs the phase speed of photons in the direction $\hat{p}$. $\delta_\gamma(\hat{p})$ depends on the k_F coefficients in the Lorentz-violating Lagrange density for the electromagnetic sector,

$$\mathcal{L}_A = -\frac{1}{4}F^{\mu\nu}F_{\mu\nu} - \frac{1}{4}k_F^{\mu\nu\rho\sigma}F_{\mu\nu}F_{\rho\sigma}. \tag{6}$$

However, the spin-dependent part of δ_γ has been shown to be extremely small using measurements of cosmological birefringence [21, 22, 23], and it can therefore be neglected. Moreover, for any one sector, the spin-independent part of δ may be defined away. We shall use this freedom to make the electromagnetic sector conventional. Our results can be adapted to a different convention with a nonzero k_F simply by making the replacement $c^{\mu\nu} \rightarrow c^{\mu\nu} - \frac{1}{2}k_{F\alpha}{}^{\mu\alpha\nu}$ in every fermion sector.

Existing cosmic ray data place bounds on the components of $\delta_p - \delta_\gamma$, which is just δ_p with our conventions. If protons of both helicities are part of the primary cosmic ray spectrum, these bounds apply to both helicities; if only one helicity is present, the bounds obviously apply only to that helicity. What is important is that any proton that is part of the primary cosmic ray flux is subject to these bounds.

Most of the details of the vacuum Cerenkov bounds are worked out in [24]. From the absence of vacuum Cerenkov radiation, there are two-sided bounds on the $-c^p_{(0j)} + sd^p_{(0j)}$ and $-c^p_{jk} + sd^p_{jk}$ coefficients appearing in δ_p—if it is assumed that the isotropic term $-c^p_{00} + sd^p_{00}$ is small compared with the others. Two-sided bounds in the presence of a generic c^p_{00} are impossible, because a large enough c^p_{00} can render the proton MAV less than one for every direction and spin. The absence of vacuum Cerenkov radiation thus cannot constrain a positive c^p_{00}. However, if $-c^p_{00} + sd^p_{00} \approx 0$, the other coefficients are all separately bounded at the 2×10^{-21} level. (If the anisotropic terms are assumed to vanish instead, there is a one-sided bound on the isotropic coefficient, $c^p_{00} - sd^p_{00} > -10^{-22}$.) There are also bounds on c^p and d^p from laboratory experiments with atomic clocks—some of which are much stronger than the cosmic ray bounds, ranging from the 10^{-20} to 10^{-25} levels [25, 26].

If the isotropic part of the proton Lorentz violation is small, the preexisting bounds justify setting $\delta_p = 0$ in (5), because the vacuum Cerenkov bounds in the proton sector are quite a bit tighter than the new bounds we shall be considering here. The remaining neutrino bounds coming from (5), which have the form $-\delta_{\nu_e}(\hat{p}) < m_n/E$, are based on precisely the same data set as the vacuum Cerenkov bounds on δ_p. So the consequences for the individual neutron c_L coefficients are (unsurprisingly) similar. If there is no isotropic neutrino Lorentz violation term, each of the neutrino coefficients $(c_L^{\nu_e})_{(0j)}$ or $(c_L^{\nu_e})_{jk}$ is bounded above and below by 10^{-10}. On the other hand, if there is exclusively isotropic neutrino Lorentz violation (but still no isotropic proton Lorentz violation), $(c_L^{\nu_e})_{00} < 2 \times 10^{-11}$. These are our primary results, and they tell us the level at which high-energy Lorentz violation in the neutrino sector is constrained.

Finally, the bounds are not limited to the electron neutrino sector. Muon neutrinos are subject to essentially the same bounds, because of the absence of the process $p^+ \rightarrow n + \mu^+ + \nu_\mu$. The bounds for ν_τ are worse by roughly a factor of 3, because the τ mass is comparable to the neutron mass and so affects the kinematics. There are also similar bounds for baryons that could possibly be produced by proton decay. Overall, high-energy cosmic rays are an excellent laboratory for constraining exotic physics.

References

[1] R. U. Abbasi, et al., Phys. Rev. Lett. **100**, 101101 (2008).
[2] K. Greisen, Phys. Rev. Lett. **16**, 748 (1966).
[3] G. T. Zatsepin, V. A. Kuz'min, JETP Lett. **4**, 78 (1966) [Pis'ma Zh. Eksp. Teor. Fiz. **4**, 114 (1966)].
[4] J. Abraham, et al., Science **318**, 938 (2007).
[5] S. Coleman, S. L. Glashow, Phys. Rev. D **59**, 116008 (1999).
[6] V. A. Kostelecký, S. Samuel, Phys. Rev. D **39**, 683 (1989).
[7] V. A. Kostelecký, R. Potting, Nucl. Phys. B **359**, 545 (1991).
[8] B. Altschul, V. A. Kostelecký, Phys. Lett. B **628**, 106 (2005).
[9] R. Gambini, J. Pullin, Phys. Rev. D **59**, 124021 (1999).
[10] J. Alfaro, H. A. Morales-Técotl, L. F. Urrutia, Phys. Rev. D **65**, 103509 (2002).
[11] I. Mocioiu, M. Pospelov, R. Roiban, Phys. Lett. B **489**, 390 (2000).
[12] S. M. Carroll, J. A. Harvey, V. A. Kostelecký, C. D. Lane, T. Okamoto, Phys. Rev. Lett. **87**, 141601 (2001).

[13] V. A. Kostelecký, R. Lehnert, M. J. Perry, Phys. Rev. D **68**, 123511 (2003).
[14] F. R. Klinkhamer, C. Rupp, Phys. Rev. D **70**, 045020 (2004).
[15] D. Colladay, V. A. Kostelecký, Phys. Rev. D **55**, 6760 (1997).
[16] D. Colladay, V. A. Kostelecký, Phys. Rev. D **58**, 116002 (1998).
[17] *Proceedings of the Fourth Meeting on* CPT *and Lorentz Symmetry*, edited by V. A. Kostelecký (World Scientific, Singapore, 2008).
[18] V. A. Kostelecký, N. Russell, arXiv:0801.0287.
[19] B. Altschul, D. Colladay, Phys. Rev. D **71**, 125015 (2005).
[20] B. Altschul, Phys. Rev. Lett. **98**, 041603 (2007).
[21] V. A. Kostelecký, M. Mewes, Phys. Rev. Lett. **87**, 251304 (2001).
[22] V. A. Kostelecký, M. Mewes, Phys. Rev. Lett. **97**, 140401 (2006).
[23] V. A. Kostelecký, M. Mewes, Phys. Rev. Lett. **99**, 011601 (2007).
[24] F. R. Klinkhamer, M. Risse, Phys. Rev. D **77**, 016002 (2008); addendum arXiv:0709.2502v5.
[25] V. A. Kostelecký, C. D. Lane, Phys. Rev. D **60**, 116010 (1999).
[26] P. Wolf, F. Chapelet, S. Bize, A. Clairon, Phys. Rev. Lett. **96**, 060801 (2006).

Journal of Physics: Conference Series **173** (2009) 012004 doi:10.1088/1742-6596/173/1/012004

Searches on Neutrino Physics with Cryogenic detectors

Ettore Fiorini[1]
Dipartimento di Fisica G.P.S.Occhialini dell' Universita' di Milano-Bicocca
and Istituto Nazionale di Fisica Nucleare, Sezione di Milano-Bicocca
Piazza della Scienza 3, 20126 Milan, Italy

E-mail: Ettore.Fiorini@mib.infn.it

Abstract. The impact in neutrino physics of the recent results on oscillations and the consequent need to measure the value of the neutrino mass are briefly discussed. The operating principle of cryogenic detectors working at low temperatures, where the small heat capacity allows to record and measure the temperature increase due to the tiny energy lost by a particle in form of heat is described. An application of these detectors is the measurement, or at least an upper constraint, of the neutrino mass in β decay. This approach is complementary and can be in the future competitive with experiments based on the spectrometric measurement of the electron energy. The presently running bolometric CUORICINO experiment searching for neutrinoless double beta decay of ^{130}Te already yields the best constraint on the absolute value of the Majorana neutrino mass. An experiment, named CUORE, for Cryogenic Underground Observatory for Rare Events, is presently in construction. With its active mass of 750 kg of natural TeO_2 it aims to reach the sensitivity in the determination of the Majorana neutrino mass suggested by the results of neutrino oscillation under the inverse hierarchy hypothesis.

1. Introduction

The existence of the neutrino was first suggested by Wolfang Pauli in a famous letter "*Liebe radioactive damen und herren*" dated December 4, 1930 and addressed to a Conference on radioactivity in Tubingen. The justification of his absence was somewhat surprising*: I cannot be with you due to a ball which is going to take place here in the night from January 6 to 7.* In this letter Pauli explains the lack of energy (see later) delivered the β decay of ^{210}Bi and the "*poor statistics*" in this decay as due to the emission of a not yet discovered neutral particle which he named *neutron* (the real neutron was not yet discovered!) A later report given by Pauli in Italy strongly impressed Enrico Fermi and prompted him to give this particle the Italian name of *neutrino* and to introduce it in his beautiful theory of weak interactions. Since the very beginning, Fermi suggested that the neutrino had to have a very low mass *"even lower than the mass of the electron"*. In fact in most later papers the neutrino was assumed to be massless. A great step forward is due to the discovery of oscillations by solar atmospheric, reactor and accelerator neutrinos (Fig.1) which is amply discussed in this Symposium [1-5]. It indicates that the masses of at least two neutrinos of different flavours are finite. The measurement of the effective neutrino mass becomes therefore mandatory and the present and

[1] To whom any correspondence should be addressed.

future developments of new type of detectors like the cryogenic ones can play an important role in this fascinating field of research.

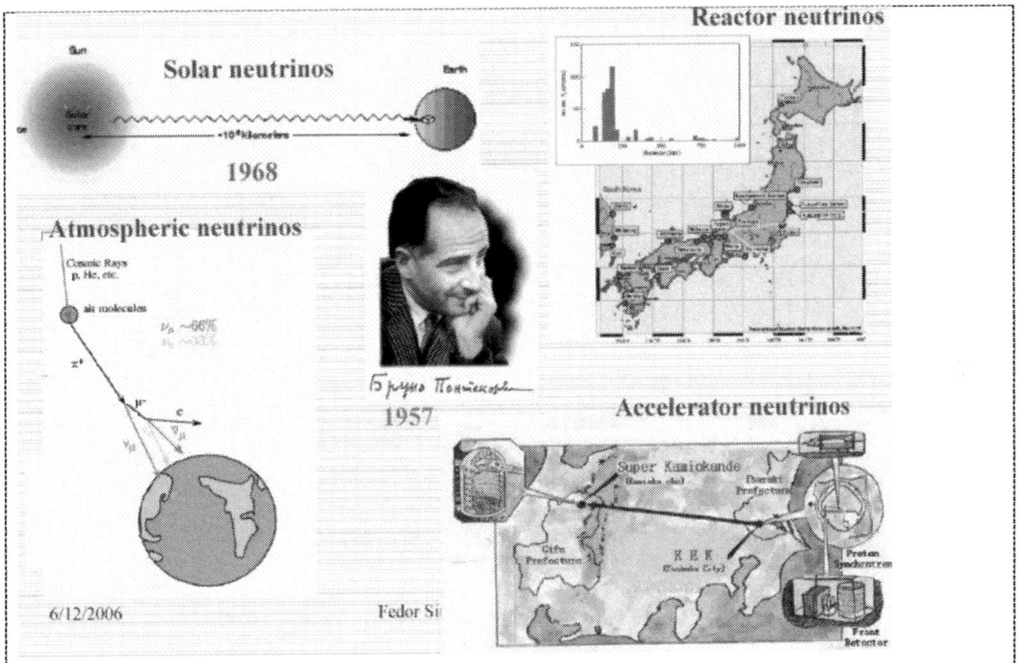

Figure 1. The discovery of oscillations of atmospheric, reactor and accelerator neutrinos.

2. The principle of cryogenic detectors

We can date the origin of these detectors [6] since 1880 when Langley first developed resistive bolometers for infrared rays from Sun, while in 1905 Curie and Laborde performed the first calorimetric measurement of radioactivity. An important step towards the discovery of the neutrino is due to Ellis and Wuster who detected a lack of heat in the radioactive decay of ^{210}Bi which prompted the above mentioned letter by Pauli. In 1935 Simon discovered that the detector sensitivity was enhanced by lowering the temperature. In a period around 1983 T. Niinikosky at CERN recorded pulses induced by cosmic rays in resistors kept at low temperature.

The first suggestions for the use of cryogenic detectors in fundamental physics date from 1984 and are due independently to two groups working on the two sides of the Atlantic. In USA [7] S.H. Moseley et al. suggested the use of cryogenic detectors mainly for astrophysics while in Europe T. Niinikosky and myself [8] proposed their use to search for rare events and in particular for double beta decay. Both groups were also interested in other subjects of physics like direct measurement of the neutrino mass. An "artist view" and a more serious scheme of the performance of a cryogenic detector is shown in Fig. 2 and 3.

Figure 2. An "artist view" of a cryogenic detector.

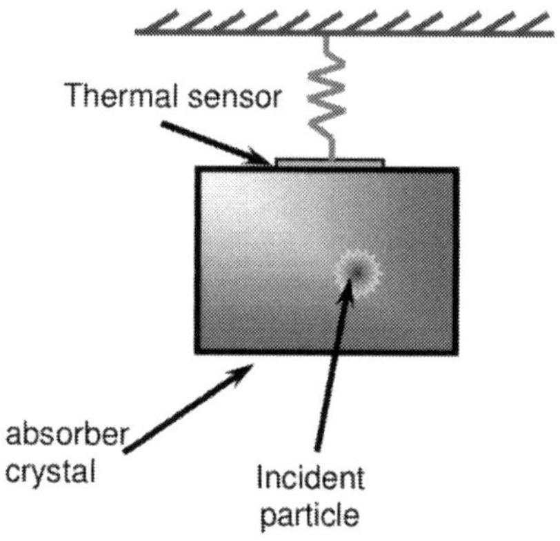

Figure 3. A more serious scheme of a cryogenic detector.

The principle of these detectors is simple. The heat capacity of an absorber, possibly of dielectric and diamagnetic material, in a suitable refrigerator is proportional to the cube of the ratio between the operating and Debye temperatures. It can therefore become so low that even the tiny energy released by a particle incident or generated in it, can induce a sizable increase of temperature which can be measured by a suitable "thermometer". The outstanding performances of these detectors with masses of the order of a milligram in the region of tens of keV is shown in Fig. 4. The resolution is now around 2 eV, more that an order of magnitude better than in any other detector. For searches for rare events larger masses are needed and the resolution is obviously poorer, but still overcomes the one of germanium detectors. This is show in fig. 5, where the sharp α line of ^{210}Po can be seen.

3. Measurement of the neutrino mass in single β decay

The discovery of neutrino oscillations has strongly stimulated searches to determine or at least to constrain the absolute value of the neutrino mass [9] by the deformation (Fig. 6) of the spectrum in beta decay. The more stringent limits, around 2 eV, were obtained in the decay of ^{3}H into ^{3}He. An improvement of an order of magnitude is aimed for by the international KATRIN experiment to be carried out in Germany.

A complementary approach can be provided by cryogenic detectors and consists in the bolometric measurement of the energy released by beta decay occurring in a micro-absorber. Even if the sensitivity of cryogenic detectors is still far from those from of spectrometers one should note that bolometric detection enables to record the total energy delivered in the decay and not only the electron one. It could also include the decay energy from excited states. A good friend of mine and an excellent physicist, the late Antonio Vitale of the Genoa University, suggested bolometric experiments on the β decay of ^{187}Re into ^{187}Os which has the lowest transition energy (~ 2.5 keV). Bolometric searches on this decay are being carried out in Genoa and Milan with absorbers of metal Re and AgReO, respectively. These experiments will be merged in the international MARE search (Fig. 7).

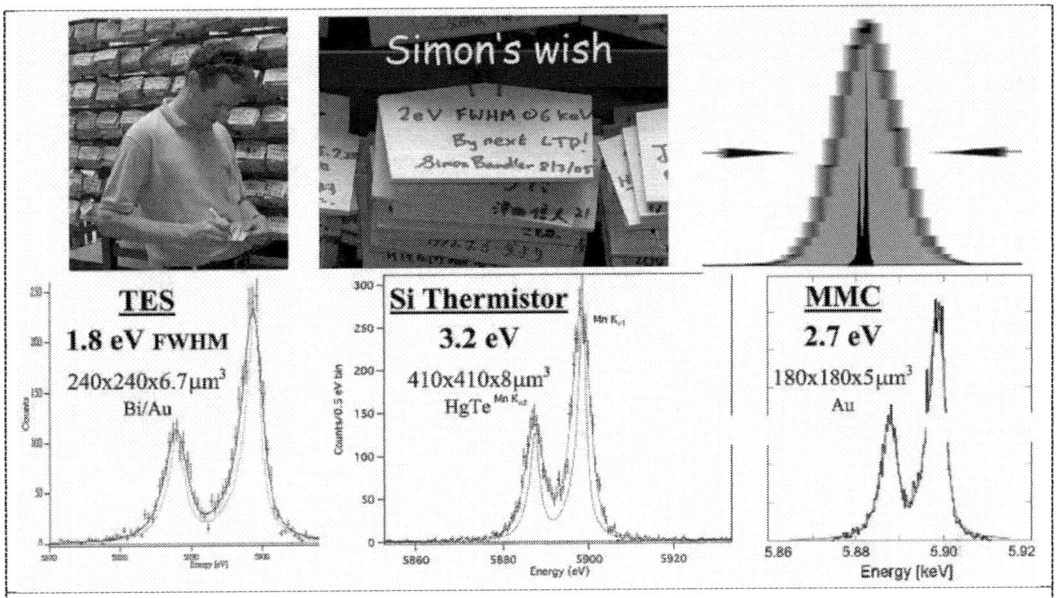

Figure 4. The resolution for X-rays with different thermistors.

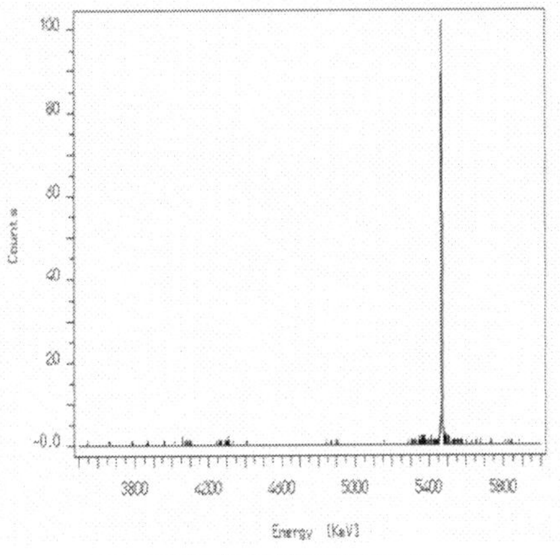

Figure 5. α particle spectrum obtained with a 760 grams bolometer. The FWHM resolution on the ^{210}Po line is 3.2 keV.

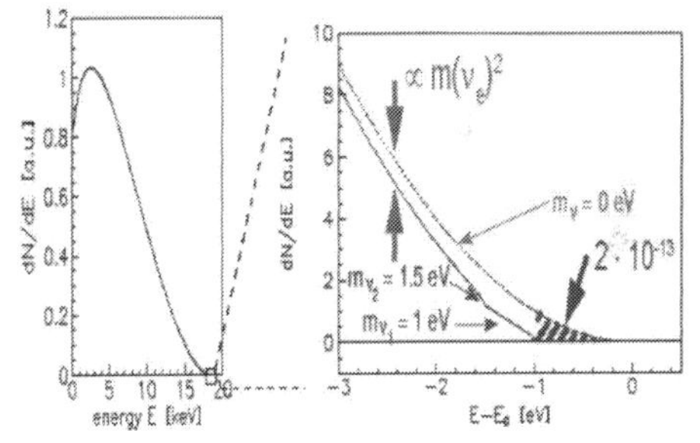

Figure 6. Modification of the β spectrum due to a massive neutrino.

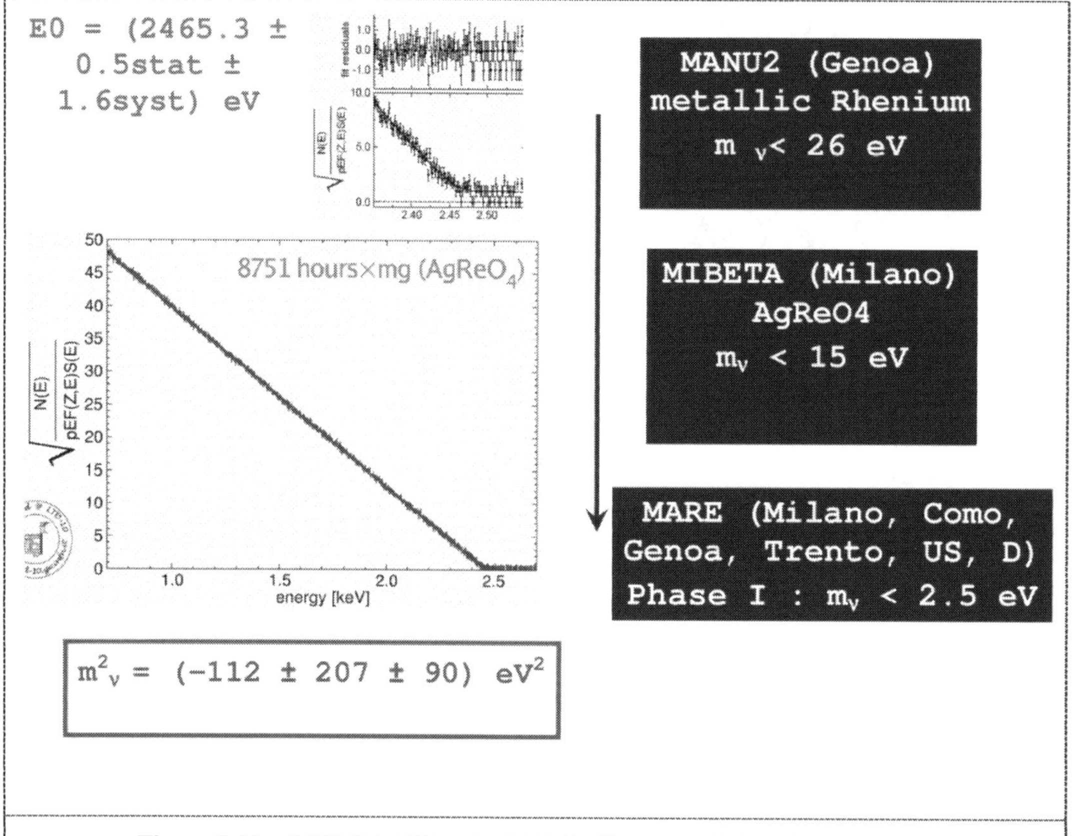

Figure 7. The MIBETA (Milano), MANU2 (Genoa) and MARE experiments.

4. Double β decay (DBD)

This very rare process, amply discussed at this Symposium [10-15] consists, in the most common case, in the transition (Fig. 8) from the isotope (A, Z) to (A, Z+2) with contemporary emission of two electrons. Several DBD mode are possible: one is the 2ν, where two electron antineutrinos are emitted. This decay mode conserves lepton number and it is allowed in the framework of Standard Model (SM) of electroweak interaction; another is the 0ν mode, which violates the lepton number and has been recognized since a long time as a powerful tool to test neutrino properties [20]. A third decay channel, is the process with the emission of one or more light neutral bosons, named Majorons. In the processes where no additional particle is emitted with the two electrons, lepton number conservation is violated (Fig. 9) and a peak appears in the sum of the two electron energies corresponding to the transition energy.

The presence of the second channel, rather improperly called neutrinoless Double Beta Decay (DBD) implies that neutrino is a Majorana [16] massive particle, and would allow to determine the absolute value of its mass $< m_\nu >$. The predicted values from the results of oscillations are of a few tens of meV or of a few meV, for the case of inverse and direct hierarchies, respectively [17].

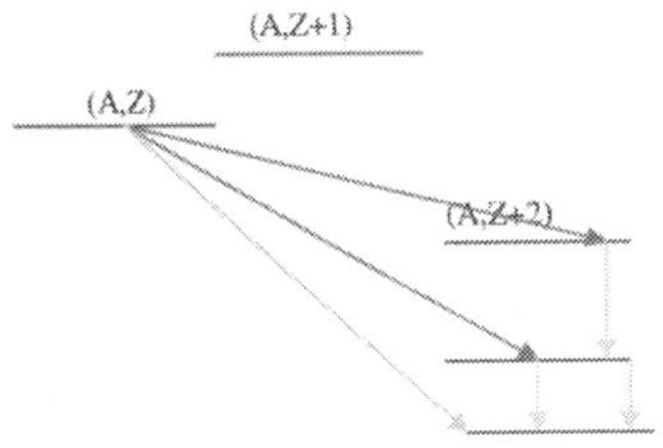

Figure 8. Double beta decay.

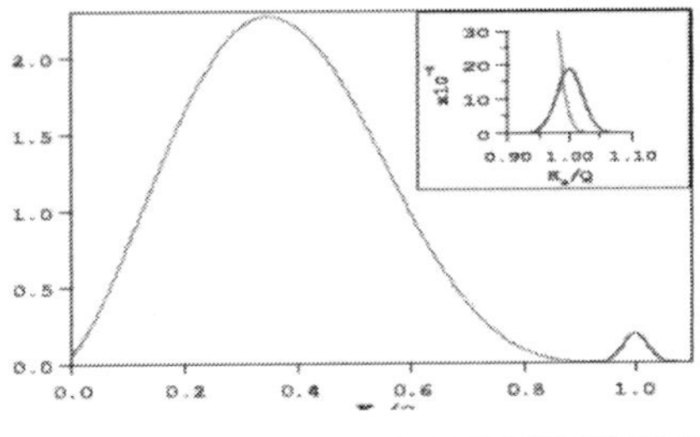

Figure 9. Spectrum of the sum of the two electron energies in two neutrino and neutrinoless $\beta\beta$ decay.

In "direct" experiments where DBD is actually detected two approaches are possible: the source = detector and the source ≠ detector as shown in Fig.10 and 11.

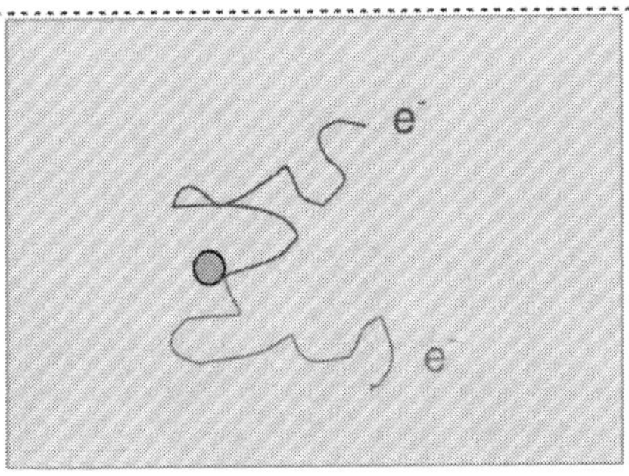

Figure 10. Source = detector.

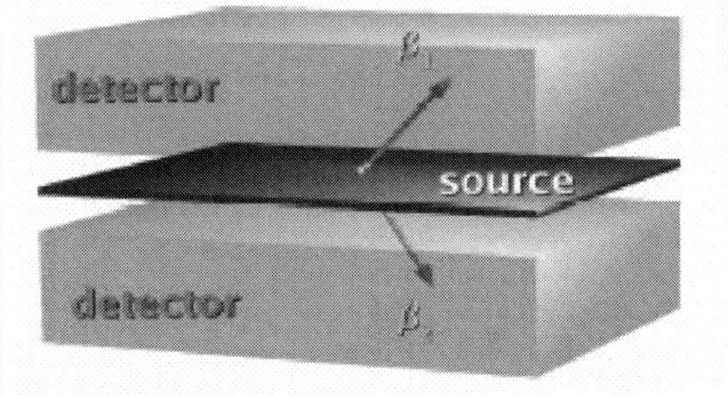

Figure 11. source ≠ detector.

No evidence has been obtained so far for neutrinoless double beta decay with the exception of the claim by a subset of the Heidelberg-Moscow collaboration [17] which indicates a value of $< m_\nu >$. around 0.5 eV. I leave to the reports by P. Vogel [18] and J. Engel [19] the discussion of the difficult calculations of the nuclear matrix elements for neutrinoless DBD.

Most of the running and planned experiments aiming to detect neutrinoless DBD with the sensitivity suggested by the results of oscillations (Fig.12) are reported to this Symposium [10-15]. Let me only mention here those, running or in construction, which are based on the use of cryogenic detectors, whose mass has constantly increased with time as shown in Fig. 13.

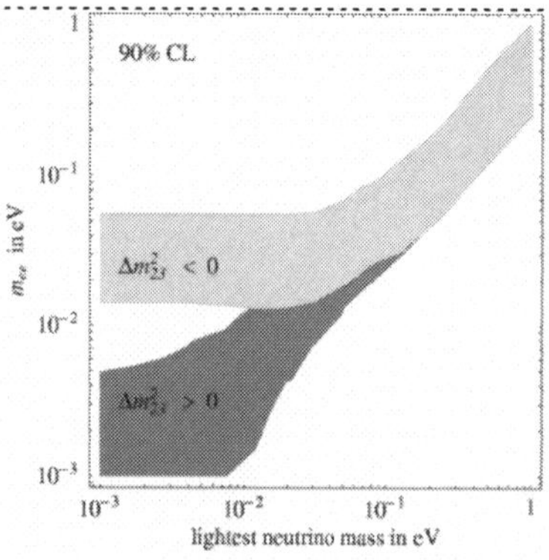

Figure 12. Values of $<m_\nu>$. expected from the inverse (upper) and direct hierarchy.

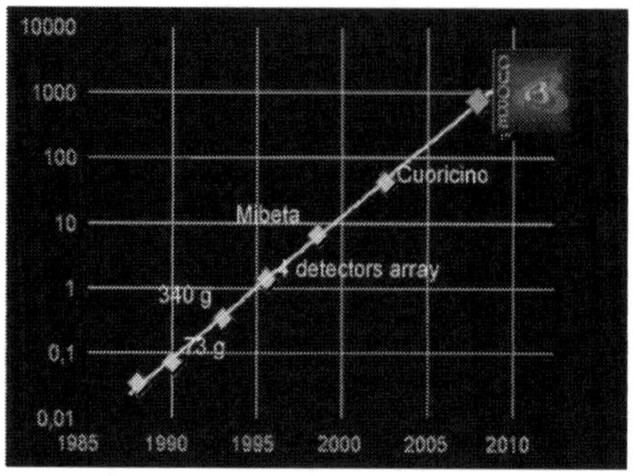

Figure 13. Increase of the obtained mass of cryogenic detectors.

The most sensitive detector of any type presently running is CUORICINO, by far the largest operating cryogenic setup in the world, shown during its mounting in Fig.14. It consists in a single tower made by absorbers of TeO2 with a total mass of ~ 41 kg. It is devoted to the search for neutrinoless DBD of 130Te, a nucleus with an isotopic abundance of ~34 % and a transition energy of ~ 2530 keV.
The CUORICINO array is made by 18 crystals of 33x6 cm3 of which two are enriched in 128Te and two in 130Te, and by 44 cubic crystals of natural Te of 5 cm side. No evidence is found for the peak indicative of neutrinoless double beta decay of 130Te, but with a 90% confidence level lower limit of 3.1 x 1024 years of DBD0v life time, the corresponding constrains on the effective neutrino mass is in the range of 0.16 - 0.84 eV using the recent calculation of Rodin et al. [18]. The corresponding values coming from the positive indication of Klapdor et al. are in the mass range 0.1-0.9 eV. As a consequence our result, while not confirming the one by Klapdor et al. cannot be used to exclude their evidence.

Journal of Physics: Conference Series **173** (2009) 012004 doi:10.1088/1742-6596/173/1/012004

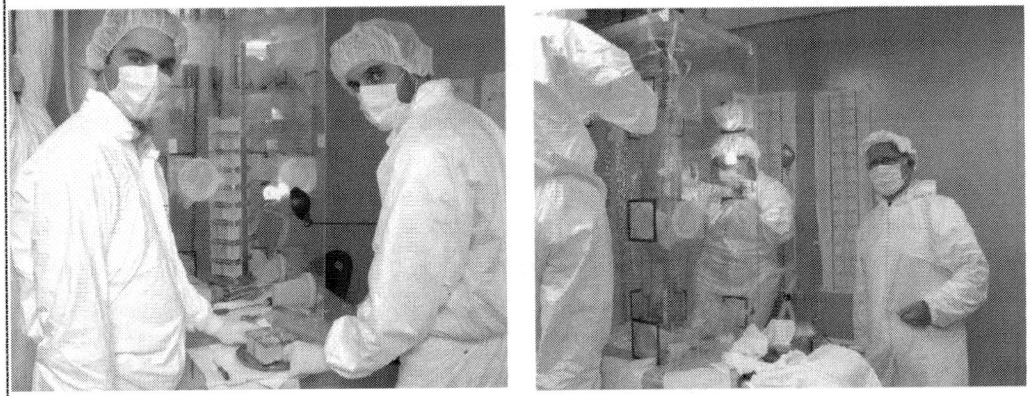

Figure 14. Mounting of CUORICINO.

The only second generation experiment approved so far is CUORE (from Cryogenic Underground Observatory for Rare Events) already in construction at the Laboratori Nazionali del Gran Sasso by a large international collaboration (fig.15). CUORE will be a giant bolometric detector made of 988 cubic crystals of natural TeO2 with a total mass of ~ 750 kg corresponding to ~ 203 kg of 130 Te. It could reach the sensitivity on the effective neutrino mass predicted by the results on neutrino oscillations under the inverse hierarchy hypothesis.

I would like to add that many other compounds can be chosen as bolometric candidates for searches on neutrinoless double beta decay. All those reported in Table 1 have been successfully tested as bolometers, the only exception being so far the compounds of 150 Nd.

Table 1. Candidates for bolometric detection of neutrino-less DBD

Compound	**I. A. (%)**	**ΔE (keV)**
^{48}**CaF**$_2$	0.0187	4212
^{76}**Ge**	7.44	2038.7
^{100}**Mo**PbO$_4$	9.63	3034
^{116}**Cd**WO$_4$	7.49	2804
^{130}**Te**O$_2$	34	25430
^{150}**Nd**F$_3$	5.64	3368
^{150}**Nd**Ga0		

Journal of Physics: Conference Series **173** (2009) 012004 doi:10.1088/1742-6596/173/1/012004

Figure 15. The CUORE collaboration.

Figure 15. CUORE.

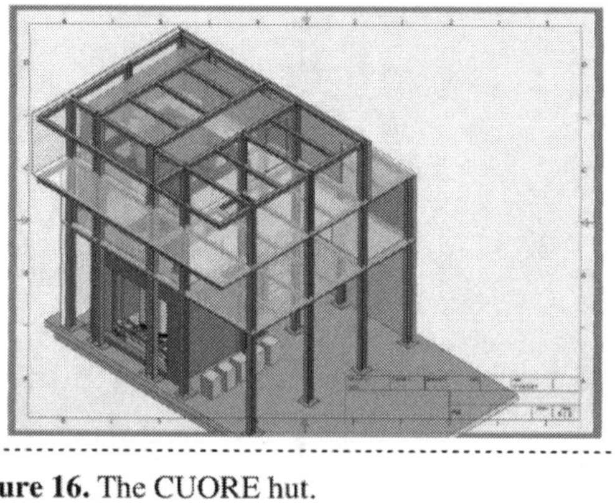

Figure 16. The CUORE hut.

5. Conclusions and acknowledgments

Thermal cryogenic detectors are already playing an important role in nuclear, subnuclear and astroparticle physics, but their impact in these fields is going to increase greatly in the future. In the field of neutrino physics bolometers constitute an approach complementary to the spectrometric measurement of the electron energy. Bolometric measurements of the β decay spectrum is an excellent model independent approach, even if its sensitivity is not yet at the level predicted by the results coming from neutrino oscillation experiments. The sensitivity in determination of the absolute value of neutrino mass from neutrinoless DBD is expected to be quite superior, even if less direct, to that from single β decay and could allow to test directly if the neutrino is a Dirac and Majorana particle. The efforts devoted with various techniques to this fascinating problem are clearly shown by the description of beautiful experiments reported here in an atmosphere of collaboration and friendship. Both in single and in double beta decay experiments, as in other fields of nuclear, subnuclear and astroparticle physics, the new approach of bolometric detection of particles is definitely going to play a relevant role.

Mine cannot be here the standard acknowledgments at the end of an invited talk. I am really moved by being here with my old friend (in fact he is young, at least younger than myself!) Frank with whom we deeply miss our great friend Peter who stimulated and helped us since the beginning of our searches on DBD. I have no words to adequately express my gratitude to organizers and to the outstanding physicists who came here to give beautiful talks honouring me much more that I deserve. I will always remember with pleasure and gratitude these wonderful days spent in this beautiful city and university.

References

[1] Turner M, *Quarks and Cosmos from Massive neutrinos to Dark Energy* , 2008 *this Symposiu.*
[2] McDonald A., *Physics at SNO and the New SNOLAB* , 2008 *this Symposium.*
[3] Lande K., *Neutrino oscillation at Homestake from Chlorine to Megadetector*, 2008 *this Symposium.*
[4] Calaprice F., *First results from the Borexino Solar Neutrino Experiment*, 2008 *this Symposium.*
[5] Hartman F., *Double Chooz and the seach for Short Range Electron Anti-neutrino oscillations*, 2008 *this Symposium.*
[6] Enns C. and McCammon D., 2008 *J. Low Temp. Phys.* **151** 5 and references therein.
[7] Moseley S.H., Mather J.C. and McCammon D., 1984 *J. Applied. Phys.* **56** 1257 .
[8] Fiorini E. and Niinikosky T., 1984 *Nucl. Instrum. Methods.* **83** 1257.
[9] Robertson R.G.H., *Direct Measurement of Neutrino Mass* , 2008 *this Symposium.*
[10] Elliott S., *The Majorana Project* , 2008 *this Symposium.*
[11] Barabash A., *NEMO-3 Double Beta Decay Experiment : Lat Results* , 2008 *this Symposium.*
[12] Ejiri H., *Double Beta Decay and Related Subjects in Japan* , 2008 *this Symposium.*
[13] Wilkerson J., *Peter Rosen and Neutrinoless Double Beta Decay* , 2008 *this Symposium.*
[14] Miley H., *A Review of the Germanium Collaboration between USC and PNNL*, 2008 *this Symposium.*
[15] Gratta G., *Measuring the Neutrino Mass One Atom at the Time*, 2008 *this Symposium.*
[16] Holstein B., *The Strange Disappearance of Ettore Majorana*, 2008 *this Symposium.*
[17] Klapdor Kleingrothaus H.V. 2008 *Int. J. Mod. Phys.* **E 17** 505.
[18] Petkov S. 2008 *Concluding Remarks.*
[18] Vogel P, *Anatomy of Nuclear Matrix Elements for Neutrinoless Double Beta Decay*, 2008 *this Symposium.*
[19] Engel J., Nuclear-Structure Corrections to the Double Beta Decay Operator, 2008 this Symposium.
[20] Furry W. H., *Phys. Rev. 56* (1939) 1184.

Carolina International Symposium on Neutrino Physics IOP Publishing
Journal of Physics: Conference Series **173** (2009) 012005 doi:10.1088/1742-6596/173/1/012005

The Deep Underground Science and Engineering Laboratory at Homestake

Kevin T. Lesko

Department of Physics, University of California Berkeley and Institute for Nuclear and Particle Astrophysics, Lawrence Berkeley National Laboratory, 1 Cyclotron Road, MS 50R5239, Berkeley, CA 94720-8156, USA

ktlesko@lbl.gov

Abstract. The US National Science Foundation and the US underground science community are well into the campaign to establish a world-class, multi-disciplinary deep underground science and engineering laboratory - DUSEL. The NSF's review committee, following the first two NSF solicitations, selected Homestake as the prime site to be developed into an international, multidisciplinary, world-class research facility. Homestake DUSEL will provide much needed underground research space to help relieve the worldwide shortage, particularly at great depth, and will develop research campuses at different depths to satisfy the research requirements for the coming decades. The State of South Dakota has demonstrated remarkable support for the project and has secured the site with the transfer of the former Homestake Gold Mine and has initiated re-entry and rehabilitation of the facility to host a modest interim science program with state funds and those from a substantial philanthropic donor. I review the scientific case for DUSEL and the progress in developing the preliminary design of DUSEL in Homestake and the initial suite of experiments to be funded along with the facility.

1. Introduction

In March 2004 the National Science Foundation (NSF) announced a three-solicitation process to revitalize its efforts to create a dedicated deep underground research laboratory. The first solicitation focused on defining the science goals for a deep underground laboratory and defining many of the laboratory's critical parameters and characteristics. The S-1 principal investigators, lead by Professor Bernard Sadoulet, produced a document, Deep Science [1] to summarize their finding of the key scientific questions to be addressed with the aid of a deep underground laboratory and many of the laboratory essential characteristics. Grouping the fundamental questions by disciplines Deep Science presents the questions to be addressed within DUSEL:

Physics:

- What is the universe made of?
- What is dark matter?
- What are neutrinos telling us?
- What happened to the antimatter?
- Are protons unstable?
- How did the universe evolve?

Biology and Geology:

- How do biology and geology interact to shape the world underground?
- How does subsurface microbial life evolve in isolation?
- Did life on earth originate beneath the surface?
- Is there life underground, as we don't know it?

Geology:

- What are the interactions among subsurface processes?
- Are underground resources of drinking water safe and secure?
- Can we reliably predict and control earthquakes?
- Can we make the earth "transparent" and observe underground processes in action?

Geology and Engineering:

- What are the mechanical properties of rock?
- What lies between the boreholes?
- How does rock respond to human activity?
- How does water flow deep underground?
- How can technology lead to a safer underground?

The report goes on to assess physical characteristics and key parameters for DUSEL including:

- Depth – the facility should access depths in excess of 7000 meters water equivalent (mwe) and support a variety of different depths for different research requirements.
- Rock type – the rock type should be geologically interesting and able to support large excavations.
- Pristine Regions – the rock mass should possess pristine and undisturbed regions for research.
- Distance from Accelerators – the facility should be sited at an appropriate distance from accelerators capable of producing neutrino beams.
- General Accessibility – the facility should be a dedicated facility supporting continuous access. The facility should support access by all research disciplines and be truly international.
- Management – the facility should be managed in a style to facilitate DUSEL's research and education goals.
- Adaptability and Expandability – the facility should be able to expandable and adaptable to support decades of research.

2. Current status of DUSEL Planning – Facility Preliminary Design

With the selection of Homestake as the site to be developed for DUSEL, a cooperative agreement between the University of California Berkeley and the NSF was established. The focus of the agreement is the development of a preliminary design for the facility and the integration of an initial suite of experiments into the facility. This preliminary design including both the facility and the suite of experiments would become the basis of application to the Major Research Equipment and Facility Construction account for the construction of DUSEL and its experiments. The requirements for the preparation of the application are defined in NSF's Large Facilities Manual [2].

2.1. Schedule of Activities. The table presented below outlines major milestones in the creation of the DUSEL Proposal including the NSF's S-4 and S-5 solicitations. The S-4 solicitation, anticipated to be $15M distributed over three years, would fund experimental collaborations to prepare conceptual and preliminary designs of candidate Initial Suite of Experiments for consideration for inclusion in the DUSEL Proposal. S-5would select the experiments to be integrated into the proposal. S-4 funding is neither a sufficient nor a necessary condition to be considered by S-5 for inclusion in the ISE. Following this schedule, the earliest construction of DUSEL could begin would be Fiscal Year 2012 (Fall of 2011).

Date	**Completed**	**Activity or Milestone**
November 2007	x	Town Meeting In Washington DC
April 2008	x	DUSEL Workshops in Lead, South Dakota
Spring/Early Summer08		NSF S-4 Solicitation to Fund Experimental Development
July 2008		Internal Review of DUSEL Project
Fall 2008		Funding of S-4 Proposals
December 2008		NSF Review of DUSEL Project
Summer 2009		Review and Selection of DUSEL's Initial Suite of Experiments (S-5 Solicitation)
Summer/Fall 2009		Integration of ISE into Facility and Preparation of Integrated Baseline Project and MREFC Proposal
Fall 2009		Completion of DUSEL Preliminary Design and Review by NSF
Winter 2010		Presentation of Proposal to and Review by the NSF
March 2010 (earliest)		Presentation to the National Science Board
FY2012		MREFC (earliest possible date capital fundingfor construction)

Table 1: Milestone Schedule for DUSEL Construction within the Major Research Equipment and Facility Construction (MREFC) Account.

3. Initial Suite of Experiments for DUSEL

As discussed above, the envisioned DUSEL proposal and presentation to the National Science Board would include a suite of experiments to be funded concurrently with the facility construction. The MREFC funding would cover normal construction costs for the experiments. Current estimates and general discussions with the agencies indicate the MREFC proposal is likely to be in the range of $500M with approximately an equal split between experiments and facility costs.

The suite of experiments would be fully multidisciplinary, with candidates anticipated from physics, geology, biology, engineering and education and public outreach. To help facilitate and coordinate the experimental collaborations the NSF working with the DUSEL project have established a coordination committee, the DUSEL Experiment Development Committee (DEDC), to serve as a point of contact, to provide coordination, and in general to assist the experimental collaborations with their interactions with DUSEL. The DEDC consists of Dr. Steve Elliott, Los Alamos National Laboratory; Professor Derek Elsworth, Penn State (Chair); Dr. Daniela Leitner, Lawrence Berkeley National Laboratory; Professor Larry Murdoch, Clemson University; Professor T.C. Onstott, Princeton University; and Professor Hank Sobel, UC Irvine.

The April workshops in Lead, South Dakota [3] was the initial meeting of the collaborations interested in creating proposals for the ISE. The working groups and disciplines presenting at this meeting are presented in the table below:

Working Group	**Group Leader(s)**
Physics	
Dark Matter	Dan Akerib, Case Western University &Rick Gaitskell, Brown University
Long Baseline Neutrinos & Nucleon Decay	Bob Svoboda, University of California at Davis
Neutrinoless Double Beta Decay	Giorgio Gratta, Stanford University
Nuclear Astrophysics	Michael Wiescher, University of Notre Dame
1-km Vertical Space	Yuri Kamyshkov, University Of Tennessee, Knoxville
Gravity Waves	Vuk Mandic, University of Minnesota
Low Background Counting and Assay	Prisca Cushman, University of Minnesota
Solar Neutrinos	Bruce Vogelaar, Virginia Tech
Studies of effects of energetic particles on electronic devices, biological systems, materials and development of structure imaging from cosmic rays.	Rob McTaggart, South Dakota State University
Geosciences, Geomicrobiology, and Engineering	
Baseline characterization and monitoring	Stephen Martel, University of Hawaii
Ambient rock deformation processes	Herb Wang, University of Wisconsin
Induced rock deformation processes	Leonid Germanovich, Georgia Tech
Ambient flow, transport, diversity andactivity	David Boutt, University of Massachusetts
Induced flow, transport and activity	Eric Sonnenthal, LBNL
Underground construction, mining and environment	Charles Fairhurst, University of Minnesota
CO_2 Sequestration	Joe Wang, LBNL
Resource extraction	Jean-Claude Roegiers, University of Oklahoma
Subsurface imaging and sensing	Steven Glaser, UC Berkeley
Ultra-deep drilling and exploration	Tom Kieft, New Mexico Tech
Self-Initiated Working Groups	
Petrogenisis, Ore Deposits, & Structures	Derric Iles, University of South Dakota
Geoneutrino Radiometric Analysis Group for Geosciences	P. Ila, MIT
Toxicology of Heavy Metals (THM) in Degenerative Diseases	Glenn I. Lykken, University of North Dakota

Table 2: April Workshop on DUSEL's Initial Suite of Experiments – Working Groups and Group Leaders

The recent release of the Particle Physics Project Prioritization Panel (P5) report [4] of the High Energy Physics Advisory Panel highlights the significance of Homestake to the future of the domestic high energy physics programs, primarily establishing a world-leading long baseline neutrino program, simultaneously creating a major nucleon-decay program, as well as focusing the US Dark Matter and Neutrinoless Double Beta Decay efforts at Homestake DUSEL. The DUSEL Project team advocates a multi-chamber approach t o creating the ~ megaton targets needed for the long baseline neutrino and nucleon decay programs.

Beginning with a single 150kt capacity cavity has been demonstrated to create a scientifically well-motivated program with significant reach in determining θ_{13}, the neutrino mass hierarchy, and even in establishing the CP phase for sufficiently large values of θ_{13}. Monte Carlo simulations for these quantities are presented in Figure 1.

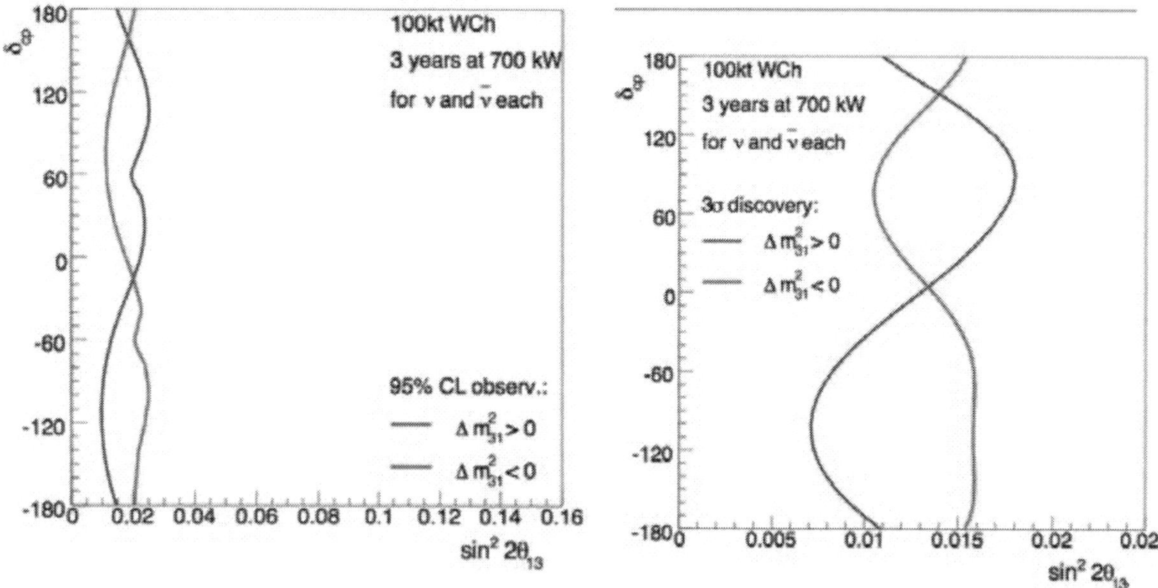

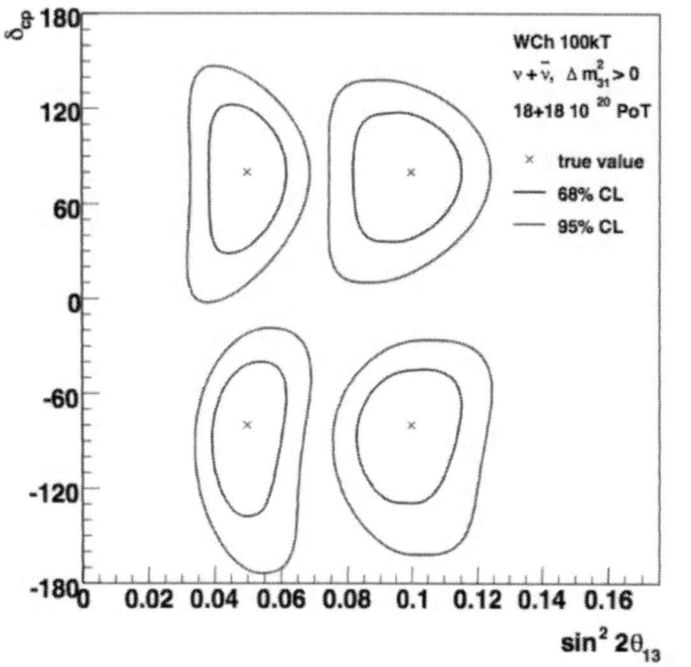

Figure 1: Discovery potential for $\sin^2 2\theta_{13}$ with 100kt of fiducial mass and 3 years of neutrino and antineutrino running each at 700kW of beam power(top left). The top-right plot demonstrates the potential for resolving the mass hierarchy issue. The sensitivity is shown as a function of the CP phase δ and $\sin^2 2\theta_{13}$. The discovery can be made on the right side of the curves with confidence level of 95% CL. The bottom plot presents simultaneous determination of δ (degrees) and $\sin^2 2\theta_{13}$ with 100kt detector and 3 years of neutrino and antineutrino running at 700kW beam power. The precision at 68% and 90%C.L. is shown. At each point we simulated the spectrum and calculated the measurement precision using the statistics of the simulated signal and background and the assumption of 10% systematic errors on the background. Calculations are from Dierckxsens, Diwan and Bishalhttp://nwg.phy.bnl.gov/~mdier/bnl/nwg/longbaseline/120GeV_0.5deg/WCC_100kt/18+18/ as part of the Fermilab – Brookhaven Long Baseline Neutrino Working Group.

Experiments developing techniques for direct detection of dark matter have made significant progress in recent years. The current state-of-the-art and anticipated increases in dark matter detection sensitivity is presented in Figure 2.

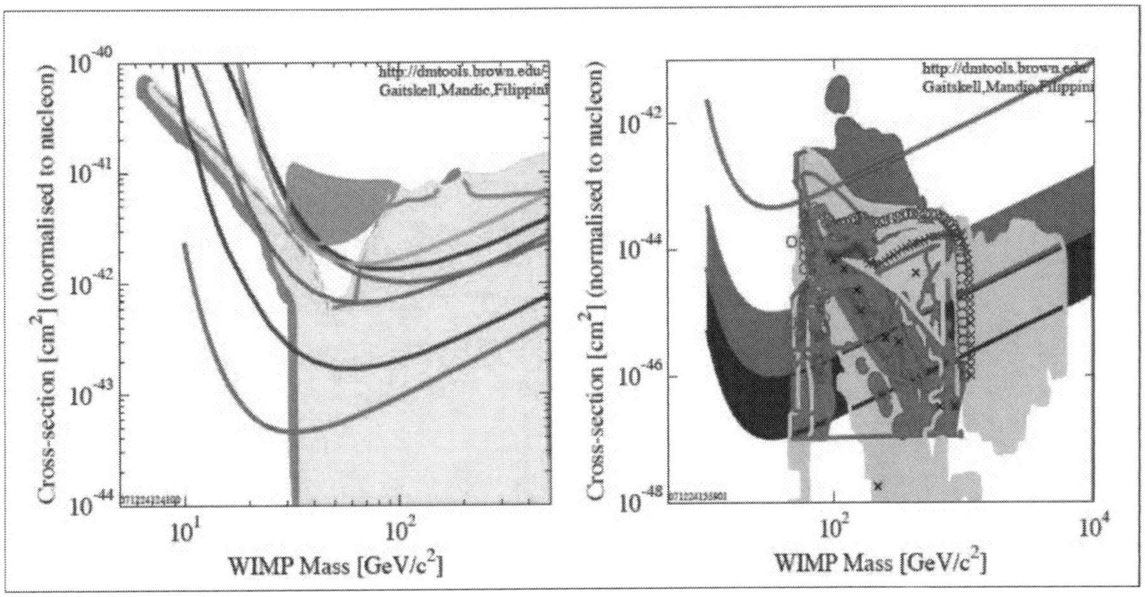

Figure 2: Plots of the elastic scattering cross section for spin-independent couplings versus WIMP mass. (a) The left panel shows the leading experimental results in which the solid curves represent experimental upper limits from the CRESST thermal and scintillation cryogenic detectors (cyan), the EDELWEISS thermal and ionization cryogenic detectors (dark red), the WARP two-phase liquid argon detector (magenta; 55 keV threshold), the ZEPLIN-II two-phase liquid xenon detector (dark green), the CDMS-II athermal-phonons and ionization detectors (blue), and the XENON-10 two-phase liquid xenon detector (red), in all cases assuming the standard halo model and nuclear-mass-number A^2 scaling. The green region shows the contested DAMA annual-modulation claim. The yellow and grey regions represent unconstrained Minimal Supersymmetric Standard Model (MSSM) predictions for low-mass WIMPs that result from relaxing the GUT-scale unification of gaugino masses. (b) The right panel displays a broad range of models. CMSSM models from are shown in dark green. Well-tempered neutralinos are shown in light red. Models within the minimal supergravity(mSUGRA) framework are shown in magenta and light blue. More specific predictions are given by split Supersymmetry models shown by blue circles and crosses (for positive and negative values of the parameter, respectively) and red circles. Black crosses show the set of representative post-LEP LHC-benchmark models. Experimental projections are shown as grey and black regions for the two ISE phases discussed in the Roadmap section of this report, at the 4850- and 7400-foot levels, respectively. The XENON-10 limit is shown again as a solid red curve for reference with the left panel. This figure is taken from and references for the shown data and predictions are included in*http://cosmology.berkeley.edu/DUSEL/Town_meeting_DC07/working_groups.html*.

Searches for neutrinoless double beta decay have been making significant technical advances as well. Figure 3 presents the sensitivity for these experiments anticipated in DUSEL's initial suite of experiments.

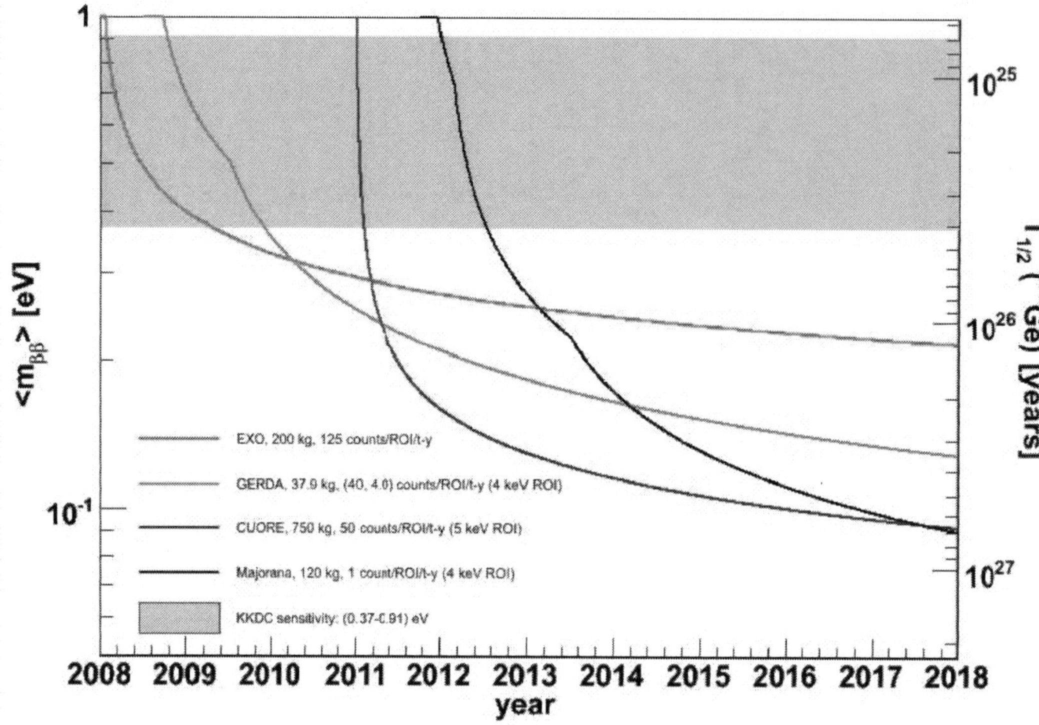

Figure 3: Majorana neutrino mass sensitivity for proposed neutrinoless double beta decay experiments. Figure courtesy of the Majorana Collaboration.

4. The Sanford Laboratory and South Dakota Supported Efforts

The State of South Dakota has demonstrated remarkable commitment to establishing DUSEL at Homestake. In October 2005, the State legislature approved an addition \$20M for a total of \$46M of state controlled funds to create an interim laboratory. In June 2006, T. Denny Sanford pledged \$70Min a notably philanthropic donation to create the Sanford Laboratory at Homestake. In 2006, the Barrick Corporation donated the 186 acres of the former Homestake gold mine site to the South Dakota Science and Technology Authority (the Authority). Currently the Authority is overseeing efforts to re-enter Homestake, refurbish the Ross and Yates shafts, and to control the water entering Homestake since the pumps were shut off in 2003. The water is currently above the 4850 Level in the facility, with rehabilitation efforts having reached the 4200 Level. Pumping has commenced with 3.5M gallons of water expelled in May. Routine pumping is anticipated by June, with access to the 4850 Level anticipated in the fall of 2008.

The State of South Dakota has expressed a great desire to participate in and to develop educational and public outreach opportunities with Homestake (both as part of the Sanford Laboratory and as DUSEL).

5. Summary

Figure 4 presents the vision for DUSEL from the Conceptual Design Report [5]. The near-surface campus at the 300 Level, major campus at 4850 Level, and deep campus at 7400 Level are illustrated. In addition several of Homestake's ~ 60 Levels of interest to biology, geology, and engineering are illustrated. In particular the 8000 Level presents geomicrobiologists with attractive opportunities for searches for exotic life forms. The DUSEL Project Team is actively advancing the design of the facility and working to develop the Initial Suite of Experiments to be paired with the facility in the MREFC application. The Project Team is working closely with the experimental collaborations to understand the experimental requirements and to fold these requirements into the facility design. The Project Team welcomes communication with any interested experiment or potential experiment that might benefit from an underground facility. While the principal avenues for funding so far have been through the NSF's solicitations, the recent close coupling of the Department of Energy, notably the High Energy Physics, efforts to DUSEL have opened additional mechanisms to fund research and development, experimental programs, and educational opportunities in and with DUSEL.

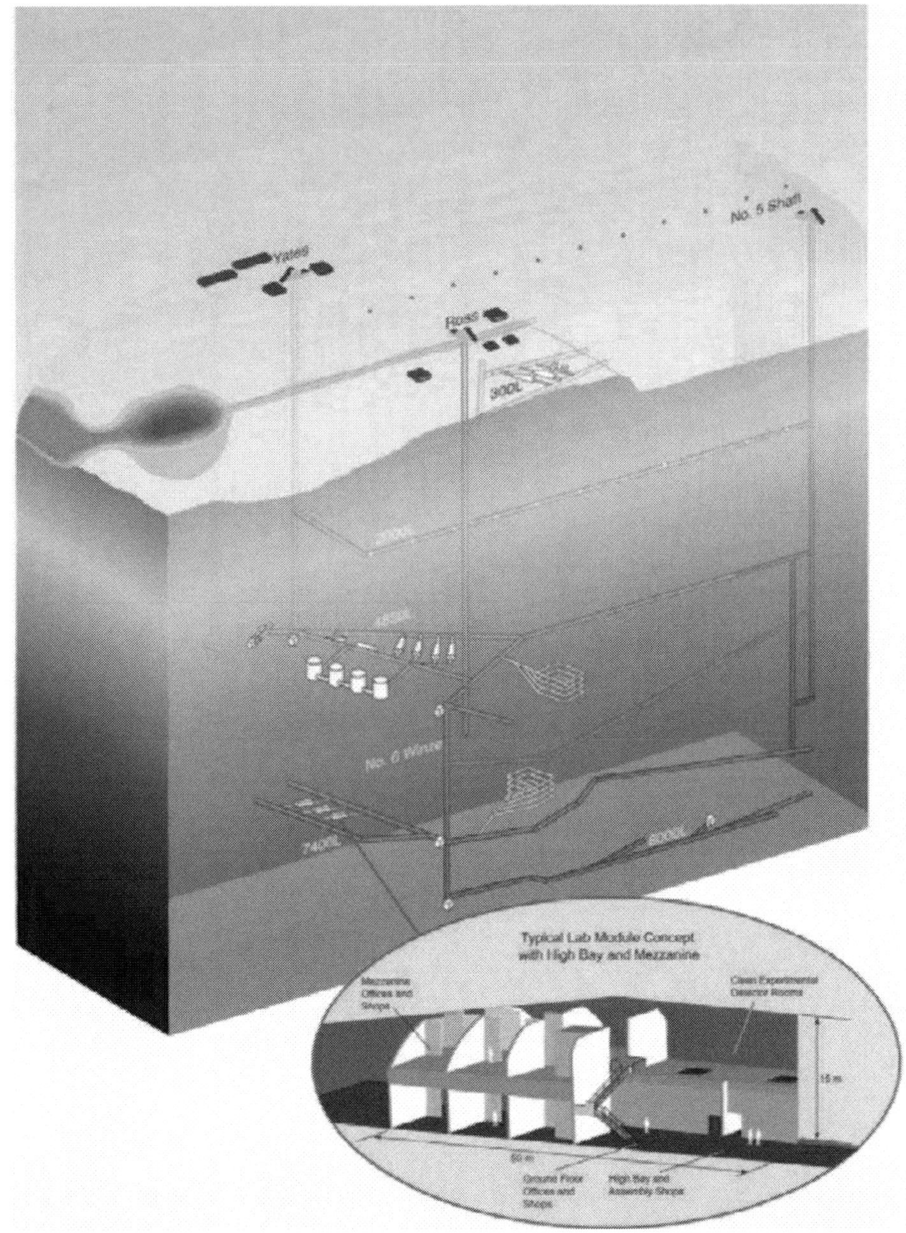

Figure 4: Concepts for underground research campuses within Homestake DUSEL. A typicalresearch module is presented in the "inset". The levels depths refer to the depth (in feet) below theshaft collar, so the 7400 Level is 7400 feet below the shaft collar.

References

[1] Deep Science, http://www.deepscience.org/
[2] NSF Large Facilities Manual, NSF 07-38, http://www.nsf.gov/pubs/2007/nsf0738/nsf0738.pdf
[3] Workshop presentations, whitepapers and summaries are available at: http://www.lbl.gov/nsd/homestake/Workshops.html
[4] http://www.science.doe.gov/hep/hepap_reports.shtm.
[5] The Homestake DUSEL Conceptual Design Report can be downloaded from: http://www.lbl.gov/nsd/homestake.

Opportunities for Neutrino Physics at the Spallation Neutron Source (SNS)

Yu Efremenko[1,2] and W R Hix[2,1]
[1]University of Tennessee, Knoxville TN 37919, USA
[2]Oak Ridge National Laboratory, Oak Ridge TN 37981, USA

E-mail: yefremen@utk.edu

Abstract. In this paper we discuss opportunities for a neutrino program at the Spallation Neutrons Source (SNS) being commissioned at ORNL. Possible investigations can include study of neutrino-nuclear cross sections in the energy range important for supernova dynamics and neutrino nucleosynthesis, a search for neutrino-nucleus coherent scattering, and various tests of the standard model of electro-weak interactions.

1. Introduction

It seems that only yesterday that we gathered together here in Columbia for the first Carolina Neutrino Symposium on Neutrino Physics. To my great astonishment I realized it was already eight years ago. However by looking back we can see that enormous progress has been achieved in the field of neutrino science since that first meeting. Eight years ago we did not know which region of mixing parameters (SMA, LMA, LOW, Vac) [1] would explain the solar neutrino deficit. We did not know whether this deficit is due to neutrino oscillations or some other even more exotic phenomena, like neutrino decay [2], or due to the some other effects [3]. Hints of oscillations of atmospheric neutrinos had not been confirmed in accelerator experiments. Double beta decay collaborations were just starting to think about experiments with sensitive masses of hundreds of kilograms. Eight years ago, very few considered that neutrinos can be used as a tool to study the Earth's interior [4] or for non-proliferation [5]. This impressive success in neutrino physics lead to increasing interest in this field. It is a growing area of research and more and more young students are entering the field of neutrino physics. In this article we discuss opportunities for a comprehensive neutrino experimental program that could start in ORNL at SNS.

2. Neutrino Production at SNS

Let us begin by describing the mechanism of neutrino production at the SNS. The Spallation Neutron Source, currently under commissioning at ORNL, will be the world's premier facility for neutron-scattering research, producing pulsed neutron beams with intensities an order of magnitude larger than any currently operating facility. When full beam power is reached in 2009, 10^{14} one GeV protons will bombard a liquid mercury target in 700 ns wide bursts with a frequency of 60 Hz. Neutrons produced by spallation reactions with the mercury will thermalize in hydrogenous and helium moderators surrounding the target and be delivered to neutron scattering instruments in the SNS experimental hall.

As a by-product, the SNS will also provide the world's most intense pulsed source of neutrinos, in the energy regime of interest for nuclear astrophysics. Interactions of the proton beam in the mercury target will produce π-mesons in addition to neutrons. These will stop inside the dense mercury target and their subsequent decay chain will produce neutrinos with a flux of ~2×10^7 ν/cm^2/s for three neutrino flavors at 20 m from the spallation target [6].

The time structure of the SNS beam is particularly advantageous for neutrino studies. Time correlations between candidate events and the SNS proton beam pulse will greatly reduce background rates and may provide sensitivity to neutral current events. As shown in the left panel of Figure 1, all neutrinos will be produced within several microseconds of the 60 Hz proton beam pulses. As a result, background events resulting from cosmic rays will be suppressed by a factor of ~2000. At the beginning of the beam spill the neutrino flux is dominated by muon neutrinos resulting from pion decay, making it possible to isolate pure neutral-current events. This time structure provides a great advantage over a continuous-beam facility such as the Los Alamos Neutron Science Centre (LANSCE), where the LSND [7] experiment was located.

The energy spectra of SNS neutrinos are shown in the right hand panel of Figure 1. These spectra are known with high accuracy because nearly all neutrinos originate from decay-at-rest processes in which the kinematics is well defined. The decay of stopped pions produces mono-energetic muon neutrinos at 30 MeV. The subsequent 3-body muon decay produces a spectrum of electron neutrinos and muon antineutrinos with energies up to 52.6 MeV. Decay chains for negative pions and muons are strongly suppressed by their quick absorption in the bulk of the mercury target.

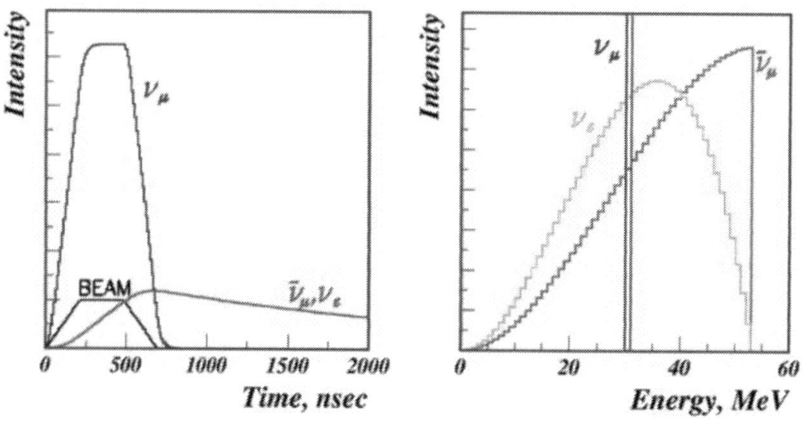

Figure 1. Time and energy distributions for different neutrino flavors produced at the SNS.

3. Neutrino Nucleus Cross Sections

As has been mentioned on many occasions [8], [9] neutrino energy spectra from SNS nicely match the neutrino spectra produced during core collapse supernovae (SN) explosions. This gives us a unique opportunity to study neutrino interactions in this important energy range, which will improve our understanding of SN dynamics and help us to design and calibrate the response of supernova neutrino detectors.

Core collapse supernovae, are among the most energetic explosions in our universe, releasing 10^{46} Joules of energy in the form of neutrinos of all flavours at a rate of 10^{57} neutrinos per second. Marking the death of a massive star (mass >8-10 solar masses) and the birth of a neutron star or black hole, core collapse supernovae serve as laboratories for physics at extremes of density, temperature, and neutronization that cannot be produced in terrestrial laboratories. The kinetic energy and the rich mix of recently synthesized elements delivered into the interstellar medium by the ejecta of each supernova make core collapse supernovae a key link in our chain of origins from the Big Bang to the formation of life on Earth. Currently, the lack of data on neutrino-nucleus interactions limits our understanding of the mechanism by which core collapse supernovae explode our understanding of the resulting nucleosynthesis, and our ability to interpret the results of neutrino astronomy.

Recent studies [10] have demonstrated unequivocally that electron and neutrino capture reactions on nuclei play a major role in dictating the dynamics of stellar core collapse, which set the stage for all of post-bounce dynamics and the formation of the supernova shock wave. Comparisons of results using modern prescriptions for electron and neutrino capture on nuclei with earlier calculations demonstrate quantitative and qualitative changes in the launch radius of the supernova shock wave after stellar core bounce and in the density, temperature, and compositional structure of the stellar core.

Nucleosynthesis in core collapse supernovae falls into three basic categories: (1) explosive nucleosynthesis that occurs as the shock wave passes through the stellar layers and causes nuclear fusion through compression and heating of the material, (2) neutrino nucleosynthesis in the ejected layers that occurs as these layers are exposed to the intense neutrino flux emerging from the proto-neutron star, and (3) *r*-process nucleosynthesis that occurs in a neutrino-driven wind emanating from the proto-neutron star after the explosion is initiated. In all cases, the final elemental abundances produced and ejected are affected through nuclear transmutations by the neutrino-nucleus interactions that occur.

The ability to detect, understand, and ultimately use the detailed neutrino "light curve" from a future core collapse supernova in our galaxy is vital to better understanding supernovae and the use of supernova models, together with detailed astronomical observations, to constrain fundamental physics that is otherwise inaccessible in terrestrial experiments. To achieve this will require an accurate normalization of the neutrino flux in a supernova neutrino detector and knowledge of the cross sections and by-products of neutrino interactions in the detector material. From Deuterium to Lead, a number of nuclei have been proposed and, in some cases, used as supernova neutrino detector materials. In all cases, accurate neutrino-nucleus cross sections are essential. Only ν+Carbon has been measured to ~10% accuracy by LSND [11,12] and KARMEN [13]. With much lower accuracy (~40%) there are data for ν+Iron [14], ν+Iodine [15], and ν+Deuterium [16].

While it is impossible to experimentally measure all of the cross sections for the hundreds of weak interaction rates needed for realistic simulations of supernovae and supernova nucleosynthesis, a strategically chosen set of measurements will validate the fundamental nuclear structure and reaction models at the foundation of the hundreds of rate computations that are input for supernova models. Total neutrino-nucleus charged-current cross sections at low energy depend strongly on the charge number of the nucleus. For example the cross section in Lead is predicted to be 300 times that of Carbon. The charged-current reaction cross section induced by ν_e is thought to scale nearly as the square of the electron energy and is particularly sensitive to the detailed structure of the induced nuclear excitation spectrum. It is therefore important to either obtain the cross sections directly from experiment and/or calibrate theoretical models so that systematic uncertainties can be estimated.

In addition to being crucially important to astrophysics, measurements of neutrino-nucleus cross sections provide an opportunity to study issues of vital interest to nuclear structure related to the weak interaction. One of these involves understanding of the axial-vector response in nuclei. That response is often characterized by a modification of the axial vector coupling constant, in comparison to its value for free nucleons. For the Gamow-Teller operator that arises as a leading term from the low-energy (long wavelength) expansion of the weak axial current, one finds empirically that the effective axial-vector coupling constant is modified (quenched) in the nuclear medium. It is unknown to what extent other operators, of higher multipolarity, arising from the other terms in the expansion of the weak axial current are analogously modified. Using neutrinos from the SNS to probe medium energy strength distributions in neutrino-nucleus scattering would open the possibility of investigating this fundamental problem. Such measurements could be performed on target materials for which low-energy charge-exchange experiments are available in order to compare low-energy excitations.

Estimations show that a ten ton active mass detector located at twenty meters from the SNS target can detect thousands of neutrino-nucleus interaction per year, therefore enabling an extensive program of low energy neutrino-nucleus interactions. Details of the detector types that can be deployed can be found elsewhere [17], [18], [19].

4. Coherent Elastic Neutral Current Neutrino-Nucleus Scattering

Coherent elastic neutral current neutrino-nucleus scattering [20], [21] has never been observed. In this process, a neutrino of any flavor elastically scatters off a nucleus with low momentum transfer such that the nucleon wavefunction amplitudes are in phase and add coherently. The cross-section for such a process can be calculated exactly in the framework of the standard model, and for spin-zero nucleus, neglecting radiative corrections, is given by [22],

$$(d\sigma / dx)_{\nu A} = \frac{G_F^2 Q_W^2}{8\pi} F^2(2ME)M[2 - \frac{ME}{k^2}]$$

where k is the incident neutrino energy, E is nuclear recoil energy, M is nuclear mass, F is the ground state elastic form factor, Q_w is the weak nuclear charge, and G_F is the Fermi Constant. The condition for coherence is largely satisfied for neutrino energies up to ~50 MeV for medium mass nuclei.

For neutrino energies up to ~50 MeV, typical values of the total coherent elastic cross-section are of the order of 10^{-39} cm^2, which is ten times larger than inverse beta decay on protons and a thousand times larger then elastic neutrino- electron scattering. In spite of its large cross-section this scattering is hard to observe due to the very low nuclear recoil energies, typically in tens of keV.

Usually nuclear power plants with their large antineutrino fluxes have been considered as prime locations for the detection of coherent elastic neutral current neutrino-nucleus scattering [23],[24],[25]. Recently [26] it has been pointed that SNS has several key advantages relative to the nuclear plants for the detection of this process. Higher neutrino energies at SNS relative to the nuclear reactors give large recoil energies. This brings detection of such reaction within the reach of the current generation of typical dark mater detectors. The large duty cycle of SNS creates favorable background suppression conditions that are not available at reactors. In addition signals from neutrons will show clear decay curves with the lifetime of the muon. This can clearly manifest detection of coherent scattering with unambiguous separation from beam related background. Details of the expected signal and rates can be found in [26] together with analysis of possible tests of the standard model that can be done via this process. Presently the CLEAR collaboration is proceeding with a proposal to deploy an ~50 kg noble liquid detector near the SNS target station.

5. Tests Of The Standard Model

The electron spectrum from the reaction $^{12}C(\nu_e,e^-)^{12}N_{gs}$ can be used to derive the original ν_e spectrum from muon decay, taking into account $E_\nu = E_e+Q$ (17.8 MeV), the detector response function, and the $(E_{\nu e} - Q)^2$ dependence of the differential cross section. It is well known that a measurement of Michel parameters in muon decay is a direct test of the Standard Model and is a method to search for manifestations of new physics since these parameters are sensitive to the Lorentz structure of the Hamiltonian of weak interactions. The Michel parameters are related to the electron spectrum shape in muon decay. The neutrino spectrum from this decay can be described in terms of similar parameters. In this manner the neutrino spectrum can provide complementary information to the set of Michel parameters, increasing the accuracy and reliability in the search for new physics in the lepton sector.

As pointed out in [27], the shape of the ν_e spectrum from μ^+ decay at rest is sensitive to scalar and tensor admixtures to pure V-A interactions due to the parameter ω_L which is analogous to the Michel parameter ρ, which determines the shape of the electron spectrum in muon decay. The ν_e spectrum can be written as:

$$dN_{\nu_e}/dx = \frac{G_F^2 m_\mu^5}{16\pi^3} Q_L^\nu (G_{0(x)} + G_1(x) + \omega_L G_2(x))$$

where m_μ is the muon mass, $x = 2E_\nu / m_\mu$ is the reduced neutrino energy, $Q^\nu{}_L$ is the probability for emission of a left-handed electron neutrino, G_0 describes pure V-A interactions, G_1 takes into account radiative corrections, and $\omega_L G_2$ includes effects of scalar and tensor components. G_1 is very small and can be neglected

In the Standard Model, ω_L is exactly zero. The KARMEN experiment [28] determined an upper limit for ω_L: $\omega_L \leq 11\%$ (90% CL). Our calculations show that at SNS, due to the much higher neutrino flux, even with a modest 20 t liquid scintillator detector the limit on ω_L can be lowered down to the few percent level.

6. Neutrino Oscillations and Lepton Flavour Number Violation in Muon Decay

Production of $\bar{\nu}_e$ is strongly suppressed at SNS relative to other neutrino species. This is due to the very fast stopping and absorption of negative pions and muons at the SNS bulk mercury target. Calculations show that ratio of electron antineutrino to electron neutrino can be as low as a few per 10,000. It was pointed out a long time ago that due to this fact SNS is extremely attractive for the search for neutrino oscillations [29], [30]. Recently this interest has been revived by part of MiniBOONE collaboration [31]. A Kiloton class detector located at the distance of 60-70 meters can easily, in one year, unambiguously test the LSND claim.

Another interesting aspect that can be explored due to the low electron antineutrino rate is lepton number violation in muon decay. This has been clearly demonstrated in the neutrino sector in the form of neutrino flavor oscillations. Many extensions to the standard model, e.g. grand unified theories [32] and left-right symmetric models [33], violate lepton flavor number in exotic decays of the muon, e.g. $\mu^+ \to e^+ + \bar{\nu}_e + \nu_\mu$, which may be present at branching ratios as high as 10^{-4} [34]. SNS provides a special opportunity to search for lepton flavor number violating processes due to the small flux of electron antineutrinos and large fluxes of electron neutrinos produced. Electron antineutrinos can be easily observed by the inverse beta decay on proton reaction in liquid scintillator detectors located in close proximity to the SNS target.

7. Conclusion

For this wealth of reasons, it is clear that a neutrino program at SNS would make an important contribution to the field of neutrino physics. Timely realization of this program will have a significant impact on Astrophysics, Nuclear Physics and High-Energy physics.

References

[1] Bahcall J, 2000 Proc. *Carolina Symposium on Neutrino Physics (Columbia)* p 56.
Bellini G, 2000 Proc. *Carolina Symposium on Neutrino Physics (Columbia)* p 171.
[2] Acker A, Anjan, and Pakvasa S, 1992 *Phys. Lett. B* ***285*** p 371.
[3] Benatti F and Floreanini 2000 *JHEP* ***0002*** p 32.
[4] Araki T *et al.* 2005 *Nature* ***436*** p 499.
[5] Bernstein A, 2002 *J. Appl. Phys.* ***91*** p 4672.
[6] Avignone F and Efremenko Yu, 1999 *Nucl Phys Proc Suppl* ***87*** p 304.
[7] C. Athanassopoulos *et al.*, 1996 *Nucl. Instrum. Meth. A* ***388*** p 149.
[8] Mezzacappa A, 2000 Proc. *Carolina Symposium on Neutrino Physics (Columbia)* p 36.
[9] Avignone F, Gudkov V, Chatterjee L, Efremenko Yu, Strayer M, 2001 *Nucl.Phys.Proc.Suppl.* ***91*** p

Journal of Physics: Conference Series **173** (2009) 012006 doi:10.1088/1742-6596/173/1/012006

113.
[10] Langanke K *et al.*, 2003, *Phys. Rev. Lett.* ***91*** p 241102.
[11] Athanassopoulos C, *et al.*, 1997 *Phys. Rev. C* ***55*** p 2078.
[12] Auerbach L B, e*t al.*, 2001 *Phys. Rev. C* ***64*** p 065501.
[13] Bodmann B, *et al.*, 1992 *Phys. Lett. B* ***280*** p 198.
[14] Maschuw R, *at al.*, 1998 *Prog. Part Phys.* ***40*** p 183.
[15] Distel J R, *et al.*, 2003 *Phys. Rev. C* ***68*** p 054613.
[16] Willis S E, *et al.*, 1980 *Phys. Lett.* ***44*** p 522.
[17] Avignone F and Efremenko Yu, *2003 J. Phys .G* ***29*** *p 2615.*
[18] Efremenko Yu, 2005 *Nucl Phys Proc Suppl* ***138*** p 343.
[19] ν-SNS collaboration 2005, Proposal for a Neutrino Facility at the Spallation Neutron Source, Oak Ridge.
[20] Freedman D, Schramm D, and Tubbs D, 1997 *Ann. Rev. Nucl. Part. Sci.* ***27*** p 167.
[21] Drukier A and Stodolsky L, 1984 *Phys. Rev. D* ***30*** p 2295.
[22] Horowitz C J, Coalkey K J, and McKinsey D N, 2003 *Phys. Rev. D* ***68*** p 023005.
[23] Cabrera B, Krauss L M, and Wilczek F, 1984 *Phys. Rev. Lett.****55*** p 25.
[24] Barbeau P, Collar J I, Miyamoto J, Shipsey I, 2003 *IEEE Trans. Nucl. Sci.* ***50*** p 1285.
[25] Hagmann C and, Bernstein A, 2003 *IEEE Trans. Nucl. Sci.* ***51*** p 2151.
[26] Scholberg K, 2006 *Phys. Rev. D* ***73*** p 033005.
[27] W. Fetshcher, 1992 *Phys. Rev. Lett.* **69**, p 2758.
[28] B. Armbruster *at al.*,1998 *Phys. Rev. Lett.* **81**, p 520.
[29] Bugg B *at al.*, 1998 *AIP Conf. Proc.* ***412*** p 141.
[30] VanDalen G J, 2003 nucle-ex/039014.
[31] Garvey G T *at al*, 2005 *Phys. Rev. D* ***72*** p 092001.
[32] Fujii H *at al*, 1994 *Phys. Rev. D* ***49*** p 559.
[33] Herczeg P and Mohapatra R N, 1992 *Phys. Lett.* ***69*** p 2475.
[34] Herczed P, 1997 *Los Alamos Science* ***25*** p 28.

Journal of Physics: Conference Series **173** (2009) 012007 doi:10.1088/1742-6596/173/1/012007

The MAJORANA Project

S.R. Elliott[1], C.E. Aalseth[2], M. Akashi-Ronquest[3,4], M. Amman[5], J.F. Amsbaugh[6], F.T. Avignone III[7,8], H.O. Back[9,4], C. Baktash[7], A.S. Barabash[10], P. Barbeau[11], J.R. Beene[7], M. Bergevin[12], F.E. Bertrand[7], M. Boswell[3,4], V. Brudanin[13], W. Bugg[14], T.H. Burritt[6], Y-D. Chan[12], T.V. Cianciolo[7], J. Collar[11], R. Creswick[8], M. Cromaz[12], J.A. Detwiler[12], P.J. Doe[6], J.A. Dunmore[6], Yu. Efremenko[14], V. Egorov[13], H. Ejiri[15], J. Ely[2], J. Esterline[16,4], H. Farach[8], T. Farmer[2], J. Fast[2], P. Finnerty[3,4], B. Fujikawa[12], V.M. Gehman[1], C. Greenberg[11], V.E. Guiseppe[1], K. Gusey[13], A.L. Hallin[17], R. Hazama[15], R. Henning[3,4], A. Hime[1], T. Hossbach[2], E. Hoppe[2], M.A. Howe[6], D. Hurley[12], B. Hyronimus[2], R.A. Johnson[6], M. Keillor[2], C. Keller[21], J. Kephart[9,4,2], M. Kidd[16,4], O. Kochetov[13], S.I. Konovalov[10], R.T. Kouzes[2], K.T. Lesko[18,19], L. Leviner[9,4], P. Luke[5], S. MacMullin[3,4], M.G. Marino[6], A.B. McDonald[20], D.-M. Mei[21], H.S. Miley[2], A.W. Myers[6], M. Nomachi[15], B. Odom[11], J. Orrell[2], A.W.P. Poon[12], G.Prior[12], D.C. Radford[7], J.H. Reeves[2], K. Rielage[1], N. Riley[11], R.G.H. Robertson[6], L. Rodriguez[1], K.P. Rykaczewski[7], A.G. Schubert[6], T. Shima[15], M. Shirchenko[13], V. Timkin[13], R. Thompson[2], W. Tornow[16,4], C. Tull[12], T. D. Van Wechel[6], I. Vanyushin[10], R.L. Varner[7], K. Vetter[12,22], R. Warner[2], J.F. Wilkerson[6]J.M. Wouters[1], E. Yakushev[13], A.R. Young[9,4], C.-H. Yu[7], V. Yumatov[10], Z.-B. Yin[21],

[1] Physics Division, Los Alamos National Laboratory, Los Alamos, NM, USA
[2] Pacific Northwest National Laboratory, Richland, WA, USA
[3] Dept. of Physics and Astron., University of North Carolina, Chapel Hill, NC, USA
[4] Triangle Universities Nuclear Laboratory, Durham, NC, USA
[5] Engineering Division, Lawrence Berkeley National Laboratory, Berkeley, CA, USA
[6] Center for Nucl. Phys. and Astrophys., Univ. of Washington, Seattle, WA, USA
[7] Oak Ridge National Laboratory, Oak Ridge, TN, USA
[8] Dept. of Physics and Astronomy, University of South Carolina, Columbia, SC, USA
[9] Department of Physics, North Carolina State University, Raleigh, NC, USA
[10] Institute for Theoretical and Experimental Physics, Moscow, Russia
[11] University of Chicago, Chicago, IL, USA
[12] Nuclear Science Division, Lawrence Berkeley National Lab., Berkeley, CA, USA

[13] Joint Institute for Nuclear Research, Dubna, Russia
[14] Dept. of Physics and Astronomy, University of Tennessee, Knoxville, TN, USA
[15] Res. Center for Nucl. Phys. & Dept. of Phys., Osaka Univ., Ibaraki, Osaka, Japan
[16] Department of Physics, Duke University, Durham, NC, USA
[17] Centre for Particle Physics, University of Alberta, Edmonton, Alberta, Canada
[18] Inst. for Nucl. & Part. Astro., Lawrence Berkeley Natl. Lab., Berkeley, CA, USA
[19] Physics Department, UC Berkeley, Berkeley, CA, USA
[20] Dept. of Physics, Queen's University at Kingston, Kingston, Ontario, Canada
[21] Dept. of Earth Science and Phys., Univ. of South Dakota, Vermillion, SD, USA
[22] Department of Nuclear Engineering, UC Berkeley, Berkeley, CA, USA

E-mail: `elliotts@lanl.gov`

Abstract. Building a $0\nu\beta\beta$ experiment with the ability to probe neutrino mass in the inverted hierarchy region requires the combination of a large detector mass sensitive to $0\nu\beta\beta$, on the order of 1-tonne, and unprecedented background levels, on the order of or less than 1 count per year in the $0\nu\beta\beta$ signal region. The MAJORANA Collaboration proposes a design based on using high-purity enriched ^{76}Ge crystals deployed in ultra-low background electroformed Cu cryostats and using modern analysis techniques that should be capable of reaching the required sensitivity while also being scalable to a 1-tonne size. To demonstrate feasibility, the collaboration plans to construct a prototype system, the MAJORANA DEMONSTRATOR, consisting of 30 kg of 86% enriched ^{76}Ge detectors and 30 kg of natural or isotope-76-depleted Ge detectors. We plan to deploy and evaluate two different Ge detector technologies, one based on a p-type configuration and the other on n-type.

1. Introduction

This is an exciting time in our quest to understand neutrinos — fundamental particles that play key roles in the early universe, in cosmology and astrophysics, and in nuclear and particle physics. Recent results from atmospheric, solar, and reactor-based neutrino oscillation experiments (Super-Kamiokande, SNO, and KamLAND)[1, 2, 3] have provided compelling evidence that neutrinos have mass and give the first indication after nearly forty years of study that the Standard Model (SM) of nuclear and particle physics is incomplete.

With the realization that neutrinos are massive, there is an increased interest in investigating their intrinsic properties. Understanding the neutrino nature (Majorana or Dirac), the neutrino mass generation mechanism, the absolute neutrino mass scale and the neutrino mass spectrum are some of the main focii of future neutrino experiments.

Lepton number, L, is conserved in the Standard Model because neutrinos are assumed to be massless and there is no chirally right-handed neutrino field. The guiding principles for extending the Standard Model are the conservation of electroweak isospin and renormalizability, which do not preclude each neutrino mass eigenstate ν_i to be identical to its anti-particle $\bar{\nu}_i$, or a "Majorana" particle. However, L is no longer conserved if $\nu = \bar{\nu}$. Theoretical models, such as the seesaw mechanism that can explain the smallness of neutrino mass, favor this scenario. Therefore, the discovery of Majorana neutrinos would have profound theoretical implications in the formation of a new Standard Model while possibly yielding insights into the origin of mass itself. If neutrinos are Majorana particles, they may fit into the leptogenesis scenario for creating the baryon asymmetry, and hence ordinary matter, of the universe. As of yet, there is no firm experimental evidence to confirm or refute this theoretical prejudice.

Many even-even nuclei are forbidden to undergo β decay, but are unstable with respect to the second order weak process of two-neutrino double beta decay ($2\nu\beta\beta$). In this process, the nucleus emits 2 β particles and 2 $\bar{\nu}$. This is an allowed process and has been observed. A similar process, neutrinoless double-beta decay ($0\nu\beta\beta$) can occur if a neutrino is exchanged between two neutrons and no neutrinos are emitted. In this case, the only leptons in the final state are the two electrons and hence the decay violates L by two units. This exchange of the neutrino however, will only occur if the neutrino is a massive Majorana particle and therefore experimental evidence of $0\nu\beta\beta$ would establish the Majorana nature of neutrinos. The science of $\beta\beta$ has been described in detail in many recent reviews [4, 5, 6, 7, 8, 9].

The most-restrictive upper limits on the $0\nu\beta\beta$ half-life come from Ge detector experiments [10, 11]. The half-life is $>1.9\times10^{25}$ y and this corresponds to an effective Majorana mass limit from $0\nu\beta\beta$ greater than about 400 meV depending on the choice of nuclear matrix element. The oscillation experiments, however, indicate that at least one of the neutrino mass eigenvalues is greater than about 45 meV. In the inverted hierarchy, this would imply an effective Majorana neutrino mass 45 meV or greater. The predicted half-life for $0\nu\beta\beta$ with an effective mass in this region is greater than 10^{27} years. An experiment will require approximately 1 tonne of isotope to be sensitive to such a long half-life, and it will require a background level of 1 count/tonne-year or better.

2. The Majorana Demonstrator

The objective of the first experimental phase of Majorana is to build a 60-kg module of high-purity Ge, of which 30 kg will be enriched to 86% in ^{76}Ge, to search $0\nu\beta\beta$. This module is referred to by the collaboration as the Demonstrator. The physics goals for this first phase are to:

(i) Probe the neutrino mass region above 100 meV

(ii) Demonstrate that backgrounds at or below 1 count/tonne-year in the $0\nu\beta\beta$-decay region of interest can be achieved that would justify scaling up to a 1 tonne or larger mass detector.

(iii) Definitively test the recent claim[12] of an observation of $0\nu\beta\beta$ decay in ^{76}Ge in the mass region around 400 meV.

The half-life of $0\nu\beta\beta$ is at least 10^{25} years, making it an extremely difficult process to observe and therefore very sensitive experiment is required. The Majorana Demonstrator will consist of ^{76}Ge detectors, deployed in multi-crystal modules, located deep underground within a low-background shielding environment. The technique will be augmented with recent improvements in signal processing and detector design, and advances in controlling intrinsic and external backgrounds. The justification for a detector mass size of 60 kg is directly linked to all three of the above-stated science goals. The enriched Ge is handled differently than natural Ge and the isotopic makeups are different. Since both these issues lead to background differences, enriched material must also be used to validate the background model.

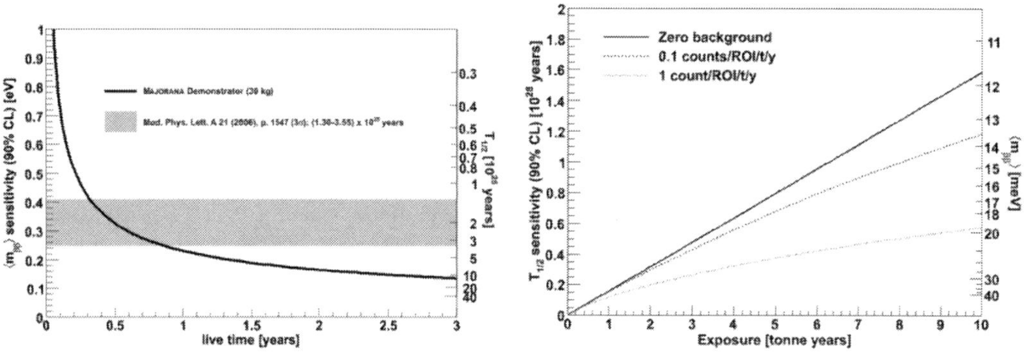

Figure 1. The left figure shows the sensitivity of the MAJORANA DEMONSTRATOR module as a function of time and indicates the range of values indicated by a recent controversial claim[12]. The right figure shows the sensitivity of a 1-ton experiment as a function of exposure for 3 different background levels. Note the factor of 1000 difference in the scales of the two panels. The matrix elements used are from Ref. [13]

The goal of the proposed MAJORANA DEMONSTRATOR is the construction of an instrument that provides sufficient sensitivity to test the recent claim, that allows a path forward to achieving a background level below 1 count/tonne-year in the 4-keV region of interest, and is scalable towards a large-scale instrument. We plan a module consisting of three cryostats each containing about 20 kg of HPGe detectors. One module will consist of 14 n-type segmented-contact (N-SC) HPGe detectors, while the other two modules will each consist of 28 smaller p-type point-contact (P-PC) HPGe detectors without segmentation. Approximately 40 of the P-PC detectors will be built of ~30 kg of enriched ^{76}Ge material. Employing these two complementary detector configurations allows us to evaluate and compare the two most promising implementations under the realistic conditions of an ultra-low background experiment. The multi-cryostat approach allows us to optimize each individual implementation, providing a fast deployment with minimum interference with already operational detectors. In addition, it allows us to separate both configurations in the data analysis. The proposed configuration is illustrated in Figure 2. Our initial emphasis will be on the first cryostat assembled with natural-Ge P-PC detectors.

The efficient commercial production of Ge detectors depends strongly on the yields for producing high-purity single crystals of Ge and for producing p-n junction diodes from these crystals. Single-crystal boules naturally favor right circular cylindrical geometries, making coaxial detectors the natural choice to efficiently use the precious enriched Ge material. To minimize the passivated surface area and maximize the sensitive volume, coaxial detectors are typically fashioned in so-called closed-end geometries with a bore hole extending ~80% along the detector axis. This bore hole, with a central electrode, is necessary to produce sufficiently high electric fields to achieve not only depletion throughout the crystal but also high drift velocity of the charge carriers to minimize

Journal of Physics: Conference Series **173** (2009) 012007 doi:10.1088/1742-6596/173/1/012007

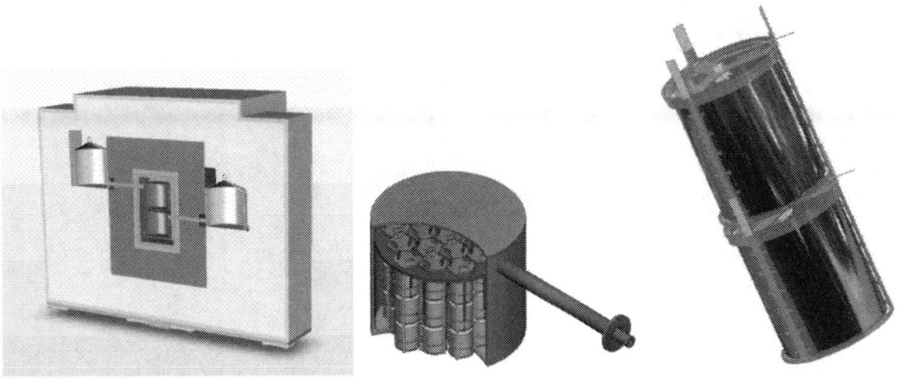

Figure 2. Left: Setup of proposed MAJORANA DEMONSTRATOR. The cryostats will be built of ultrapure electroformed copper. The inner passive shield will also be constructed of electroformed copper surrounded by lead, which itself is surrounded by an active muon veto and neutron moderator. Middle: A conceptual design of a 3x19 detector arrangement of Ge detectors mounted in one cryostat, envisioned as a building block for a tonne-scale experiment. Right: A conceptual design of a two N-SC detector string.

charge trapping effects.

Previous-generation $0\nu\beta\beta$ decay searches in ^{76}Ge favored detectors made of p-type material primarily because p-type crystals can be grown more efficiently to larger dimensions than n-type crystals. In addition, with p-type detectors trapping effects are reduced, and generally slightly better energy resolution is obtained. P-type detectors are typically fabricated with a p+ B-implanted contact on the inner bore hole and an n+ Li-diffused contact on the outside electrode to obtain efficient and full depletion from the outside. While the inside B contact is very thin (typically <1 μm) the Li-diffused contact is typically more then 100 μm thick and can be up to 1 mm or more depending on the fabrication process and any subsequent annealing or prolonged storage at room temperatures. The thickness of the outside-Li contact provides an advantage since this dead layer absorbs alpha-particle radiation from surface contamination that may otherwise generate background for ^{76}Ge $0\nu\beta\beta$ decay.

While the pulse-shape obtained at the central contact in coaxial detectors[14] can provide radial separation of multiple interactions in any implementation, pulse-shapes obtained in segmented detectors[15, 16] provide improved sensitivity in the radial separation and, more importantly, in complementary directions as well. A high-degree of segmentation, such as a 6x6-fold segmentation, enables the full reconstruction of γ-ray interactions within the detectors. GRETINA in particular has provided very encouraging analyses that events can be reconstructed with a position uncertainty of $\sim$2 mm [17]. Even without absolute event vertex reconstruction, events with multiple energy depositions can be identified and rejected. Preliminary results from GRETINA and other highly-segmented HPGe arrays indicate that a minimum separation of 4 mm will be achievable. However, these advanced capabilities come at the price of a

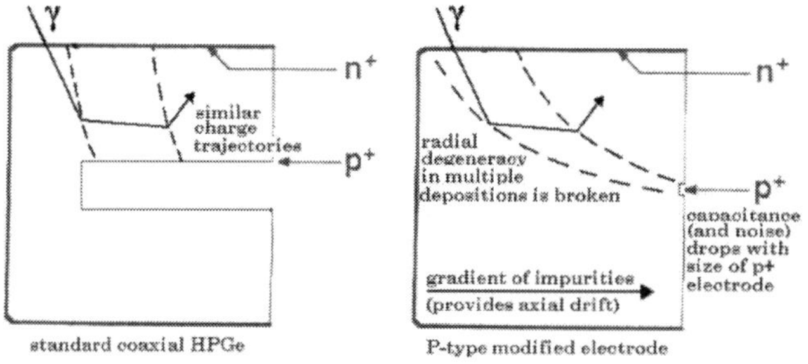

Figure 3. Closed-end coaxial p-type Ge detector (left) versus the p-type point contact Ge detector implementation (right). Figure is adapted from Ref. [19]

proportionally larger number of small parts such as cables near the detector, and their selection must be optimized against their ability to overcome added backgrounds.

An alternative right circular cylindrical detector design has been developed in which the bore hole is removed and in its place a point-contact, either B implanted or Li diffused, is formed at the center of the intrinsic (open end) detector surface [18]. The changes in the electrode structure result in a drop in capacitance to $\sim$1 pF, reducing the electronics noise component and enabling sub-keV energy thresholds. This point-contact configuration also has lower electric fields throughout the bulk of the crystal and a weighting potential that is sharply peaked near the point contact. (Earlier attempts employing n-type material exhibited poor energy resolution at high energies due to increased charge trapping.) This in turn results in an increased range of drift times and a distinct electric signal marking the arrival of the charge cloud at the central electrode. Figure 3 illustrates the p-type point-contact (P-PC) implementation in contrast to the conventional, closed-end coaxial detector approach. Instrumented with a modern FET, a P-PC detector was recently demonstrated by MAJORANA collaborators to provide low noise, resulting in a low energy trigger threshold and excellent energy resolution, as well as excellent pulse-shape capabilities to distinguish multiple interactions [19]. This detector was developed with the goal of detecting the very soft (sub-keV) recoils expected from coherent neutrino-nucleus scattering in a reactor experiment.

The P-PC detector in [19] has a diameter of 5 cm and a length of 5 cm resulting in a total mass of 0.5 kg. While it may be possible to expand the radial dimension of these detectors, it will be more difficult to expand the axial dimension due to the requirement of a significant impurity gradient between the detector faces. Without an impurity concentration gradient, the electric fields can be sufficiently weak that the detector is susceptible to excessive charge trapping, even in the p-type configuration. It is interesting to note that increasing the voltage, *e.g.* beyond full depletion, will not significantly increase the electric field in the detector bulk away from the point contact. In contrast, the presence of space charge due to an appropriate impurity concentration and gradient is able to increase the electric field throughout the crystal. The first attempt to make a P-PC detector with a flat impurity concentration failed. The detector in [19] resulted from a second attempt with a factor of two difference in impurity concentration between the detector faces. The collaboration has recently purchased and received several additional P-PC detectors from various vendors.

Although the detector mounts and feedthroughs require different designs, the overall dimensions of the cryostats can be similar, reducing the design effort and simplifying production. Fabricating multiple, similar cryostats will provide an opportunity to assess several risks associated with a large-scale modular deployment for a 1-tonne experiment. These risks include the reproducibility of the mechanical and thermal integrity of the cryostats, as well as the ability to consistently achieve and maintain radiopurity throughout the module production and assembly chain. Figure 2 shows the conceptual design of a larger, 57-detector cryostat currently envisioned for a tonne-scale modular deployment of ^{76}Ge detectors.

Following the selection of two different detector designs, *i.e.* the N-SC and P-PC configurations, the mechanical components within the cryostat will be optimized independently. However, the overall dimensions and wall thicknesses can be the same since each cryostat will contain about 20 kg of Ge detectors. Our reference design is a cylindrical vessel with a thick cold plate at the top from which the detectors hang. A thermal shroud mounted to the cold plate provides radiative cooling for the detector array. The cryostat, cold plate and thermal shroud will be fabricated using ultra-low-background, relatively-thick electroformed copper to eliminate any concern about collapse of the vacuum vessel. The coldplate shields the germanium detectors from backgrounds originating in the front-end electronics; thus, a thicker coldplate is favored. The cryostats themselves are situated within a thick layer of electroformed Cu comprising the innermost passive shielding layers, so extra cryostat wall and coldplate thickness is not a concern from the standpoint of backgrounds. However, thicker cryostat walls do reduce the potential for inter-cryostat granularity rejection. A combination of thermal, mechanical and background analyses will be used to determine the preferred thickness of these components for the cryostats in the DEMONSTRATOR with the upper limit being set by considerations of the copper electroforming process (time and material quality).

The detector mounts for each configuration will be different, as will the number of cables and feedthroughs. While the assembly and readout for the central channel in the N-SC configuration will be similar to the P-PC approach, all electronics components for the segment signals will be located outside the cryostat, at the maximum distance possible to minimize radioactive background in the detectors. This distance is constrained by the bandwidth and noise requirements for the pulse-shape analysis to achieve sufficent position sensitivity. Using warm FETs for the segments also reduces the heat load to the cryostat. Nevertheless, beyond the feedthroughs for high voltage, test input, feedback, ground, and signal for the central channel for each detector, N-SC detectors require an additional feedthrough for each outer segment of each crystal, resulting in an additional thermal burden to the cryostat and background burden to the detectors. Assuming, for comparison, 28 750-g P-PC detectors arranged in 4 layers of 7 detectors for each of the two cryostats, each P-PC cryostat will have 140 feedthroughs. In contrast, a cryostat with 14 1.5-kg N-SC 36-segment detectors arranged in 2 layers of 7 detectors will have 574 feedthroughs.

In both configurations, the detectors will be mounted in a string-like arrangment as shown in Figure 2 for the N-SC configuration. This reference design consists of a thick copper "lid", a copper support frame, and low-mass Teflon support trays or standoffs. Each individual detector is mounted in a separate frame built of electroformed Cu that eases handling of individual detectors during shipping, acceptance testing and string assembly. The assembled detector string is simply lowered through a hole in the cryostat cold plate until the string lid sits on the cold plate. This allows us to easily mount and dismount individual detector strings from the top. Cables are run vertically from the detector contacts through a slit in the string lid, above which low-background front-end electronics packages are mounted to read out the central contact. For N-SC detectors, HV blocking capacitors will also be mounted above the lid. The string lid provides some shielding between these components and the detectors, and allows the bandwidth to be maximized by placing the central contact FETs near the germanium detectors. The front-end and the outer contact leads are routed out of the detector along the cold finger.

The MAJORANA DEMONSTRATOR cryostats will be enclosed in a graded passive shield and an active muon veto to eliminate external backgrounds. Shielding reduces signals from γ rays originating in the experiment hall, cosmic-ray μ's penetrating the shielding, and cosmic-ray μ-induced neutrons. The strategy is to provide extremely low-activity material for the inner shield. Surrounding this will be an outer shield of bulk γ-ray shielding material with lower radiopurity. This high-Z shielding enclosure will be contained inside a gas-tight Rn exclusion box made of stainless-steel sheet. Outside this bulk high-Z shielding will be a layer of hydrogenous material, some of which will be doped with a neutron absorber such as boron, intended to reduce the neutron flux. Finally, active cosmic-ray anti-coincidence detectors will enclose the entire shield. The MAJORANA collaboration plans to site the DEMONSTRATOR deep underground in the Sanford Laboratory at the Homestake gold mine in Lead, South Dakota, USA.

The data acquisition software system will be constructed using the Object-oriented Real-time Control and Acquisition (ORCA) [20] application to achieve the goal of providing a general purpose, highly modular, object-oriented, acquisition and control system that is easy to develop, use, and maintain. The object-oriented nature of ORCA enables a user to configure it at run-time to represent different hardware configurations and data read-out schemes by dragging items from a catalog of objects into a configuration window.

Monte Carlo (MC) radiation transport simulation models have been developed for MAJORANA using MaGe [21], an object-oriented MC simulation package based on ROOT [22] and the Geant4 [23, 24] toolkit and optimized for simulations of low-background germanium detector arrays. MaGe is being jointly developed by the MAJORANA and GERDA [25] collaborations in consultation with collaborators from the National Energy Research Scientific Computing Center (NERSC) at LBNL. MaGe defines a set of physics processes, materials, constants, event generators, etc. that are common to these experiments, and provides a unified framework for geometrical definitions, database access, user interfaces, and simulation output schemes in an effort to reduce repetition of labor and increase code scrutiny.

3. Conclusion

The MAJORANA collaboration is carrying out R&D to develop a 1-tonne $0\nu\beta\beta$ experiment based on enriched ^{76}Ge. The Ge detectors will be housed in a classic cryostat design that is constructed from ultra low-radioactivity materials. The cryostats will be contained within a passive shield, and active veto system and operated deep underground. The project should demonstrate that this technology will be low enough in background to verify the feasibility of constructing a 1-tonne Ge-based $0\nu\beta\beta$ experiment.

Acknowledgments

This conference was organized to celebrate the 75th birthdays of Frank Avignone, Etorre Fiorini and Peter Rosen. The entire field of $\beta\beta$ research has benefitted from the long and storied efforts of Frank and Etorre. I admire their scientific leadership at this time as we celebrate their careers. The contributions to this field by Peter Rosen are also paramount. It is with fondness that we remember his legacy, but with sadness that we can't also share his 75th birthday at this event.

The author wishes to acknowledge the contributions of the MAJORANA Collaboration. Our efforts have also benefitted from a close cooperation with the GERDA collaboration. The preparation of this manuscript was supported in part by Laboratory Directed Research and Development at LANL.

References

[1] Ashie Y *et al.* 2004 *Phys. Rev. Lett.* **93** 101801
[2] Ahmed S N *et al.* 2004 *Phys. Rev. Lett.* **92** 181301
[3] Araki T *et al.* 2005 *Phys. Rev. Lett.* **94** 081801
[4] Elliott S R and Vogel P 2002 *Ann. Rev. Nucl. Part. Sci.* **52** 115
[5] Elliott S R and Engel J 2004 *J. Phys. G: Nucl. Part. Phys.* **30** R 183
[6] Barabash A S 2004 *Phys. Atom. Nucl.* **67** 438
[7] Ejiri H 2005 *J. Phys. Soc. Jap.* **74** 2101
[8] Avignone F T III, King G S III and Zdesenko Y 2005 *New Journal of Physics* **7** 6
[9] Avignone F T III, Elliott S and Engel J 2008 *Rev. Mod. Phys.* **80** 481 (*Preprint* `arXiv:0708.1033`)
[10] Baudis L *et al.* 1999 *Phys. Rev. Lett.* **83** 41
[11] Aalseth C E *et al.* (IGEX) 2002 *Phys. Rev. D.* **65** 092007
[12] Klapdor-Kleingrothaus H V and Krivosheina I V 2006 *Mod. Phys. Lett. A* **21** 1547
[13] Rodin V A, Faessler A, Šimkovic F and Vogel P 2006 *Nucl. Phys. A* **766** 107 erratum arXiv:0706.4304
[14] Aalseth C E, Avignone F T, Brodzinski R L, Miley H S and Reeves J H 1998 *J. Radioanal. Nucl. Chem.* **233** 119
[15] Elliott S, Gehman V, Kazkaz K, Mei D M and Young A 2006 *Nucl. Instrum. and Meth. A* **588** 504
[16] Abt I *et al.* 2006 (*Preprint* `nucl-ex/0701005`)
[17] Vetter K *et al.* 2000 *Nuclear Instruments and Methods in Physics Research A* **452** 105–114
[18] Luke P, Goulding F, Madden N and Pehl R 1989 *IEEE trans. Nucl. Sci.* **36** 926
[19] Barbeau P, Collar J and Tench O 2007 *JCAP* **09** 009
[20] Howe M, Cox G, Harvey P, McGirt F, Rielage K, Wilkerson J and Wouters J 2004 *IEEE Transactions on Nuclear Science* **51** 878–883
[21] Y-D Chan *et al.* 2008 (*Preprint* `arXiv:0802.0860v1`)
[22] Brun R and Rademakers F 1997 *Nuclear Instruments and Methods in Physics Research A* **389** 81–86
[23] Agostinelli S *et al.* (GEANT4) 2003 *Nuclear Instruments and Methods in Physics Research A* **506** 250–303
[24] Allison J *et al.* 2006 *IEEE Transactions on Nuclear Science* **53**
[25] Abt I *et al.* (GERDA) 2004 (*Preprint* `LNGS LOI 35/04; hep-ex/0404039`)

NEMO 3 double beta decay experiment: latest results

Alexander Barabash (on behalf of the NEMO Collaboration)
Institute of Theoretical and Experimental Physics, B. Cheremushkinskaya 25, 117218 Moscow, Russia
E-mail: barabash@itep.ru

Abstract. The double beta decay experiment NEMO 3 has been taking data since February 2003. The aim of this experiment is to search for neutrinoless decay and investigate two neutrino double beta decay in seven different enriched isotopes (^{100}Mo, ^{82}Se, ^{48}Ca, ^{96}Zr, ^{116}Cd, ^{130}Te and ^{150}Nd). After analysis of the data corresponding to 693 days, no evidence for $0\nu\beta\beta$ decay in ^{100}Mo and ^{82}Se was found. The half-life limits at 90% C.L. are $5.8 \cdot 10^{23}$ y and $2.1 \cdot 10^{23}$ y, respectively. Using $\sim$ 940 days of data, preliminary results were obtained for ^{48}Ca, ^{96}Zr and ^{150}Nd. For two neutrino decay half-life values were measured, $T_{1/2}(^{48}\mathrm{Ca}) = [4.4^{+0.5}_{-0.4}(stat) \pm 0.4(syst)] \cdot 10^{19}$ y, $T_{1/2}(^{96}\mathrm{Zr}) = [2.3 \pm 0.2(stat) \pm 0.3(syst)] \cdot 10^{19}$ y and $T_{1/2}(^{150}\mathrm{Nd}) = [9.2^{+0.25}_{-0.22}(stat) \pm 0.62(syst)] \cdot 10^{18}$ y. Also for $0\nu\beta\beta$ decay the following limits at 90% C.L. were obtained, $> 1.3 \cdot 10^{22}$ y for ^{48}Ca, $> 8.6 \cdot 10^{21}$ y for ^{96}Zr and $> 1.8 \cdot 10^{22}$ y for ^{150}Nd.

1. Introduction

Interest in neutrinoless double-beta decay has seen a significant renewal in recent years after evidence for neutrino oscillations was obtained from the results of atmospheric, solar, reactor and accelerator neutrino experiments (see, for example, the discussions in [1, 2, 3]). These results are impressive proof that neutrinos have a non-zero mass. However, the experiments studying neutrino oscillations are not sensitive to the nature of the neutrino mass (Dirac or Majorana) and provide no information on the absolute scale of the neutrino masses, since such experiments are sensitive only to the difference of the masses, Δm^2. The detection and study of $0\nu\beta\beta$ decay may clarify the following problems of neutrino physics (see discussions in [4, 5, 6]): (i) lepton number non-conservation, (ii) neutrino nature: whether the neutrino is a Dirac or a Majorana particle, (iii) absolute neutrino mass scale (a measurement or a limit on m_1), (iv) the type of neutrino mass hierarchy (normal, inverted, or quasidegenerate), (v) CP violation in the lepton sector (measurement of the Majorana CP-violating phases).

The currently running NEMO 3 experiment is devoted to the search for $0\nu\beta\beta$ decay and to the accurate measurement of two neutrino double beta decay ($2\nu\beta\beta$ decay) by means of the direct detection of the two electrons. This tracking experiment, in contrast to experiments with ^{76}Ge, detects not only the total energy deposition, but other parameters of the process. These include the energy of the individual electrons, angle between them, and the coordinates of the event in the source plane. Since June of 2002, the NEMO 3 detector has operated in the Fréjus Underground Laboratory (France) located at a depth of 4800 m w.e. Since February 2003, after

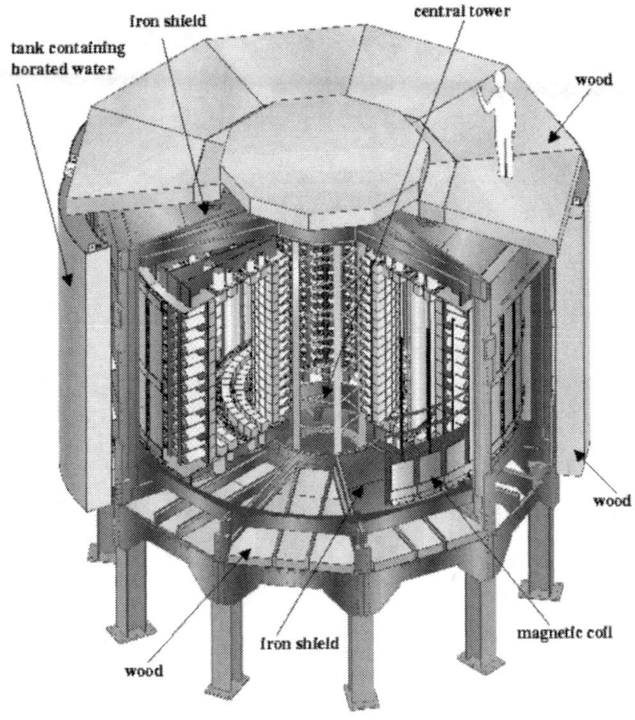

Figure 1. A schematic view of the NEMO 3 detector.

the final tuning of the experimental set-up, NEMO 3 has been taking data devoted to double beta decay studies. The first obtained results with ^{100}Mo and ^{82}Se were published in [7, 8, 9, 10].

2. The NEMO 3 detector

The NEMO 3 detector has three main components, a foil consisting of different sources of double beta decay isotopes and copper, a tracker made of Geiger wire cells and a calorimeter made of scintillator blocks with PMT readout, surrounded by a solenoidal coil. The detector has the ability to discriminate between events of different types by positive identification of charged tracks and photons. A schematic view of the NEMO 3 detector is shown in Fig. 1.

The NEMO 3 detector is cylindrical in design and is composed of twenty equal sectors. The external dimensions of the detector with shields are about 6 m in diameter and 4 m in height. NEMO 3 is based on the techniques tested on previous incarnations of the experiment NEMO 1 [11], and NEMO 2 [12].

The wire chamber is made of 6180 open octagonal drift cells which operate in Geiger mode (Geiger cells). A gas mixture of $\sim$ 95% helium, 4% ethyl-alcohol, 1% argon and 0.15% water at 10 mbar above atmospheric pressure is used as the filling gas of the wire chamber. Each drift cell provides a three-dimensional measurement of the charged particle tracks by recording the drift time and the two plasma propagation times. The transverse position is determined from the drift time, while the longitudinal position is deduced from the difference between the plasma propagation times at both ends of the cathode wires. The average vertex position resolution for the two-electron events is $\sigma_t = 0.5$ cm in the transverse plane of the detector and $\sigma_l = 0.8$ cm in the longitudinal plane. The Geiger counters information is treated by the track reconstruction program based on the cellular automaton algorithm, described in [13].

The calorimeter, which surrounds the wire chamber, is composed of 1940 plastic scintillator

Table 1. Investigated isotopes with NEMO 3 [15].

Isotope	^{100}Mo	^{82}Se	^{130}Te	^{116}Cd	^{150}Nd	^{96}Zr	^{48}Ca
Enrichment, %	97	97	89	93	91	57	73
Mass of isotope, g	6914	932	454	405	36.6	9.4	7.0

blocks coupled by light-guides to very low-radioactivity photomultiplier tubes (PMTs) developed by Hammamatsu. The energy resolution FWHM of the calorimeter ranges from 14.1 to 17.6% for 1 MeV electrons, while the time resolution is of 250 ps at 1 MeV.

The apparatus accommodates almost 10 kg of different double beta decay isotopes (see table 1). Most of these isotopes are highly enriched and are shaped in the form of thin metallic or composite foils with a density of 30-60 mg/cm^2. Three sectors are also used for external background measurement and are equipped respectively with pure Cu (one sector, 621 g) and natural Te (1.7 sectors, 614 g of ^{nat}TeO$_2$). Some of the sources, including ^{100}Mo, have been purified in order to reduce their content of ^{208}Tl (from ^{232}Th and ^{228}Th, and from ^{228}Ra with a half-life of 5.75 y) and ^{214}Bi (from ^{226}Ra with a half-life of 1600 y) either by a chemical procedure [14], or by a physical procedure [15]. The foils are placed inside the wire chamber in the central vertical plane of each sector. The majority of the detector, 12 sectors, is mounted with 6.9 kg of ^{100}Mo.

The detector is surrounded by a solenoidal coil which generates a vertical magnetic field of 25 Gauss inside the wire chamber. This magnetic field allows electron-positron identification by measuring the curvature of their tracks. The ambiguity of the e^+/e^- recognition based on the curvature reconstruction is 3% at 1 MeV.

The whole detector is covered by two types of shielding against external γ-rays and neutrons. The inner shield is made of 20 cm thick low radioactivity iron which stops γ-rays and slow neutrons. The outer shield is comprised of tanks filled with borated water on the vertical walls and wood on the top and bottom designed to thermalise and capture neutrons.

At the beginning of the experiment, the radon inside the tracking chamber, and more precisely the its decay product ^{214}Bi present in its radioactive chain was found to be the predominant background. Radon is present in the air of the laboratory and originates from the rock surrounding. It can penetrate the detector through small leaks. A tent coupled to a radon-free air factory was installed around the detector in October 2004 in order to decrease the presence of radon inside the traker.

Since February 2003, after the final tuning of the experimental set-up, NEMO 3 has routinely been taking data devoted to double beta decay studies. The calibration with radioactive sources is carried out every 6 weeks. The stability of the calorimeter is checked daily with a laser based calibration system [15].

The advantage of the NEMO 3 detector rests in its capability to identify the two electrons from $\beta\beta$ decay and the de-excitation photons from the excited state of the daughter nucleus. The NEMO 3 calorimeter also measures the detection time of the particles. The use of appropriate time-of-flight (TOF) cuts, in addition to energy cuts, allows an efficient reduction of all backgrounds.

A full description of the detector and its characteristics can be found in [15].

3. Experimental results

A candidate for a $\beta\beta$ decay is a two-electron event which is defined with the following criteria: two tracks coming from the same vertex in a source foils, the curvature of the tracks corresponds to a negative charge, each track has to be associated with a trigged scintillator, and the time-

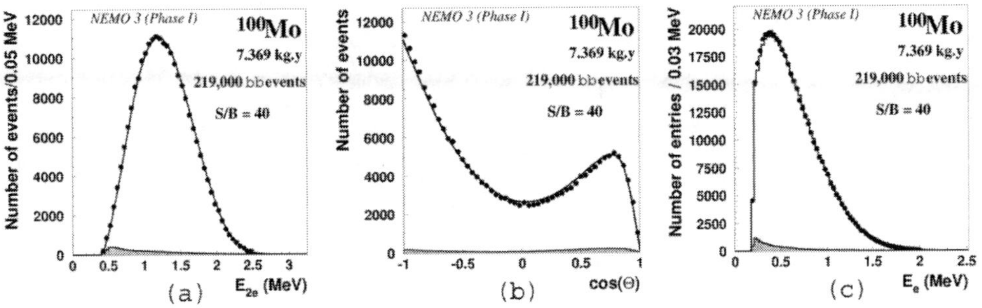

Figure 2. (a) Energy sum spectrum of the two electrons, (b) angular distribution of the two electrons and (c) single energy spectrum of the electrons, after background subtraction from ^{100}Mo with of 7.369 kg·years exposure (Phase I+II) [8]. The solid line corresponds to the expected spectrum from $2\nu\beta\beta$ simulations and the shaded histogram is the subtracted background computed by Monte-Carlo simulations.

of-flight has to correspond to the case of two electrons emitted at the same time from the same vertex. The energy deposited in the counter is required to exceed 200 keV, except with ^{82}Se where the threshold was selected to be 300 keV. In order to suppress backgrounds from ^{214}Bi decay inside the tracking detector, that is followed by ^{214}Po α-decay, it is required that there is no delayed Geiger cell hit close to the event vertex.

A complete study of backgrounds has been performed to date. The level of each background has been directly measured from the data. The first run period, from February 2003 to September 2004, with high level of radon is Phase I. The radon level inside NEMO 3 during the second run after the installation of the radon trapping facility, Phase II (since November 2004 up to now), has been reduced by a factor of six.

3.1. Results for ^{100}Mo ($Q_{2\beta}$ = 3.034 MeV)

The $2\nu\beta\beta$ decay of ^{100}Mo has been measured with high accuracy in NEMO 3. Fig. 2a displays the spectrum of $2\nu\beta\beta$ events for ^{100}Mo that were collected over 389 days (Phase I) [8]. The angular distribution (Fig. 2b) and single electron spectrum (Fig. 2c) are also shown.

The total number of events exceeds 219,000 which is much greater than the total statistics of all of the preceding experiments with ^{100}Mo (and even greater than the total statistics of all previous $2\nu\beta\beta$ decay experiments!). It should also be noted that the background is as low as 2.5% of the total number of $2\nu\beta\beta$ events. The half-life obtained for Phase I data (389 days) is $T^{2\nu}_{1/2} = [7.11 \pm 0.02(stat) \pm 0.54(syst)] \cdot 10^{18}$ y and the single electron energy spectrum is in favour of the Single State Dominance (SSD) mechanism (see [16]). The $T_{1/2}$ value is in agreement with previous measurements [17] but it has higher precision.

The $2\nu\beta\beta$ decay of ^{100}Mo to the excited 0^+ state of ^{100}Ru (1130 keV) has also been measured, $T^{2\nu}_{1/2} = [5.7^{+1.3}_{-0.9}(stat) \pm 0.8(syst)] \cdot 10^{20}$ y [10]. The result is in good agreement with previous measurements [18, 19, 20].

Fig. 3c shows the tail of the two-electron energy sum spectrum in the $0\nu\beta\beta$ energy window for ^{100}Mo (Phase I+II; 690 days of measurement). One can see that the experimental spectrum is in good agreement with the calculated spectrum, which was obtained taking into account all sources of background. Using a maximum likelihood method, the following limit on neutrinoless double beta decay of ^{100}Mo (mass mechanism; 90% C.L.) has been obtained:

$$T_{1/2}(^{100}\text{Mo}; 0\nu) > 5.8 \cdot 10^{23} \text{ y} \tag{1}$$

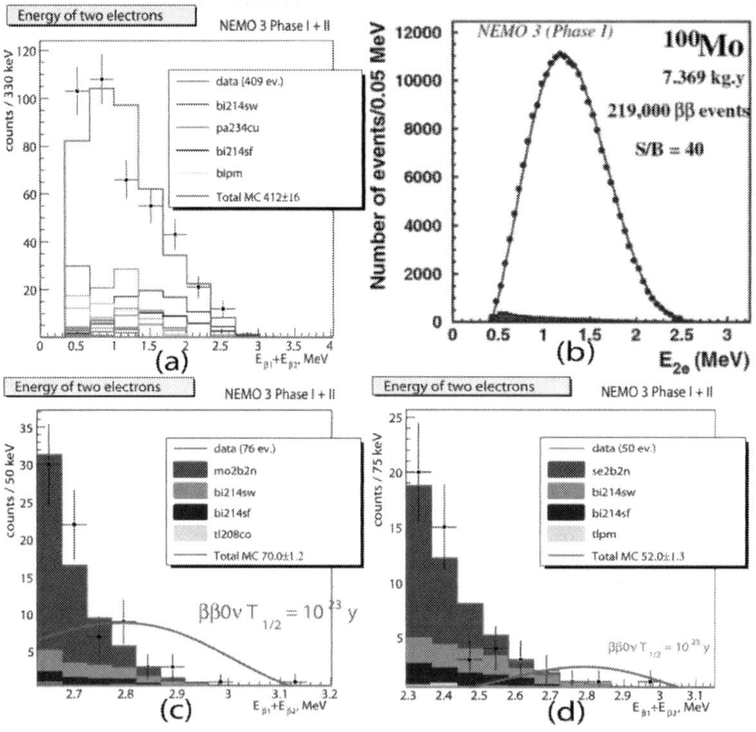

Figure 3. NEMO 3 spectra for ^{100}Mo, ^{82}Se and Cu for 690 days of data. (a) Background $\beta\beta$ spectrum in Cu foil (Phase I + Phase II); (b) ^{100}Mo $2\nu\beta\beta$ spectrum (Phase I); (c) ^{100}Mo $\beta\beta$ spectrum at $Q_{\beta\beta}$ (Phase I + Phase II); (d) ^{82}Se $\beta\beta$ spectrum at $Q_{\beta\beta}$ (Phase I + Phase II).

Using NME values from [21, 22] the bound on $\langle m_\nu \rangle$ gives 0.61-1.26 eV for ^{100}Mo. In this experiment the best present limits on all possible modes of double beta decay with Majoron emission have also been obtained [9].

3.2. Results for ^{82}Se ($Q_{2\beta}$ = 2.995 MeV)

Fig. 4 displays the spectrum of $2\nu\beta\beta$ events for ^{82}Se that were collected over 389 days (Phase I) [8]. The signal contains 2,750 $2\nu\beta\beta$ events and the signal-to-background ratio is 4. The measured half-life value is $T^{2\nu}_{1/2} = [9.6 \pm 0.3(stat) \pm 1.0(syst)] \cdot 10^{19}$ y and is also in good agreement with previous measurements [17].

Fig. 3d shows the tail of the two-electron energy sum spectrum in the $0\nu\beta\beta$ energy window for ^{82}Se (Phase I+II; 690 days of measurement). One can see that the experimental spectrum is in good agreement with the calculated spectrum, which was obtained taking into account all sources of background. Using a maximum likelihood method, the following limits on neutrinoless double beta decay of ^{82}Se (mass mechanism; 90% C.L.) has been obtained:

$$T_{1/2}(^{82}\mathrm{Se}; 0\nu) > 2.1 \cdot 10^{23}\ \mathrm{y} \qquad (2)$$

In addition very strong limits on different types of decay with Majoron emission were obtained [9].

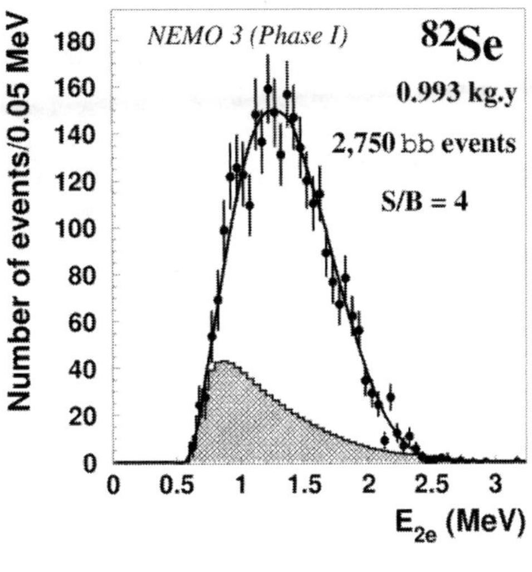

Figure 4. Energy sum spectrum of the two electrons after background subtraction from ^{82}Se with 0.993 kg·years exposure (same legend as Fig. 2) [8].

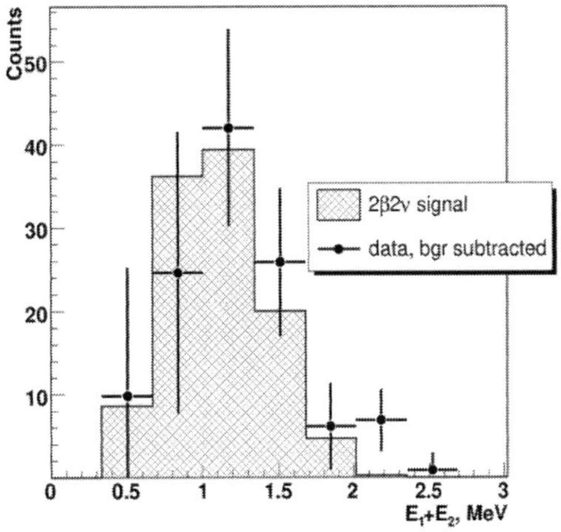

Figure 5. The energy sum spectrum of ^{130}Te $\beta\beta_{2\nu}$ events with background subtracted, Phase II data (534 days).

3.3. Results for ^{130}Te ($Q_{2\beta}$ = 2.529 MeV)

A preliminary measurement of ^{130}Te half-life is reported here. Due to the extremely low decay rate, only the low background data (Phase II) was used. In total, 534 days of data were processed. The number of events observed from the Te source foil is 607, while the predicted background is 492 two-electron events. A binned maximum likelihood method was used to analyse the data. A $2\nu\beta\beta$ signal equal to 109 ± 21.5(stat) events was found, Fig. 5. This corresponds to a half-life of $T^{2\nu}_{1/2}(^{130}\text{Te}) = [7.6 \pm 1.5(\text{stat}) \pm 0.8(\text{syst})] \cdot 10^{20}$ y.

This result is an improvement over the previous attempt at the direct measurement of ^{130}Te

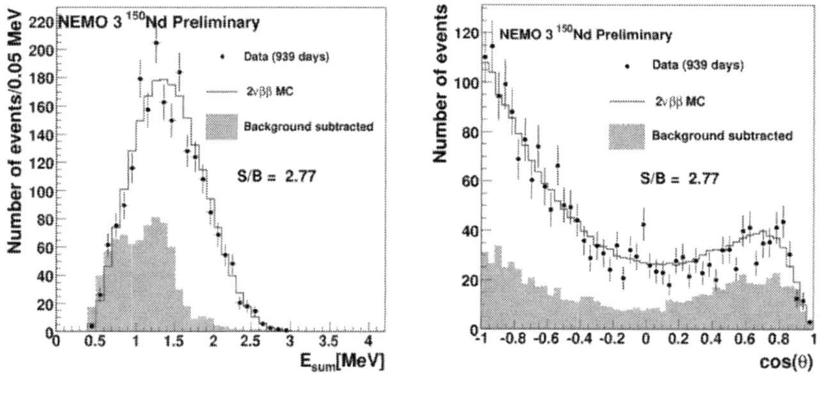

Figure 6. The energy sum spectrum (left) and the angular distribution (right) of ^{150}Nd $2\nu\beta\beta$ events with background subtracted Phase I+II data (939 days).

decay [23]. It is also in agreement with predictions based on the geochemical ^{82}Se/^{130}Te ratio and the present ^{82}Se decay rate from counting experiments [17].

Thus it is of particular interest, because there is a disagreement between geochemical measurements. One group found $T_{1/2} \approx 8 \cdot 10^{20}$ y [24], while the other gives $T_{1/2} \approx (2.5 - 2.7) \cdot 10^{21}$ y [25]. Also it was noted that a smaller $T_{1/2}$ value was obtained in the experiments with "young" ores (younger than 100 million years). This led to the hypothesis that the differences can be accounted for by variations of the Fermi constant G_F with time [26]. The NEMO 3 measurement reported does not contradict this hypothesis, and suggests that the possible G_F variation should be tested using geochemical methods for other $\beta\beta$ decaying nuclei.

3.4. Results for ^{150}Nd ($Q_{2\beta}$ = 3.367 MeV)

A preliminary measurement of the half-life of ^{150}Nd was obtained for 46.6 g of $^{150}Nd_2O_3$ (enrichment of $(91 \pm 0.5)\%$ and the weight of ^{150}Nd is 36.6 g) from data collected between February 2003 and December 2006 corresponding to 939 days of data during the Phases I and II of the experiment. The 2828 $\beta\beta$ type events were observed with a signal-over-background ratio of 2.8. The distribution of the energy sum of the electrons in $2\nu\beta\beta$ type events and their angular distribution are shown in Fig. 6. The background subtracted data and the $2\nu\beta\beta$ signal expectation obtained from Monte Carlo calculations are in good agreement. The $2\nu\beta\beta$ selection efficiency is 7.2%. The measured half-life is $T^{2\nu}_{1/2}(^{150}\mathrm{Nd}) = [9.20^{+0.25}_{-0.22}(\mathrm{stat}) \pm 0.62(\mathrm{syst})] \cdot 10^{18}$ y. This value is between the two previous results obtained from experiments with time projected chambers, $T^{2\nu}_{1/2} = [18.8^{+6.6}_{-3.9}(stat) \pm 1.9(syst)] \cdot 10^{18}$ y [27] and $T^{2\nu}_{1/2} = [6.75^{+0.37}_{-0.42}(stat) \pm 0.68(syst)] \cdot 10^{18}$ y [28], but has much better statistical error.

A $0\nu\beta\beta$ decay signal would correspond to an excess of $\beta\beta$ type events in the distribution of the energy sum of the electrons around the energy of the transition $Q_{\beta\beta}$. No excess of events in the distribution of the energy sum of the electrons from $\beta\beta$ type events originating from the ^{150}Nd sample was observed during the 939 days of data collection. A limit on the half-life of the $0\nu\beta\beta$ process is subsequently set using the CL_s method [29] for $E_{sum} > 2.5$ Mev. The limit on the half-life is $T_{1/2}(0\nu) > 1.8 \cdot 10^{22}$ y at a 90% C.L., which translates into an upper limit $\langle m_\nu \rangle < 1.7 - 2.4$ eV using the NME from QRPA calculations (deformation is not taken into account) [30] or $\langle m_\nu \rangle < 4.8 - 7.6$ eV using the NME from pseudo-SU(3) model (developed for deformed nuclei) [31]. The limit on the half-life was improved when compared to the previous

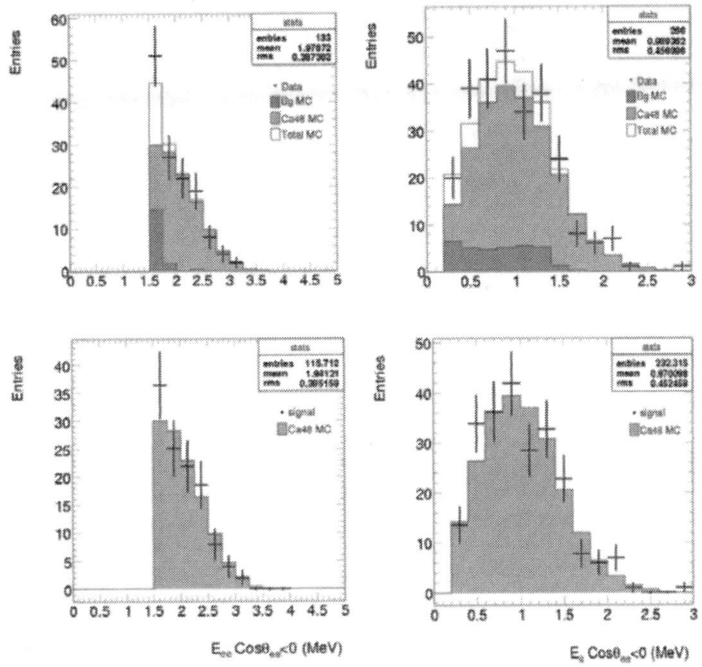

Figure 7. Spectra obtained with ^{48}Ca (Phase I+II; 943.16 days). Energy sum spectrum of the two electrons without background subtraction (top-left); single energy spectrum of the two electrons without background subtraction (top-right); energy sum spectrum of the two electrons after background subtraction (bottom-left); single energy spectrum of the two electrons after background subtraction (bottom-right).

result, $T_{1/2}(0\nu) > 1.7 \cdot 10^{21}$ y (95% C.L.) [32].

In the assumption of a $0\nu\beta\beta$ process involving right currents (V+A), the limit is found to be $T_{1/2}(0\nu) > 1.27 \cdot 10^{22}$ y at a 90% C.L. For a $0\nu\beta\beta$ process with Majoron emission (spectral index n = 1) the obtained limit is $T_{1/2}(0\nu\chi^0) > 1.55 \cdot 10^{21}$ y (90% C.L.).

3.5. Results for ^{48}Ca ($Q_{2\beta}$ = 4.272 MeV)

A preliminary measurement of the half-life of ^{48}Ca was obtained for 17.5 g of $^{48}CaF_2$ (enrichment (73.2 ± 1.6)% and the weight of ^{48}Ca is 7 g) from the data corresponding to 943.16 days of data collection during the Phases I and II.

The sample of ^{48}Ca inside NEMO 3 is known to be contaminated with ^{90}Sr, which is the main background source in this case. The activity for the ^{90}Sr contamination as measured by the NEMO 3 detector is 1699 $\pm$ 3(stat) mBq/kg. To suppress the ^{90}Sr background contribution $\beta\beta$ events with a sum energy > 1.5 MeV and $\cos\theta < 0$ were selected. Fig. 7 shows the two electron sum spectrum and single electron spectrum after these cuts. After 22636 h of data collection, a total of 133 data events were selected, and background was estimated as 17.13 events. The efficiency from MC is 3.3%. The obtained half-life value is, $T^{2\nu}_{1/2} = [4.4^{+0.5}_{-0.4}(stat) \pm 0.4(syst)] \cdot 10^{19}$ y. This value is in good agreement with two previous measurements [33, 34], and has much higher precision. The half-life value obtained is in good agreement with the Shell Model prediction of $T^{2\nu}_{1/2} = 3.9 \cdot 10^{19}$ y [35].

In the search for the $0\nu\beta\beta$ decay of ^{48}Ca, the CL_s method [29] was employed. The range $E_{ee} > 2.0$ MeV was used for the limit calculation. The limit Obtained is $T_{1/2}(0\nu) > 1.3 \cdot 10^{22}$ y

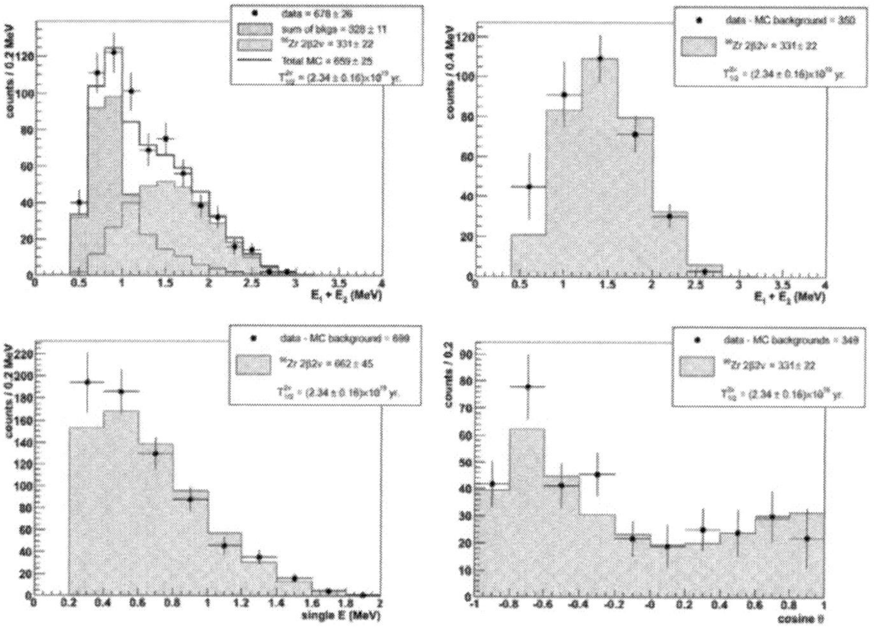

Figure 8. Spectra obtained with ^{96}Zr (Phase I+II; 924.67 days). The energy sum spectrum of the two electrons without background subtraction (top-left); energy sum spectrum of the two electrons after background subtraction (top-right); single energy spectrum of the electrons, after background subtraction (bottom-left); angular distribution of the two electrons after background subtraction (bottom-right)

(90% C.L.).

Using the NME from [36], the corresponding limit on the effective Majorana neutrino mass is $\langle m_\nu \rangle < 29.7$ eV. The limit obtained is approximately the same as the best previous result, $T_{1/2}(0\nu) > 1.4 \cdot 10^{22}$ y (90% C.L.) [37].

3.6. Results for ^{96}Zr ($Q_{2\beta} = 3.350$ MeV)

A preliminary measurement of the half-life of ^{96}Zr was obtained for a 22 g sample of $^{96}ZrO_2$ (enrichment $(57.3 \pm 1.4)\%$ and the weight of ^{96}Zr is 9.4 g) from the data corresponding to 924.67 days of data collection during the Phases I and II. Fig. 8 shows the two electron sum spectrum, single electron spectrum and angular distribution after subtraction of background. These were 331 $2\nu\beta\beta$ events selected (best fit). The efficiency from MC is 7.6%. The half-life obtained is, $T^{2\nu}_{1/2} = [2.3 \pm 0.2(stat) \pm 0.3(syst)] \cdot 10^{19}$ y. This value is in good agreement with the previous NEMO 2 measurement [38], but again has much higher precision.

For neutrinoless decay (mass mechanism) the limit is $T_{1/2}(0\nu) > 8.6 \cdot 10^{21}$ y (90% C.L.) and the corresponding limit on the effective Majorana neutrino mass is $\langle m_\nu \rangle < 7.4 - 20.1$ eV (using NME values from [21, 22]). This limit is in 8.6 times better than the previous NEMO 2 result [38].

4. Conclusion

The NEMO 3 detector has been operating within the target performance specifications since February 2003. The $2\nu\beta\beta$ decay has been measured for a seven isotopes with high statistics and greater precision than previously. The ^{100}Mo $2\nu\beta\beta$ decay to the 0_1^+ excited state of ^{100}Ru has also been measured. No evidence for $0\nu\beta\beta$ decay was found for all seven isotopes. The best limits have been obtained with ^{100}Mo ($> 5.8 \cdot 10^{23}$ y at 90% C.L.) and ^{82}Se ($> 2.1 \cdot 10^{23}$ y at 90% C.L.). After five years of data collection the expected sensitivity at the 90% C.L. will be $\sim 2 \cdot 10^{24}$ y for ^{100}Mo, $\sim 8 \cdot 10^{23}$ y for ^{82}Se and $\sim 10^{22} - 10^{23}$ y for the other isotopes.

5. Acknowledgments

This work was supported by the Russian Federal Agency for Atomic Energy and by the Federal Agency for Science and Innovations (contract No. 02.516.11.6099). A portion of this work was supported by grants from RFBR (No. 06-02-72553).

References

[1] Valle J W F 2006 *Preprint* hep-ph/0608101.
[2] Bilenky S M 2007 *J. Phys.* A **40**, 6707.
[3] Mohapatra R N and Smirnov A Y 2006 *Ann. Rev. Nucl. Part. Sci.* **56** 569.
[4] Pascoli S, Petcov S T and Rodejohann W 2003 *Phys. Lett.* B **558** 141.
[5] Mohapatra R N *et al* 2005 *Preprint* hep-ph/0510213.
[6] Pascoli S, Petcov S T and Schwetz T 2006 *Nucl. Phys.* B **734** 24.
[7] Arnold R *et al* 2004 *JETP Lett.* **80** 377.
[8] Arnold R *et al* 2005 *Phys. Rev. Lett.* **95** 182302.
[9] Arnold R *et al* 2006 *Nucl. Phys.* A **765** 483.
[10] Arnold R *et al* 2007 *Nucl. Phys.* A **781** 209.
[11] Dassie D *et al* 1991 *Nucl. Instr. Meth.* A **309** 465.
[12] Arnold R *et al* 1995 *Nucl. Instr. Meth.* A **354**, 338.
[13] Kisel I *et al* 1997 *Nucl. Instr. Meth.* A **387** 433.
[14] Arnold R *et al* 2001 *Nucl. Instr. Meth.* A **474** 93.
[15] Arnold R *et al* 2005 *Nucl. Instr. Meth.* A **536** 79.
[16] Shitov Yu A 2006 *Phys. At. Nucl.* **69** 2090.
[17] Barabash A S 2006 *Czech. J. Phys.* **56** 437.
[18] Barabash A S *et al* 1995 *Phys. Lett.* B **345** 408.
[19] Barabash A S *et al* 1999 *Phys. At. Nucl.* **62** 2039.
[20] De Braeckeleer L *et al* 2001 *Phys. Rev. Lett.* **86** 3510.
[21] Kortelainen M and Suhonen J 2007 *Phys. Rev.* C **76** 024315.
[22] Simkovic F *et al* 2008 *Phys, Rev.* C **77** 045503.
[23] Arnaboldi C *et al* 2003 *Phys. Lett.* B **557** 167.
[24] Manuel O K 1991 *J. Phys. G* **17** 221; Takaoka N and Ogata K 1966 *Z. Naturforsch* **21a** 84; Takaoka N, Motomura Y and Nagano K 1996 *Phys. Rev.* C **53** 1557.
[25] Kirsten T *et al* 1986 *Proc. Int. Symp. Nuclear Beta Decay and Neutrino (Osaka'86)* (Singapur: World Scientific) 81; Bernatowicz T *et al* 1993 *Phys. Rev.* C **47** 806.
[26] Barabash A S 2000 *Eur. Phys. J.* A **8** 137.
[27] Artemiev V *et al* 1995 *Phys. Lett.* B **345** 564.
[28] De Silva A *et al* 1997 *Phys. Rev.* C **56** 2451.
[29] Junk T 1999 *Nucl. Instr. Meth.* A **434** 435.
[30] Rodin V A *et al* 2007 *Nucl. Phys.* A **793** 213.
[31] Hirsch J G *et al* 1995 *Nucl. Phys.* A **582** 124.
[32] Klimenko A A *et al* 1986 *Nucl. Instr. Meth.* B **17** 445.
[33] Balysh A *et al* 1996 *Phys. Rev. Lett.* **77** 5186.
[34] Brudanin V B *et al* 2000 *Phys. Lett.* B **495** 63.
[35] Caurier E *et al* 1999 *Nucl. Phys.* A **654** 973a.
[36] Caurier E *et al* 2008 *Phys. Rev. Lett.* **100** 052503.
[37] Ogawa I *et al* 2004 *Nucl. Phys.* A **730** 215.
[38] Arnold R *et al* 1999 *Nucl. Phys.* A **658** 299.

Journal of Physics: Conference Series **173** (2009) 012009 doi:10.1088/1742-6596/173/1/012009

The MOON project and DBD matrix elements

H. Ejiri

RCNP, Osaka University, Ibaraki, Osaka, 567-0047, Japan
National Institute of Radiological Sciences, Chiba, 263-8555, Japan
Czech Technical University in Prague, Faculty of Nuclear Sciences and Physical Engineering, Brehova, Prague, Czech Republic

E-mail: ejiri@rcnp.osaka-u.ac.jp

Abstract. This is a brief report on experimental studies of double beta decays (DBD) in Japan, the MOON project for spectroscopic studies of neutrino-less DBD ($0\nu\beta\beta$) and on experimental studies of DBD nuclear matrix elements. Experimental DBD studies in Japan were made by geochemical methods on ^{130}Te,^{128}Te and ^{96}Zr and by a series of ELEGANT(EL) counting methods, EL III on ^{76}Ge, EL IV, V on ^{100}Mo, ^{116}Cd, and EL VI on ^{48}Ca. Future counter experiments are MOON, CANDLES, XMASS and DCBA. The MOON project, which is based on EL V, aims at studies of the Majorana nature of the neutrino (ν) and the ν-mass spectrum by spectroscopic $0\nu\beta\beta$ experiments with the ν-mass sensitivity of $< m_\nu^m >=$ 100–30 meV. The MOON detector is a super ensemble of multi-layer modules, each being composed by PL scintillator plates and position-sensitive detector planes. DBD nuclear matrix elements have been studied experimentally by using charge exchange reactions. The 2-neutrino DBD matrix elements are expressed by successive single-β matrix elements through low-lying intermediate states.

1. Introduction

Double beta decay (DBD) have been studied experimentally by indirect geochemical methods and direct counting methods. Geochemical experiments are integral measurements of both 2-neutrino double beta decay ($2\nu\beta\beta$) and neutrino-less double beta decays ($0\nu\beta\beta$). In fact $2\nu\beta\beta$ rates are in most cases much larger than $0\nu\beta\beta$ rates, and thus observed rates by geochemical methods are $2\nu\beta\beta$ rates. On the other hand direct counting experiments are exclusive measurements for individual $2\nu\beta\beta$ and $0\nu\beta\beta$ rates.

The $0\nu\beta\beta$ is a sensitive and realistic probe for studying the Majorana nature of the neutrino (ν), the lepton number non-conservation, the ν-mass spectrum and the absolute mass scale. They are discussed in recent review articles and references therein [1, 2, 3, 4, 5, 6].

Nuclear matrix elements for $2\nu\beta\beta$ and $0\nu\beta\beta$ are very interesting, as discussed in review articles and references therein [7, 8]. Experimental $2\nu\beta\beta$ rates give $2\nu\beta\beta$ matrix elements, which are used to evaluate $0\nu\beta\beta$ matrix elements, while $0\nu\beta\beta$ matrix elements are crucial for extracting neutrino properties from $0\nu\beta\beta$ rates.

The present paper reports briefly (1) geochemical and counter DBD experiments in Japan, (2) the outline of the MOON (Majorana/Mo Observatory Of Neutrinos) project for spectroscopic $0\nu\beta\beta$ experiment, and (3) experimental studies of DBD nuclear matrix elements.

DBD experiments in Japan were carried out by the geochemical methods for ^{130}Te, ^{128}Te, and ^{96}Zr since the 1960's [9, 10, 11], and also by direct counter experiments with a series of

ELEGANT(ELEctron GAmma rays and Neutrino Telescope EL) detectors since the 1980's. They are EL III on ^{76}Ge [12, 13, 14], EL IV, V on ^{100}Mo, ^{116}Cd [15, 16, 17, 18, 19, 20, 21] and EL VI on ^{48}Ca [22]. Meanwhile, DBD theories have been developed by the theory group in Japan [23]. The counter experiments have been extended to MOON [24, 25, 26, 27] and CANDLES [28] in Osaka. XMASS and DCBA are future DBD plans at ICRR and KEK [29, 30].

The MOON project is based on the ELEGANT V experiments with $\beta\beta$ sources $\neq$detectors [24, 25, 26]. The MOON detector sensitivity is enlarged by orders of magnitudes in order to study the ν-mass in the QD-IH mass region and to observe the low-energy solar-ν as well. MOON is a project for spectroscopic studies of $\beta\beta$ rays from ^{100}Mo, ^{82}Se and other $\beta\beta$ isotopes.

Nuclear matrix elements are very sensitive to nuclear spin-isospin correlations and nuclear structures, and thus theoretical evaluations of them are hard [7, 8]. Experimentally, properties of intermediate states are studied by charge exchange reactions, and are used to evaluate nuclear matrix elements [8, 25, 31, 32]. In fact, $2\nu\beta\beta$ matrix elements are well expressed by using experimental single-β matrix elements through low-lying intermediate states [1, 8, 33, 34].

This report is based partly on the recent reviews on MOON and DBD experiments [26, 27].

2. DBD experiments in Japan

DBD geochemical experiments in Japan were carried out first in the 1960's by Takaoka and Ogata at Osaka Univ. [9]. They measured the integrated half-life of ^{130}Te. The obtained value of $T_{1/2} = 8.2 \pm 0.6\ 10^{20}$ y agrees with later measurements. In the 1990's, Takaoka et al. and Kawashima et al. measured by geochemical methods integrated half-lives for ^{128}Te and ^{96}Zr [10, 11].

Direct counting experiments were made first by the DBD pioneer, Prof. E. Fiorini and his collaborator in the early 1970's [35]. They studied ^{76}Ge $\beta\beta$ decays by using Ge detectors.

Counter experiments in Japan were started in the early 1980's by the Ejiri's group at Osaka, Extensive experiments on $0\nu\beta\beta$ and $2\nu\beta\beta$ were performed by a series of ELEGANT (EL) detectors. EL III, which was a high-purity Ge detector surrounded by NaI detectors, was used to study the ground and excited state $0\nu\beta\beta$ decays from ^{76}Ge [12, 13, 14]. Meanwhile, the ambitious experiment IGEX with enriched ^{76}Ge detector was proposed by Prof. F. Avignone and his collaborator, which gave a very stringent limit on the $0\nu\beta\beta$ halflife [36].

EL IV was composed of multilayer Si disks and ^{100}Mo foils, with NaI active shields, and was used to study $2\nu\beta\beta$ from ^{100}Mo.

EL V is a detector complex of drift chambers for 2 β ray tracking, PL plates for β ray energies and times, and NaI detector arrays for γ and X-rays [16]. It started operation in 1989, and gave the first results of the $2\nu\beta\beta$ decays from ^{100}Mo and ^{116}Cd [15, 19], and most stringent limits on $0\nu\beta\beta$ decays from them [20, 21].

EL VI, which consists of CaF detectors surrounded by CsI detectors, is for studying the $0\nu\beta\beta$ from ^{48}Ca. It gave the most stringent limits on $0\nu\beta\beta$ from ^{48}Ca [22].

It is noted that the neutrino masses to be studied using the present detectors are limited by their mass sensitivity of 0.2–1 eV. Among them, CUORICINO [37] and NEMO III [38] are expected to reach the mass region of 0.2–0.5 eV in the near future.

Experimental proposals for future $\beta\beta$ experiments with the QD-IH mass sensitivity have been made on $\beta^-\beta^-$ isotopes. Some of the future experiments in Japan are as given below.

NOON is a spectroscopic experiment with multi-layer scintillation plates and tracking detector planes [24, 25, 26]. It is based on the ELEGANT V [15, 16]. It aims at studies of QD-IH mass region. Details are given in section 3.

CANDLES (CAlcium fluoride for studies of Neutrinos and Dark matter by Low Energy Spectrometer) is an array of CaF_2 crystals in liquid scintillator to study $\beta\beta$ decays of ^{48}Ca and dark matter [28]. It is based on ELEGANT VI with CsI active shields [22]. Water buffers are used as passive shields. It is noted that the natural abundance of the ^{48}Ca isotope is only 0.2 % and the enrichment process is very difficult.

XMASS uses a large liquid Xe scintillator to study ^{136}Xe double beta decays and search for DM [29]. A prototype detector is now under progress. A one year run with 10 kg will cover the 0.1 - 0.2 eV QD region.

DCBA (Drift Chamber Beta-ray Analyzer) uses a tracking chamber to study ^{150}Nd $\beta\beta$ decays [30]. Tests are under progress by using DCBA-T2 and DCBA-T3. The energy resolution is crucial to reduce the background (BG) contribution from $2\nu\beta\beta$. The half-life sensitivities are 10^{25} y and 10^{26} y with natural and enriched ^{150}Nd sources, respectively.

3. MOON for high-sensitivity $0\nu\beta\beta$ spectroscopy

MOON is a $\beta\beta$ spectroscopic experiment to study the Majorana ν-mass in the QD–IH mass region of 100–30 meV, which corresponds to the half-life of 10^{26-27} y for ^{82}Se and ^{100}Mo [24, 25, 26]. The MOON-type spectroscopic experiment has several unique features.

i. MOON uses external $\beta\beta$ sources, which are separated from $\beta\beta$ detectors. Thus the $\beta\beta$ nuclides are selected in views of the large signal rate (i.e. the large phase space volume, the large nuclear matrix element, and the large number of the $\beta\beta$ isotopes) and the small BG rate (i.e. the large $Q_{\beta\beta}$ value and the small $2\nu\beta\beta$ rate). ^{82}Se, ^{100}Mo and ^{150}Nd are possible candidates.

ii. The $\beta\beta$ event can be confirmed by measuring at least two $\beta\beta$ isotopes with different nuclear matrix elements and different $0\nu\beta\beta$ peak energies. Energy and angular correlations for two β-rays are measured. The $0\nu\beta\beta$ decays to both the ground and excited 0^+ states are also measured. These measurements are important for identifying the ν-mass process among other processes (right-handed week currents, heavy ν, SUSY and others).

iii. Using $0\nu\beta\beta$ isotopes with large $Q_{\beta\beta} \approx 3$ MeV, the signal can be placed well above most RI $\beta - \gamma$ rays to avoid the RI BGs. Two β-ray spectroscopy by tracking detectors make it possible to get required S/N (signal to BG) ratios.

iv. Major BGs from the high energy tail of the $2\nu\beta\beta$ spectrum in the $0\nu\beta\beta$ window can be reduced to the level required for the mass sensitivity by using detectors with realistic energy resolution of around $\sigma = 2.0$ %.

v. The multi-layer structure of detector modules makes the detector quite compact. It is possible to expand a small-scale detector for the QD mass study to a large-scale one for the IH mass study by increasing the number of modules. Actually MOON is an extension of the ELEGANT experiments with two detector layers [16]. Accordingly the basic concept and detector components have been well proved.

3.1. MOON and Majorana neutrino mass

The light Majorana ν-mass sensitivity of the $0\nu\beta\beta$ experiment is given by the minimum effective ν-mass to be observed. It is given as

$$< m_\nu^m >= (S_N)^{-1/2} \; (N_{\beta\beta})^{-1/2} \; \delta^{1/2}, \tag{1}$$

where $N_{\beta\beta}$ is the effective number of $\beta\beta$ isotopes and δ is the number of $0\nu\beta\beta$ signals required for the 90 % CL. Using practical units, $S_N = 6 \cdot 10^{15} \; G^{0\nu} \; |M^{0\nu}|^2 \; (A)^{-1} \; m_e^{-2}$ with the phase space volume of $G^{0\nu}$ in units of $10^{-14} y^{-1}$ and the mass number A. $N_{\beta\beta} = \epsilon^{0\nu} \; N \; T$ with $\epsilon^{0\nu}$: the signal efficiency, N: the weight of $\beta\beta$ isotopes in units of ton, T: the run live-time in units of year. δ is given as $\delta \approx 1.6 + 1.7 \; (BT)^{1/2}$ with B being the BG rate per ton per year.

The nuclear matrix element $M^{0\nu}$ is crucial for the mass sensitivity as discussed in reviews [1, 5, 7, 8] and references [39, 40, 41]. It is found that nuclear matrix elements by recent QRPA calculations [40, 41] are approximately expressed as $M^{0\nu} \approx 18A^{-1/3}$. By using this approximate value, one gets the nuclear sensitivity coefficient in equ. (1) as $(S_N)^{-1/2} \approx 10$ meV for ^{82}Se and ^{100}Mo. The sensitivity coefficient for the excited 0^+ state in ^{100}Mo is around 20 meV.

The MOON detector with $N \approx 0.1, B \approx 4, \epsilon^{0\nu} \approx 0.25$ is required to get the QD mass sensitivity of around 100 meV in the $T = 2$ year run, and that with $N \approx 0.5, B \approx 1.5, \epsilon^{0\nu} \approx 0.25$ for the IH mass sensitivity of around 30 meV in the $T = 4$ year run.

3.2. MOON detector

The MOON detector is a high-sensitivity hybrid detector ensemble to measure individual $\beta\beta$ ray energies, their emission points and their angles as well as γ rays [24, 25, 26]. The MOON detector consists of multi-layer detector modules, as shown in Fig. 1. One unit of the detector consists of 17 modules. Each module is composed by 3 kinds of detector layers and the source films.

i. Six plastic scintillator plates (PL) for $\beta\beta$ energy and time. Scintillation photons are collected by photo-multiplier tubes (PMT) around the plastic-scintillator plate.

ii. Five sets of up and down thin position-sensitive detector (PL-fiber or Si-strip) planes for $\beta\beta$ vertex position and the emission angle. Each set of the position-sensitive detector planes is inserted between the PL plates.

iii. One thick NaI detector array for X and γ rays.

iv. Five thin $\beta\beta$ source films are interleaved between the position-sensitive detector planes.

Two β-rays from a $\beta\beta$ source film are measured in coincidence by up and down position-sensitive detector planes and up and down PL plates. All other position detector planes and PL plates in the module are used as active shields to reject BG γ rays and BG neutrons. The thick NaI plate is for measuring γ rays following $\beta\beta$ decays to the 0^+ excited states. All other modules are used as active shields.

Each PL plate is around $1.25\,\mathrm{m} \times 1.25\,\mathrm{m} \times 0.015\,\mathrm{m}$ and each PL/Si-detector plane is around $0.9\,\mathrm{m} \times 0.9\,\mathrm{m} \times 0.3\,\mathrm{mm}$, while each source film is around $0.85\,\mathrm{m} \times 0.85\,\mathrm{m}$ with $0.05\,\mathrm{gr/cm^2}$. The total $\beta\beta$ source is 0.36 kg per film, 1.8 kg per module and 30 kg per unit of the detector.

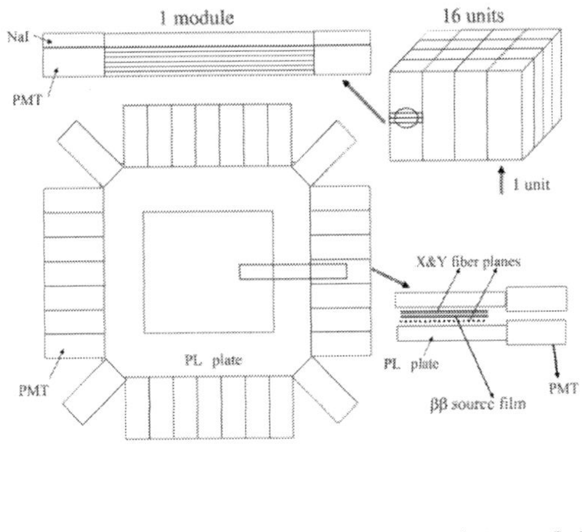

Figure 1. MOON detector. One unit with 17 modules, and 1 module with 6 PL plates and 5 sets of up and down position-sensitive detector planes.

The energy resolution is crucial for reducing the BG contribution from the high energy tail of the $2\nu\beta\beta$ spectrum into the $0\nu\beta\beta$ window. So far $\sigma \approx 2.1\,\%$ is achieved at 3 MeV ($Q_{\beta\beta}$ value of ^{100}Mo) for a small PL plate with $6\,\mathrm{cm} \times 6\,\mathrm{cm} \times 1\,\mathrm{cm}$. Good resolution is expected also for larger PL plates by correcting for the position dependence of the light collection in PL plates. This resolution is just what is required to reduce the $2\nu\beta\beta$ BG contribution and to achieve the

IH mass sensitivity of the order of 50–30 meV. Improvement of the resolution to $\sigma \approx 1.7\,\%$ is under progress for the higher sensitivity experiment by using better PL plates and PMTs.

The position-sensitive detector planes are used for identifying the $\beta\beta$ vertex point within 10-20 mm^2, and also for particle identification, i.e. selection of β rays and rejection of γ and α rays. Two options of the position-sensitive detector planes are under consideration, one is a PL fiber plane and the other is a Si-strip plane.

The $\beta\beta$ source film is interleaved between the X-type and Y-type detector planes, each with 90 cm $\times$ 90 cm $\times$ 300 μm. The PL scintillator fibers with 300 μm square cross section are stretched to X direction in case of the X-type plane and to Y direction in case of the Y-type plane. Photons from the fibers are collected by multi-anode photomultipliers.

The overall E-resolution of the PL plate and the fiber plane is expected to be $\sigma \approx 2.3\,\%$. This is good enough to achieve the ν- mass sensitivity to study the ν-mass in the QD and IH mass regions. In case of the Si-strip detector plane, one expects a similar resolution.

The multi-layer module structure with good energy and position resolutions is very powerful for selecting $0\nu\beta\beta$ signals and for rejecting RI-background signals. This makes it realistic to build a compact detector of the order of 0.4 m^3 per kg $\beta\beta$ isotopes. The detector is smaller in volume than the SuperNEMO by orders of magnitude [42].

The enriched ^{82}Se and/or ^{100}Mo isotopes are used for the $\beta\beta$ source. Enrichment to 85 % of each isotope is made by using the centrifugal separation. The ^{100}Mo isotope separation is made by the centrifugal separation of MoF_6 gas. Using 6000 centrifuges and 40 separation steps, ^{100}Mo isotopes are obtained with a rate of 350 gr per day, and a half ton in 5 years.

Isotope separation and purification of ^{82}Se and ^{100}Mo isotopes are done as in NEMO III [38]. The purity of the enriched isotopes is required to be around 300–100 mBq per ton or less for U and Th radioactive isotopes in order to achieve the QD-IH mass sensitivity.

3.3. Selection of DBD signals and the Majorana mass sensitivity

The $0\nu\beta\beta$ signal is identified by observing two β rays by a set of the up and down position-sensitive ΔE-detector planes and PL E-detector plates, each at the same point within 0.1-0.2 cm^2. All PL detector plates and position-sensitive detector planes other than the relevant set of the up and down detector plates and detector planes are used as active shields. No pre- and post-decay signals before and/or after the signal are observed at the same position of the detector set.

The energy spectra around the $0\nu\beta\beta$ peak region are evaluated by Monte Carlo simulations [43], as shown in Fig. 2 for ^{82}Se and ^{100}Mo with 50 mg/cm. Counting rates equivalent to 0.6 t y run with the signal condition mentioned above are shown, while the simulation was made for 10^6 events to get good statistics. The $0\nu\beta\beta$ efficiency is obtained as $\epsilon^{0\nu} \approx 0.25$.

The low energy tail of the $0\nu\beta\beta$ spectrum reflects the dependence of the energy loss on the emission angle. Thus the $0\nu\beta\beta$ signal efficiency decreases as the source thickness increases.

It is very important to note that the high energy slope of the $2\nu\beta\beta$ spectrum for the 50 mg/cm^2 is same as that for the 10 mg/cm^2. Thus the contribution of the $2\nu\beta\beta$ tail into the $0\nu\beta\beta$ window is mainly due to the energy resolution and the source thickness effect is not appreciable.

Natural RI BGs at the $0\nu\beta\beta$ E-window are mainly from ^{208}Tl and ^{214}Bi isotopes with $Q_\beta > Q_{\beta\beta} \approx 3$ MeV. BGs rate from ^{208}Tl and ^{214}Bi impurities in the Mo source films were evaluated by Monte Carlo simulations [43].

There are various kinds of backgrounds from cosmic ray origins even at underground laboratories. Charged muons are rejected by passive shields and multi-layer PL plates and PL-fiber/Si-strip planes. The cosmic muon BG is less than 1/t y.

Cosmic neutrons, which are produced by cosmic-μ capture reactions around the detector, are major BG sources. Energetic γ rays with $E > 3$ MeV, which follow inelastic scatterings of neutrons and neutron capture reactions, are potential BG sources at the $0\nu\beta\beta$ window. They are evaluated by Monte Carlo simulations [43]. Actually, the neutrons are also rejected by the active

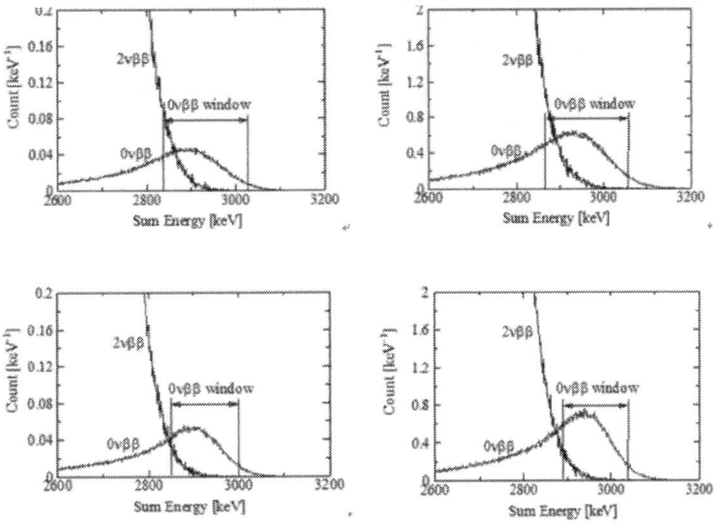

Figure 2. Sum energy spectra of $0\nu\beta\beta$ and $2\nu\beta\beta$ rays in the $0\nu\beta\beta$ peak region for 0.6 ton year run with 50 mg/cm^2 $\beta\beta$ sources. Left: ^{82}Se with $< m_\nu >= 75$ meV. Right: ^{100}Mo with $< m_\nu >= 250$ meV. Top: E-resolution $\sigma = 2.2$ %. Bottom: $\sigma = 1.7$ %. Ref. Shima[43].

shields of the PL plates, and the experiment is carried out at a deep underground laboratory of around 2500 m water equivalent (w.e). Then the rate is estimated to be $B = 1$/t y.

The high energy part of the $2\nu\beta\beta$ spectrum in the $0\nu\beta\beta$ window is the major BG in the MOON type detector with moderate energy resolution. The $2\nu\beta\beta$ BG rate is given as

$$B(2\nu\beta\beta) = 4.2\ \epsilon^{2\nu}\ (t_{1/2}\ 0.01\ A)^{-1}, \tag{2}$$

where $t_{1/2}$ is the $2\nu\beta\beta$ half-life in units of 10^{20} y, $\epsilon^{2\nu}$ is the $2\nu\beta\beta$ efficiency (probability of $2\nu\beta\beta$ events at the $0\nu\beta\beta$ window) in units of 10^{-7}. The $2\nu\beta\beta$ efficiency was evaluated for the 3σ E-window as $\epsilon^{2\nu} = 0.8$ and 0.3 in units of 10^{-7} for $\sigma = 2.2$ % and 1.7 %, respectively.

The MOON experiment will be carried out in three phases, phase I with 1 detector unit (0.03 ton $\beta\beta$ source) to search for the Majorana ν-mass in the 150 meV QD mass region and phase II with 4 units (0.12 ton) for the ν-mass in the lower QD mass region of around 100–70 meV. Phase III with 16 units (0.48 ton) is for the IH mass region of around 30–40 meV.

The E-resolution of $\sigma = 2.2$ % and 1.7 % and the source impurities of 60 and 20 mBq for ^{208}Tl isotopes and 300 and 100 mBq for ^{214}Bi isotopes are assumed for the phase I-II and the phase III experiments, respectively, at the underground laboratory of 2450 m w.e. The ^{208}Tl and ^{214}Bi impurities for the phase I-II are the values in the NEMO III experiment [38] and those for the phase III are 1/3 of the values of NEMO III. The BG rates per ton year (/t y) are 4.2 and 1.6 for the $2\nu\beta\beta$ contributions, 0.9 and 0.3 for ^{208}Tl and 0.03 and 0.01 for ^{214}Bi, respectively, for the phase I-II and the phase III experiments. The BG rates from natural and cosmogenic RI isotopes are much less than the $2\nu\beta\beta$ rate, and well below 1/t y.

The $0\nu\beta\beta$ half-life sensitivity and the ν-mass sensitivity for each phase are evaluated for the

^{82}Se and ^{100}Mo by using $M^{0\nu} = 18\ A^{-1/3} = 4.2$ and 3.9 for ^{82}Se and ^{100}Mo. The evaluated half-lives and the ν-mass sensitivities are given in Table 1.

Table 1. Half-life sensitivities ($t_{1/2}$) and ν- mass sensitivities ($< m_\nu^m >$) for the phases I, II, and III experiments with N=1, 4,and 16 detector units, respectively.

Phase	$N_{\beta\beta}$	$\Sigma N_{\beta\beta}T$	$t_{1/2}$ ^{82}Se	$t_{1/2}$ ^{100}Mo	$< m_\nu^m >^{82}$Se	$< m_\nu^m >^{100}$Mo
I	0.03 t	0.06 t y	0.32 10^{26} y	0.15 10^{26} y	134 meV	161 meV
II	0.12 t	0.30 t y	1.12 10^{26} y	0.41 10^{26} y	70 meV	98 meV
III	0.48 t	1.92 t y	5.9 10^{26} y	2.0 10^{26} y	31 meV	44 meV

3.4. Detector research and development

Photon responses of PL and NaI scintillators to be used for MOON were studied by using the detector test bench CROSS (Correlation Response Observatory for Scintillation Signals) at NIRS (National Institute for Radiological Sciences) [44].

One PL scintillator plate with 6 cm × 6 cm × 1 cm was set at CROSS. The scintillation photons were collected by 2 inch ϕ H6410 photomultiplier tubes with 25 % photo-electron efficiency at the 4 sides of the plate. The energy resolution was found to be given by $\sigma \approx 3.7E^{-1/2}$ %, and the expected resolution at $Q_{\beta\beta} \approx 3$ MeV is $\sigma = 2.1$ %.

A prototype MOON-1 detector was constructed to prove feasibility of MOON with the IH mass sensitivity [26, 45, 46, 47]. It consists of 6-layer plastic-scintillator (PL) plates, each with $53 \times 53 \times 1$ cm^3. RP-408 (BC-408 equivalent) PL.

The 94.5 % enriched ^{100}Mo films with 40 mg/cm^2 are interleaved between the PL plates in the MOON-1 detector. The 6-layer PL plates are viewed by 56 6×6 cm^2 square-type photo-multiplier tubes (PMT) R6236-01-KMOD provided by Hamamatsu Photonics. One PMT collects photons from 3 PL plates and the hit PL plate is identified by the PMT hit pattern.

Photon response for the same PL plate as used for the MOON-1 detector was studied using the 976 keV K conversion electron from a ^{207}Bi RI source at the centre of PL. The number of total photo-electrons is 1830 ± 35 for the 976 keV electron line. The energy resolution is found to be $\sigma = 4.8 \pm 0.2$ %, which corresponds to $\sigma = 2.7$ % at 3 MeV.

The MOON-1 detector was set in the active and passive shields of ELEGANT V [16]. The measured energy resolution is found to be well reproduced by $\sigma = 5.0\ E^{-1/2}$ % with E in units of MeV. This leads to the energy resolution of $\sigma = 2.9$ % at 3 MeV, as required for the half-life sensitivity of 2.3 10^{26} y (QD-IH mass of 50 meV) for ^{82}Se. The resolution of around $\sigma = 2.2$ % is expected by correcting for the position dependence of the photon collection.

4. DBD matrix elements

Nuclear $\beta\beta$ responses are evaluated by using relevant β matrix elements for intermediate states. The $0\nu\beta\beta$ matrix element is expressed in terms of the single-β matrix elements as

$$M^{0\nu}(\tau\sigma) \approx \Sigma_J \left[\frac{M_S(J)M_{S'}(J)}{\Delta_S(J)}\right], \tag{3}$$

where $M_S(J)$ and $M_{S'}(J)$ are single-β matrix elements through the low-lying (single particle-hole) states $|S_J>$ in the intermediate nuclei [8, 33]. Extensive studies of charge exchange reactions have been made to get GT(1$^+$) responses with $\tau^{\pm}\sigma$. The reactions studied are (p,n), (n,p), (d,^{2}He), (^{3}He, t) (t,^{3}He), (^{7}Li, ^{7}Be), and others [8, 25, 48, 49].

Recently it has been pointed out that low energy photons are used to study nuclear weak responses relevant to $\beta\beta$ decays. Nuclear matrix elements from intermediate 1^+ and 1^- states to the final 0^+ state are studied by investigating M1 and E1 photon absorptions into IAS (Isobaric Analog States) of the intermediate states [25].

One direct way to get the weak (neutrino) nuclear responses is to use ν-beams. Low energy ν-beams with $E \leq 100\,\mathrm{MeV}$ can be obtained from pion decays. Intense pions are produced by nuclear Macro 6interaction with GeV protons. The J-PARC booster synchrotron provides high energy 3 GeV protons with $1.2 \cdot 10^{15}$ per sec, and the expected ν intensity is $3 \cdot 10^{14}$ per sec. These low energy neutrinos are very interesting for studying neutrino nuclear response [31]. NuSNS uses low energy neutrinos from stopped pions produced by intense SNS protons in order to study neutrino nuclear responses [50, 51].

Then $M_S(J)$ and $M_{S'}(J)$ are obtained experimentally from the charge exchange reactions and/or single β decay rates [8]. Then $2\nu\beta\beta$ matrix elements derived from $2\nu\beta\beta$ half-lives are well reproduced by using single-β GT matrix elements through low-lying 1^+ states in the intermediate nucleus [33, 34]. Thus $0\nu\beta\beta$ matrix elements may be evaluated from single-β matrix elements through low-lying $0^\pm$, $1^\pm$, $2^\pm$, $3^\pm$, ... in the intermediate nucleus.

5. Concluding remarks

DBD experiments in Japan have been carried out by using the geochemical methods at the Ogata lab. Osaka Univ. since the 1960's, and by the ELEGANT series counter experiments at the Ejiri lab. Osaka Univ. since the early 1980's. The counter experiments have been extended to the future MOON experiments and others.

MOON is the high-sensitivity spectroscopic $0\nu\beta\beta$ experiment with detector $\neq \beta\beta$ source. ^{82}Se and/or ^{100}Mo with the large $Q_{\beta\beta}$ are used. Energy and angular correlations for 2–3 kinds of $\beta\beta$ nuclei and for both the ground and excited states are useful to identify the light ν-mass $0\nu\beta\beta$ process.

The MOON detector consists of multi-layer scintillator plates and position sensitive detector planes. Good energy and position resolutions, together with the compact multi-layer structure of the detector, make it realistic to study the QD and IH mass regions.

DBD matrix elements are given in terms of single-β matrix elements through low-lying states in intermediate nuclei. Charge exchange reactions are used to get the single-β matrix elements.

It is noted that MOON can be used for real-time studies of solar and supernova neutrinos since ^{100}Mo has large charged-current responses for both solar and supernova neutrinos [8, 52].

Low-energy solar neutrinos are studied by observing the inverse β decays from ^{100}Mo in delayed coincidence with the successive β decay from ^{100}Tc to reduce BGs. The good position resolution is crucial to reduce the $2\nu\beta\beta$ accidental coincidence rate. All Mo isotopes have large responses for supernova neutrinos. Thus a MOON-type detector can be used to study supernova neutrinos by using thick natural Mo plates of the order of 100 tons in place of thin ^{100}Mo foils.

6. Acknowledgments

The author is honored to present this talk at CISNP in honor of Prof. Frank Avignone, Prof. E. Fiorini, and late Prof. Peter Rosen. On behalf of our DBD colleagues in Japan, we would like to congratulate them for their great achievements in neutrino and dark matter physics, and thank them for their leadership in progress of nuclear particle physics. The author would like to extend his special personal thanks to Frank and Ettore for the valuable and constructive discussions and nice collaborations since the early 1980's. I have been always inspired and impressed by their pioneering works. They are also key persons for promoting international collaborations for future DBD experiments [53]. Thus we certainly do hope to enjoy their wonderful activities all through the 21st century. We wish to thank Peter for his enlightening discussions at various conferences and symposia, and miss him for ever. The author thanks the CISNP organizers for this wonderful symposium, and for MOON collaborator for valuable discussions.

References

[1] Ejiri H 205 *J. Phys. Soc. Jap.* **74** 2101.
[2] Ejiri H 2006 *Prog. Part. Nucl. Phys.* **57** 153.
[3] Vergados J 2002 *Phys. Rept.* **361** 1.
[4] Elliott S and Vogel P 2002 *Ann. Rev. Nucl. Part. Sci.* **52** 115.
[5] Simkovic F and Fassler A 2002 *Prog. Part. Nucl. Phys.* **48** 201.
[6] Avignone F, Elliott S and Engel J 2008 *Rev. Mod. Phys.* **80** 481.
[7] Suhonen J D and Civitarese O 1998 *Phys. Rept.* **300** 123.
[8] Ejiri H 2000 *Phys. Rept.* **338** 265.
[9] Takaoka N and Ogata K 1966 *Naturforsch* A **21** 84.
[10] Takaoka N, Motomura Y and Nagao K 1996 *Phys. Rev.* C **53** 1557.
[11] Kawashima A, Takahashi K, and Masuda A 1993 *Phys. Rev.* C **47** 2452.
[12] Ejiri H, et al. 1986 *Nuclear Physics* A **448** 271-279.
[13] Kamikubota N, Ejiri H, et al. 1986 *Nucl. Instr. Methods* A **245** 379-392.
[14] Ejiri H, et al. 1987 *J. Phys. G Nucl. Phys.* **13** 839-846.
[15] Ejiri H et al. 1991 *Nucl. Instr. Methods* A **302** 304-314.
[16] Ejiri H et al. 1991 *Phys. Lett.* B **258** 17-23.
[17] Kudomi N, et al. 1992 *Phys. Rev.* C **46** R2132-R2135.
[18] Tanaka J and Ejiri H 1993 *Phys. Rev.* C **48** 5412.
[19] Ejiri H et al. 1995 *J. Phys. Soc.Japan* **64** 339-343.
[20] Ejiri H et al. 1996 *Nucl. Phys.* A **611** 85-95.
[21] Ejiri H et al. 2001 *Phys. Rev.* C **63** 065501 1-7.
[22] Ogawa I et al. 2004 *Nucl. Phys.* A **730** 215-223.
[23] Doi M, Kotani T and Takasugi E 1985 *Prog. Theor. Phys. Supp.* **83** 1.
[24] Ejiri et al. 2000 *Phys. Rev. Lett.* **85** 2917.
[25] Ejiri H 2006 Czech. J. Phys. B **54** 317; *ibid* B **56** 459.
[26] Ejiri H 2007 *Mod. Phys. Lett.* A **22** 1277.
[27] Ejiri H et al. 2008 *Proc Spin Conference SPIN07 (Prague), European Phys. J* **102** 239.
[28] Umehara S et al. 2003 *Proc. Neutrinos Dark Matter NDM03, Nara 2003, eds. Ejiri H and Ogawa I.*
[29] Moriyama S et al. (2001) *Workshop XENON01, Tokyo*, Ueshima K 2008 *arXiv0803.2888.*
[30] Ishihara N et al. 2000 *Nucl. Instr. Meth.* A **443**, 101.
[31] Ejiri H 2003 *Nucl. Instr. Methords* **503** 276.
[32] Amos K, Faessler A and Rodin V 2007 *Phys. Rev.* C **76** 014604.
[33] Ejiri H and Toki H 1996 *J. Phys. Soc. Japan Lett.* **65** 7.
[34] Ejiri H, (2009) *J. Phys. Soc. Japan* **78** No7, to be published.
[35] Fiorini E et al. 1973 *Nuovo Cimento* A **13** 747.
[36] Aalseth C. E, et al. 2003 it Phys. Rev. D **65** 092007.
[37] Arnaboldi C et al. 2004 *Phys. Lett.* B **584** 260; *hep-exp (2008) arXiv:0802.3439.*
[38] Arnold R et al. 2005 *Phys. Rev. Lett.* **95** 182302.
[39] Civitarese O and Suhonen J 2003 *Nucl. Phys.* A **729** 867.
[40] Rodin V A, et al. 2006 *Nucl. Phys.* A **766** 107; 2007 *Nucl. Phys.* A **793** 213.
[41] Kortelainen M and Suhonen J 2007 *Phys. Rev.* C **75** 051303; **76** 024315.
[42] Piquemal F 2007 *Proc. Int. Conf. TAUP 2007, Sendai, Sep. 2007.*
[43] Shima T 2007 *Proc. TAUP07 Sendai, Sep. 2007; Private communication.*
[44] Nakamura H, et al. 2008 *to be submitted.*
[45] Doe P, et al. 2003 *Nucl. Phys.* A **721** C517.
[46] Nomachi M, et al. 2005 *Nucl. Phys. Proc. Suppl.* B **138** 221.
[47] Nakamura H, et al. (2007) *J. Phys. Soc. Japan* **76** 114201.
[48] Dohmann H, et al., (2008) *Phys. Rev.* C **78** 041602.
[49] Zegers R, et al., (2007) *Phys. Rev. Letters* **99** 2501.
[50] Avignone F 2000 *Workshop Neutr. Nucl. Phys. Stopped $\pi\mu$ facility, Oak Ridge 2000.*
[51] Hungerford Ed, *Proc. NNR05 workshop ed. H. Ejiri, CAST/Spring-8, Dec. 2005.*
[52] Ejiri H, Engel J and Kudomi N 2002 *Physics Letters* B **530** 27-32.
[53] Ejiri H 2002 *http://www.rcnp.osaka-u.ac.jp/ ejiri/DBD-Lett.*

Anatomy of double beta decay nuclear matrix elements

Petr Vogel

Kellogg Radiation Laboratory 106-38
Caltech. Pasadena, CA 91125, USA

E-mail: pxv@caltech.edu

Abstract. The necessary ingredients for a realistic evaluation of the $0\nu\beta\beta$ nuclear matrix elements are reviewed. It is argued that the short range nucleon correlations, nucleon finite size, and higher order nuclear currents need to be included in the calculation, even though a consensus on the best way to treat all of these effects has not been reached. Another positive development is the realization that the two alternative and complementary methods, the Quasiparticle Random Phase Approximation and the Nuclear Shell Model, agree on many aspects of the calculation, in particular on the competition, or canclation, between the contribution of nuclear pairing on one hand, and the other pieces of interaction that result in admixtures of broken pairs or higher seniority states on the other hand. The relatively short range ($r \lesssim$ 2-3 fm) of the effective $0\nu\beta\beta$ operator found in both methods is a consequence of that competition.

1. Introduction

Thanks to the success of neutrino oscillation experiments we know that neutrinos have a finite mass and that the lepton flavor is not a conserved quantity. Future oscillation studies will aim at refining our knowledge of the oscillation parameters, at determining the as yet unknown mixing angle θ_{13} and the so-called hierarchy (i.e. the sign of the mass square difference Δm^2_{atm}), and at searching for the CP violation in the neutrino sector.

However, oscillation experiments will not help in fixing the absolute values of the neutrino masses, and even more importantly, will tell us nothing about the charge conjugation properties of neutrinos. Different approaches are needed for that task.

Regarding the mass scale, we can combine the oscillation results and the results of tritium beta-decay experiments to conclude that the masses of at least two (out of the total of three active) neutrinos are bracketted by 10 meV $\leq m_\nu \leq$ 2 - 3 eV. Therefore, neutrino masses are six or more orders of magnitude smaller than the masses of the other fermions and their mass pattern, i.e. the mass ratios of neutrinos, is clearly rather different than the pattern of masses of the charged fermions. This suggests that, perhaps, the origin of the neutrino mass is different from the origin (presumably coupling to the Higgs) of the masses of the other charged fermions.

The most popular explanation of the smallness of neutrino mass is the see-saw mechanism which implies, among other things, that neutrinos are Majorana particles, and consequently that the total lepton number should not be conserved. Hence tests of lepton number conservation acquire a fundamental importance. Among the possible tests of lepton number conservation the study of $0\nu\beta\beta$ decay is by far the most sensitive.

Observing the $0\nu\beta\beta$ decay would guarantee that neutrinos are massive Majorana fermions. This qualitative statement (or theorem), however, does not in general allow us to deduce the magnitude of the neutrino mass once the rate of the $0\nu\beta\beta$ decay have been determined. It is important to stress, however, that quite generally an observation of **any** total lepton number violating process, not only of the $0\nu\beta\beta$ decay, would necessarily imply that neutrinos are massive Majorana fermions.

In the following I will leave aside the important issue of the $0\nu\beta\beta$ mechanism. Instead, I will assume that the only way the decay can occur is through the exchange of a virtual light, but massive, Majorana neutrino (the mass eigenstate neutrinos ν_1, ν_2 and ν_3 of the oscillation theory) between the two nucleons undergoing the transition, and that these neutrinos interact by the standard left-handed weak currents. Observation of the $0\nu\beta\beta$ decay, independently of its mechanism, would constitute a proof that the lepton number is not conserved and hence that neutrinos are massive Majorana particles.

Moreover, if we accept the above assumption about the mechanism of the decay, we can relate the $0\nu\beta\beta$-decay rate to a quantity containing information about the absolute neutrino mass. That relation is given by the well known formula

$$\frac{1}{T_{1/2}^{0\nu}} = G^{0\nu}(Q,Z)|M^{0\nu}|^2 \langle m_{\beta\beta}\rangle^2 , \tag{1}$$

where $G^{0\nu}(Q,Z)$ is an easily calculable phase space factor that depends on the transition Q value and on the nuclear charge Z. The Z dependence is caused by the Coulomb interaction of the emitted electrons. $M^{0\nu}$ is the nuclear matrix element that can be evaluated in principle, although with a considerable uncertainty and is discussed in detail later, and finally the quantity $\langle m_{\beta\beta}\rangle$ which is the effective neutrino Majorana mass.

The magnitude of $\langle m_{\beta\beta}\rangle$, in turn, is constrained by the known parameters governing the neutrino oscillation phenomena (mixing angles and Δm^2) for the two possible mass ordering (hierarchies) and by the absolute mass scale. Depending on that mass scale, one can divide the range of the quantity $\langle m_{\beta\beta}\rangle$ into three regions:

- The degenerate mass region where all $m_i \gg \sqrt{\Delta m^2_{atm}}$. In that region $\langle m_{\beta\beta}\rangle \geq 0.1$ eV, corresponding crudely to 0ν half-lives of 10^{26-27} years. To explore it (in a realistic time frame), $\sim$ 100 kg of the decaying nucleus is needed. Several experiments aiming at such sensitivity are being built and should run very soon and give results within the next $\sim$ 3 years. Moreover, this mass region (or a substantial part of it) will be explored, in a similar time frame, by the study of ordinary β decay and by the observational cosmology. These latter techniques are independent of the Majorana nature of neutrinos.
- The so-called inverted hierarchy mass region where $20 < \langle m_{\beta\beta}\rangle < 100$ meV and the $0\nu\beta\beta$ half-lives are about 10^{27-28} years. (The name is to some extent a misnomer. In that interval one could encounter not only the inverted hierarchy but also a quasi-degenerate but normal neutrino mass ordering. Successful observation of the $0\nu\beta\beta$ decay will not be able to distinguish these possibilities.) To explore this mass region, $\sim$ ton size sources would be required. Proposals for the corresponding experiments exist, but none has been funded as yet, and presumably the real work will begin depending on the experience with the various $\sim$ 100 kg size sources. Timeline for exploring this mass region is $\sim$ 10 years.
- Normal mass hierarchy region where $\langle m_{\beta\beta}\rangle \leq$ 10-20 meV. To explore this mass region, $\sim$ 100 ton sources would be required. There are no realistic proposals for experiments of this size at present.

Obviously, relating the measured decay halflife to $\langle m_{\beta\beta}\rangle$ requires a knowledge of the nuclear matrix elements $M^{0\nu}$. Any error or uncertainty in their value will be reflected in the corresponding error and uncertainty in $\langle m_{\beta\beta}\rangle$. Moreover, planning a $0\nu\beta\beta$-decay experiment, and comparing the results of experiments with different nuclei also relies on the knowledge of the nuclear matrix elements $M^{0\nu}$.

2. What is needed in order to evaluate $M^{0\nu}$ nuclear matrix elements

The matrix elements $M^{0\nu}$ must be evaluated theoretically; they cannot be extracted from any other nuclear experiments. Since all $0\nu\beta\beta$-decay candidate nuclei are relatively heavy, the corresponding many body problem cannot be solved without approximations. The fundamental issue is, in that case, how to estimate the corresponding uncertainty. But even aside from this problem, in order to perform the calculation, and do a good job, one needs to use a proper effective transition operator. In the past, various authors used different formulations, and consequently got different results. Only recently has a consensus emerged in the community of practitioners, and most recent evaluations use the same basic ingredients.

Ordinary beta decay, and also the $2\nu\beta\beta$ decay, are low momentum transfer processes. The long wavelength approximation is valid in that case and the relevant operators are just the Fermi ($\tau^{\pm}$) and Gamow-Teller ($\vec{\sigma}\tau^{\pm}$) operators. In analogy, the early evaluations of the $0\nu\beta\beta$-decay matrix elements used transition operators based on the two-nucleon generalization of these basic operators. Since the $0\nu\beta\beta$-decay involves the exchange of a virtual Majorana neutrino, its propagator should be also included. In the momentum representation the transition operator is then

$$\Omega = \tau^+\tau^+ \frac{-g_V^2 + g_A^2\sigma_{12}}{q(q + E_m - (M_i + M_f)/2)} \,, \quad \sigma_{12} = \vec{\sigma}_1 \cdot \vec{\sigma}_2 \,. \tag{2}$$

In order to obtain the coordinate representation one makes the Fourier transform and arrives at the "neutrino potential"

$$H(r, E_m) = \frac{R}{2\pi^2 g_A^2} \int \frac{d\vec{q}}{\omega} \frac{1}{\omega + A_m} e^{i\vec{q}\cdot\vec{r}} = \frac{2R}{\pi r g_A^2} \int_0^{\infty} dq \frac{q \sin(qr)}{\omega(\omega + A_m)} \,, \tag{3}$$

where the nuclear radius $R = 1.2A^{1/3}$ fm has been added as an auxiliary factor so that H becomes dimensionless. (ω is just the magnitude of q and A_m is the excitation energy of the virtual intermediate state with respect to $(M_i + M_f)/2$.)

With the operator (2) one can proceed in two ways. Either the dependence on the energy of the intermediate state E_m is taken into account explicitly (this is often done when the Quasiparticle Random Phase Approximation (QRPA) is used), or one can use the closure approximation, neglecting the dependence on E_m and thus avoiding the necessity to treat the virtual intermediate states (this is usually done in the Nuclear Shell Model (NSM)). Since the typical values of $q \equiv \omega$ are quite high, as I will show next, the closure approximation is quite adequate and the corresponding error is small (few % at most).

What are, then, the characteristic values of the neutrino momentum q? Clearly, since the virtual neutrino connects two nucleons in the nucleus, its momentum is constrained by the uncertainty relation $q \geq 1/R$. Already this argument shows that the long wavelength approximation is not valid. Very recently the calculation within QRPA [1] and NSM [2] showed that nuclear structure effects cause cancelation between the contribution of lower values of q or equivalently higher values of r, the distance between the neutrons that undergo the transition. In Figure 1 the dependence of $M^{0\nu}$ on r is shown. Note that only $r \leq$ 2-3 fm contribute, i.e. distances considerably smaller than the nuclear radii R. The lower panel, which will be explained below, shows the cancelation between two opposing tendencies that is almost perfect for $r \geq$ 2-3 fm. A very similar result have been found in NSM [2].

The relatively short range of the effective operator (or equivalently relatively high q values) has important consequences. The following effects that are usually not relevant in applications

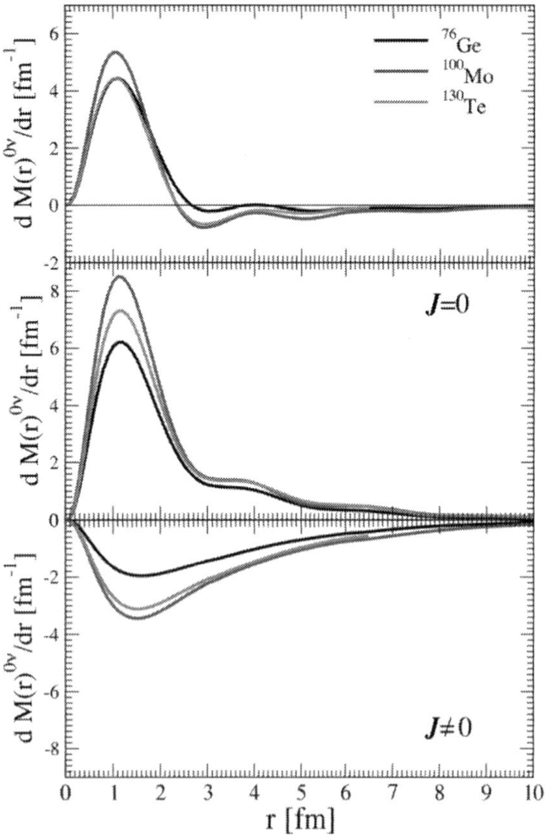

Figure 1. The dependence on r of $M^{0\nu}$ for ^{76}Ge, ^{100}Mo and ^{130}Te. The upper panel shows the full matrix element, and the lower panel shows separately 'pairing' ($\mathcal{J} = 0$ of the two participating neutrons) and 'broken pair' ($\mathcal{J} \neq 0$) contributions. The QRPA has been used.

of nuclear structure theory dealing with low-lying nuclear states and transitions between them need to be considered when the matrix elements $M^{0\nu}$ are evaluated;

- Higher order nuclear currents, in particular induced pseudoscalar and weak magnetism. These currents are proportional to q/m_π and q/M_p and thus suppressed in ordinary β decay, but need to be included when $0\nu\beta\beta$-decay is considered.
- Nucleon-nucleon short range repulsion that causes correlations between nucleons which are not included when the nuclear wave functions are constructed from products of the single-nucleon mean field wave functions.
- Nucleon composite nature or finite size. At high momentum transfer weak transitions (both the charge changing or charge conserving) that leave a nucleon intact are suppressed. This is a consequence of the finite size of the nucleon. That effect is usually taken into account by introducing nucleon form factors with the characteristic cut-off $\Lambda_{cut} \sim 1$ GeV.

In the past various authors neglected some or all of these effects. That is one of the reasons (albeit not the main one) for the variation between the published values of matrix elements $M^{0\nu}$ stressed forcefully by Bahcall and collaborators [3] and interpreted as uncertainty. Clearly, the above effects need to be included, and the proper discussion must concentrate on the most rigorous way of including them, not on the need to consider them or not.

The main issue, naturally, is the treatment of nuclear structure; in other words, how the wave functions of the ground states of the initial and final nuclei are evaluated. To understand the difficulties, let us recall that the nuclear matrix elements for both modes of the $\beta\beta$ decay (the 2ν

as well as the 0ν) depend sensitively on the competition of two opposing tendencies related to the two important components of the effective nucleon-nucleon interaction. The pairing interaction is responsible for the extra binding of even-even nuclei and thus also responsible for the existence of $\beta\beta$ decay itself. If pairing was the only relevant component of the nuclear interaction, all like nucleons would reside in Cooper-like pairs with 0^+. The matrix elements $M^{0\nu}$ would be large and positive in our phase convention (the $\mathcal{J} = 0$ part in Figs. 1 and 2).

However, that is not so in reality. The nuclear interaction contains other important parts, in particular the neutron-proton force. The presence of these components of the interaction means that the ground states of the initial and final nuclei contain admixtures of pairs with $\mathcal{J} \neq 0$. Such "broken pairs" typically give a negative contribution to $M^{0\nu}$ as also shown in Figs. 1 and 2. Moreover, the positive and negative contributions are of similar magnitude thus increasing the sensitivity of the final value of $M^{0\nu}$ to the corresponding interaction strengths. (That competition was first noted, it seems, in Ref. [4] and stressed again recently in [1] within the QRPA and in [5] within the NSM.)

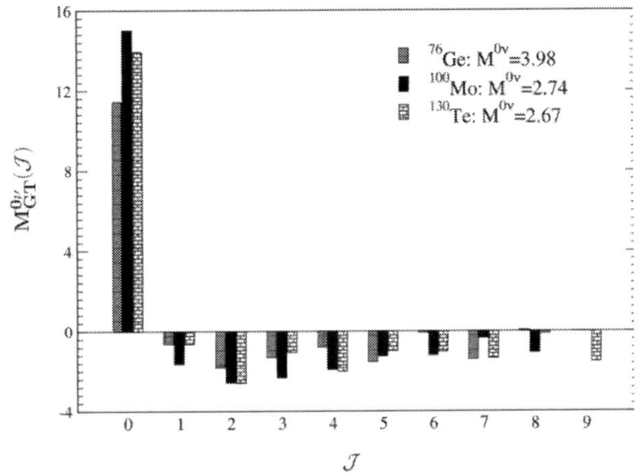

Figure 2. Contributions of different angular momenta $\mathcal{J}$ associated with the two decaying neutrons to $M^{0\nu}$ in ^{76}Ge, ^{100}Mo and ^{130}Te. The QRPA is used, with the interaction strength g_{pp} adjusted so that the $2\nu\beta\beta$ lifetime is correctly reproduced

In our calculations based on QRPA (and RQRPA, renormalized QRPA) calculations [1, 6, 7] we adjust the neutron-proton particle-particle interaction coupling constant, responsible for the "broken pairs" contribution, using the known $2\nu\beta\beta$-decay lifetimes or, equivalently, the corresponding $M^{2\nu}$ matrix elements. We based this procedure on the fact that the Gamow-Teller strength, the contribution of the 1^+ virtual intermediate states that are fully responsible for the $M^{2\nu}$, is the quantity most sensitive to the corresponding parameter, usually denoted as g_{pp}. The nominal value, corresponding to the G-matrix based on the realistic nucleon-nucleon force, is $g_{pp} = 1$. The renormalized g_{pp} has values between 0.8 and 1.2.

In the NSM there is no analog to the adjustment of the g_{pp} parameter. The hamiltonian (primarily its so-called monopole part) is adjusted so that a set of nuclear spectroscopic data is optimally reproduced. There is no attempt to reproduce specifically the $2\nu\beta\beta$-decay lifetime, but the agreement with experiment is, in most cases, acceptable [2, 5, 8].

Within QRPA one can trace the main reasons for the differences between the calculated values of $M^{0\nu}$ matrix elements by different authors noted in [3] to the different ways the effective interaction was chosen, and adjusted to the chosen set of the included single particle orbits. The advantage of the above quoted procedure (choose the coupling constant g_{pp} so that the known $2\nu\beta\beta$-decay lifetimes are reproduced) is that the dependence on the other chosen input parameters is greatly reduced.

3. Short range correlations

One of the items listed in the previous section is the inclusion in the calculation of the effect of the short-range nucleon-nucleon repulsion known from scattering experiments. Two nucleons strongly repel each other at distances $r \leq 0.5 - 1.0$ fm, i.e. distances very relevant to the evaluation of the $M^{0\nu}$ as shown in Fig. 1. The nuclear wave functions used in QRPA and NSM, which are products of the mean field single-nucleon wave functions, do not take into account the influence of this repulsion that is irrelevant in most standard nuclear structure theory applications. The most natural way to include this effect is to modify the radial dependence of the effective $0\nu\beta\beta$ operator so that the effect of short distances (small values of r) is reduced. However, no consensus on how to do this properly has been reached as yet.

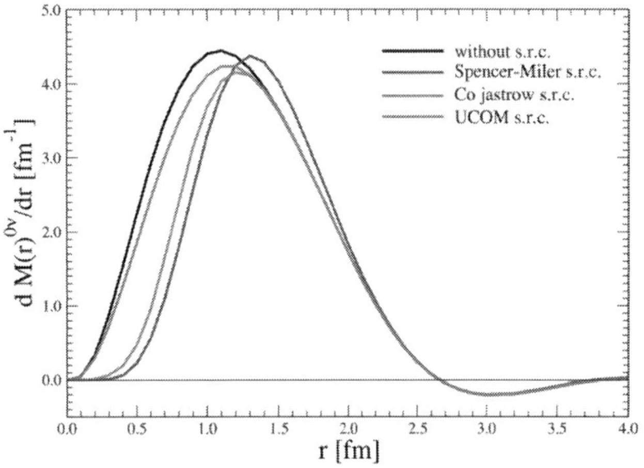

Figure 3. The r dependence of $M^{0\nu}$ in ^{76}Ge. The four curves show the effects of different treatments of short-range correlations. The resulting $M^{0\nu}$ values are 5.32 when the effect is ignored, 5.01 when the UCOM transformation [9] is applied and 4.14 when the treatment based on the Fermi hypernetted chain [10] and 3.98 when the phenomenological Jastrow function is used. [11]. The QRPA has been used.

This simplest way, used traditionally and in recent works [1, 2], includes the effect of short range correlations by multiplying the neutrino potential by the square of a Jastrow-like function first derived in [11] and in a more modern form in [10]. That phenomenological procedure reduces the magnitude of $M^{0\nu}$ by 20-25% as illustrated in Fig. 3. Recently, another procedure, based on the Unitary Correlation Operator Method (UCOM) has been proposed [9]. That procedure, still applied not fully consistently, reduces the $M^{0\nu}$ much less, only by about 5% [12]. It is prudent even though not really satisfactory, as suggested in [1], to include these two possibilities as extremes and the corresponding range as systematic error.

Another effect, also already mentioned, that needs to be taken into account is the nucleon finite size. That is included, usually again phenomenologically, by introducing the dipole form of the nucleon form factor

$$f_{V,A} = \left(\frac{1}{1 + q^2/\Lambda_{V,A}^2} \right)^2 , \tag{4}$$

where the cut-off parameters $\Lambda_{V,A}$ have values (deduced in the reactions of free neutrinos with free or quasifree nucleons) ~ 1 GeV. This corresponds to a nucleon size of $\sim 0.5 - 1.0$ fm. Note that in our case we are dealing with bound nucleons and neutrinos far off mass shell, hence it is not obvious that the above form factors are applicable. It turns out, however, that once the short range correlations are properly included (by either of the procedures discussed above) the $M^{0\nu}$ becomes essentially independent of the adopted values provided $\Lambda_{V,A} \geq 1$ GeV is used. However, in the past various authors neglected the effect of short range correlations, and in that case a proper inclusion of nucleon form factor (or their neglect) again causes variations in the calculated $M^{0\nu}$ values.

4. Discussion

Within the QRPA (and RQRPA) the full range of the calculated nuclear matrix elements $M^{0\nu}$ is presented in Table I in Ref. [1] separated into entries with the two alternative ways of treating the short range correlations (Jastrow and UCOM). The quoted ranges reflect the differences in the $M^{0\nu}$ values evaluated in different sizes of the included single-particle basis (usually two, three or four oscillator shells), differences between QRPA and RQRPA, as well as the full range obtained with the unquenched strength of the axial current weak interaction ($g_A = 1.25$) and its quenched analog ($g_A = 1.0$). The effect of the uncertainty in the experimental $2\nu\beta\beta$ decay lifetime (even though usually small) is also included. Note that these analogs of error bars are highly correlated; if the true value for one element is, say, near the upper edge, it will be near the upper edge for the other candidate nuclei as well.

The most recent NSM results [2], obtained with Jastrow type short range correlation corrections, are noticeably lower than the QRPA values. That difference is particularly acute in the lighter nuclei ^{76}Ge and ^{82}Se. While the QRPA and NSM agree on many aspects of the problem, in particular on the role of the competition between "pairing" and "broken pairs" contributions and on the r dependence of the matrix elements, the disagreement in the actual values remains to be explained.

When one compares the experimentally determined 2ν matrix elements and the calculated 0ν matrix elements one can note the fast variation in $M^{2\nu}$ when going from one nucleus to another while $M^{0\nu}$ changes only rather smoothly, in both QRPA and NSM. This is presumably related to the high momentum transfer (or short range) involved in $0\nu\beta\beta$. The smooth variation with Z and A of the $M^{0\nu}$ matrix elements makes the comparison of results obtained in different nuclei easier and more reliable. Altogether, we see that while a substantial progress has been achieved and calculations by different authors agree substantially better with each other now than before, we are still somewhat far from being able to evaluate the $0\nu\beta\beta$ nuclear matrix elements confidently and accurately.

Acknowledgments

It has been a great pleasure to participate in this workshop celebrating the contribution of Frank Avignone and Ettore Fiorini to neutrino physics. I look forward to many more years of collaboration and friendship with both of them. The original results reported here were obtained in collaboration with Fedor Šimkovic, Vadim Rodin, Amand Faessler and Jonathan Engel. The fruitful collaboration with them is gratefully acknowledged.

5. References

[1] Šimkovic F, Faessler A, Rodin VA, and Vogel P, *Phys. Rev.* C**77**, 045503 (2008).
[2] Menendez J, Poves A, Caurier E, and Nowacki F, arXiv:0801.3760.
[3] Bahcall JN, Murayama H, and Pena-Garay C, *Phys. Rev. D* **70**, 033012 (2004).
[4] Engel J, Vogel P and Ji X, *Phys. Lett. B* **225**, 5 (1989).
[5] Caurier E, Menendez J, Nowacki F and Poves A, *Phys. Rev. Lett.* **100**, 052503 (2008).
[6] Rodin VA, Fässler A, Šimkovic F, and Vogel P, *Phys. Rev. C* **68**, 044302 (2003).
[7] Rodin VA, Fässler A, Šimkovic F, and Vogel P, *Nucl. Phys.* **A766**, 107 (2006); erratum **A793**, 213 (2007).
[8] Retamosa J, Caurier E, and Nowacki F, *Phys. Rev. C* **51**, 371 (1995).
[9] H. Feldmeier, T. Neff, R. Roth and J. Schnack, *Nucl. Phys.* **A632**, 61 (1998).
[10] A. Fabrocini, F. A. de Saavedra and G. Co', *Phys. Rev. C* **61**, 044302 (2000), and G. Co', private communication.
[11] G. A. Miller and J. E. Spencer, *Ann. Phys.* **100**, 562 (1976).
[12] Kortelainen M and Suhonen J, *Phys. Rev. C* **75**, 051303 (2007); ibid *Phys. Rev. C* **76**, 024315 (2007); Kortelainen M, Civitarese O, Suhonen J, and Toivanen J, *Phys. Lett.* **B647**, 128 (2007).

Carolina International Symposium on Neutrino Physics IOP Publishing
Journal of Physics: Conference Series **173** (2009) 012011 doi:10.1088/1742-6596/173/1/012011

Nuclear-structure corrections to the double-beta-decay operator

Jonathan Engel

Dept. of Physics and Astronomy, University of North Carolina, Chapel Hill, NC, 27516-3255, USA

E-mail: engelj@physics.unc.edu

Abstract. In order to extract an effective neutrino mass from a measurement of neutrinoless double-beta decay, one must calculate nuclear matrix elements that help determine the decay rate. QRPA and shell-model calculations of the matrix elements are contrasted, and the possibility is raised of improving the shell-model calculations by incorporating missing configurations into an effective decay operator. Preliminary results from a perturbative treatment of the missing configurations are presented.

1. Introduction

Neutrinoless double beta decay is a very slow lepton-number-violating nuclear transition that occurs if neutrinos have mass (which they do) and are their own antiparticles. An initial nucleus (Z, A), with proton number Z and total nucleon number A, decays to $(Z+2, A)$, emitting two electrons in the process [1].

The neutrino masses and mixing matrix figure prominently in the decay. The rate, assuming that the process is mediated by the exchange of a light virtual neutrino, is

$$[T_{1/2}^{0\nu}]^{-1} = G_{0\nu}(Q, Z)\,|M_{0\nu}|^2\,\langle m_{\beta\beta}\rangle^2 \,, \tag{1}$$

where Q is the energy difference between the initial and final nuclei, Z is the charge of the initial nucleus, $G_{0\nu}(Q, Z)$ is a tabulated phase-space measure, $M_{0\nu}$ is the all-important nuclear matrix element to which we return shortly, and $m_{\beta\beta}$ is a linear combination of neutrino masses:

$$m_{\beta\beta} \equiv |\sum_{k=1}^{3} m_k U_{ek}^2| \,. \tag{2}$$

Here, m_k is the mass of the $k^{\rm th}$ neutrino (these mass eigenstates are linear combinations of the electron-, mu- and tau-neutrinos) and U_{ek} is the element of the unitary mixing matrix that connects that neutrino to the electron neutrino. The quantity $m_{\beta\beta}$ is what we want to extract from the decay rate. We cannot do so, however, without knowing the matrix element $M_{0\nu}$.

A precise expression for the matrix element is complicated, but with a few approximations it can be written to within an accuracy of about 30% as

$$M_{0\nu} \approx M_{0\nu}^{GT} - \frac{g_V^2}{g_A^2} M_{0\nu}^{F} \tag{3}$$

with g_V and g_A the vector and axial-vector coupling constants, and

$$M_{0\nu}^F = \langle f|\sum_{a,b} H(r_{ab},\overline{E})\tau_a^+\tau_b^+|i\rangle \ , \quad (4)$$

$$M_{0\nu}^{GT} = \langle f|\sum_{a,b} H(r_{ab},\overline{E})\vec{\sigma}_a\cdot\vec{\sigma}_b\tau_a^+\tau_b^+|i\rangle \ . \quad (5)$$

Here $|i\rangle$ and $|f\rangle$ are the initial and final nuclear ground states, a, b label particles, $\overline{E}$ is an average excitation energy, and H is a "neutrino potential", given by

$$H(r,\overline{E}) \approx \frac{2R}{\pi r}\int_0^\infty dq \frac{\sin qr}{q+\overline{E}-(E_i+E_f)/2} \ . \quad (6)$$

R is the nuclear radius, inserted to make the matrix element dimensionless. Equation (3) is the form we work with in what follows.

2. Two ways to calculate $M_{0\nu}$

Most calculations of $M_{0\nu}$ are in the neutron-proton Quasiparticle Random Phase Approximation (QRPA); the rest are in the shell model. The two methods have complementary virtues. The QRPA includes many single-particle levels outside a relatively small "inert" core, but limits itself to a particular kind of correlation. The shell model includes arbitrary complicated correlations, but only in a few single-particle levels outside a much larger inert core. The current predictions of the two models, after a recent shaking out period, are related as shown in figure 1 below.

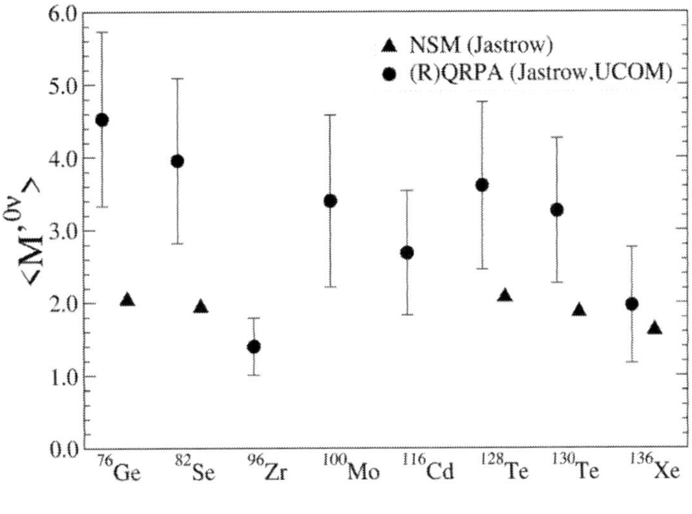

Figure 1. Shell-model (triangles, from reference [2]) and QRPA (circles with error bars, from reference [3]) 0ν matrix elements.

The error bars in the figure come from the use of two or more prescriptions for treating short-range correlations and for the effective value of the axial-vector coupling constant g_A in nuclei. The shell-model calculations simply use one of the prescriptions for each (short-range correlations treated via the method of reference [4] and an effective g_A of 1.25), and include an estimate for the relatively small corrections from forbidden terms in the weak current. Another difference between the two calculations is in the Hamiltonian. The QRPA Hamiltonian is not as realistic as possible, but adjusted to reproduce two-neutrino double-beta decay. The shell-model Hamiltonian is fixed by a combination of theory and fitting to spectra. When all is said and done, the shell-model results are pretty uniformly smaller than those of the QRPA, sometimes by a factor of 2 or more.

Which set of calculations is closer to the truth? To find out, one has to correct one or the other for the physics it omits. Although it is possible to add missing correlations to the QRPA, it is not easy to do so because one would have to undo the simplified treatment of correlations (e.g., BCS pairing) first. By contrast, there is a systematic procedure for adding the effects of missing single-particle levels to a shell-model space [5, 6]. While the procedure is perturbative and not always reliable for that reason, it often works and we try it here.

3. Effective-Operator Formalism in the Shell Model

We give the briefest of introductions to the theory of shell-model effective operators. The formalism is based on the division of the many-body Hilbert space into the shell-model space (always called P) and the rest (Q). One begins by defining operators with the same names that project onto these spaces:

$$P = \sum_{i \,\epsilon P} |i\rangle\langle i| \qquad\qquad Q = \sum_{\text{other } i} |i\rangle\langle i| \,. \tag{7}$$

Here the sums are over orthogonal sets, so that P and Q are true projectors. Next we define an effective Hamiltonian that when acting on the P-space projection of an eigenstate Ψ_a gives back that projection with the correct eigenvalue:

$$H_{\text{eff}}(E_a)P|\Psi_a\rangle = E_a P|\Psi_a\rangle \,. \tag{8}$$

Similarly, for any "bare" operator $\mathcal{M}$ (where "bare" means "acting in the full model space"), we can define an effective operator $\mathcal{M}^{\text{eff}}$ that acts in the P space with matrix elements related to those of the bare operator by:

$$\frac{\langle\Psi_a|P\mathcal{M}^{\text{eff}}P|\Psi_b\rangle}{\sqrt{\langle\Psi_a|P|\Psi_a\rangle\langle\Psi_b|P|\Psi_b\rangle}} = \langle\Psi_a|\mathcal{M}|\Psi_b\rangle \tag{9}$$

Next, through a series of steps, one can develop a perturbative expansion for the matrix elements of $\mathcal{M}^{\text{eff}}$. The starting point is the Bloch-Horowitz equation (or, more precisely, its extension to the wave function) relating an arbitrary eigenstate $\Psi(E)$ to its P-space projection:

$$\Psi(E) = \mathcal{N}\left[P|\Psi\rangle + \frac{1}{E - QH}QHP|\Psi\rangle\right] \tag{10}$$

Here E is the energy of the eigenstate and $\mathcal{N}$ a normalization coefficient.

The perturbative expansion based on equation (10) is similar to the Bethe-Breuckner-Goldstone expansion for nuclear matter. Here, things are a little more complicated; we are computing a model-space interaction rather than the energy, so the diagrams look a bit different, and the isolation of a model space that contains empty levels as well as occupied levels leads to "folded diagrams", the complications of which we need not go into here. The first steps for obtaining an effective Hamiltonian H_{eff}, however, are almost exactly the same as in nuclear matter. One defines a G matrix as the sum over all "ladder diagrams", as in figure 2 below. The railed lines in the figure indicate states well above the shell-model single-particle space (e.g., more than six or seven shells above).

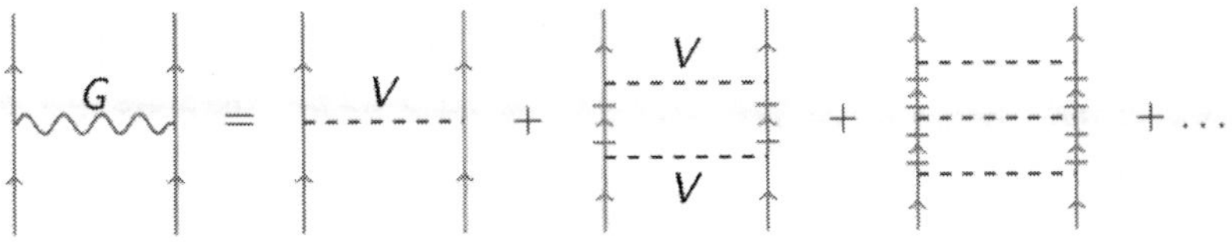

Figure 2. Diagrammatic construction of a G-matrix, the first step in an effective interaction

Although here are an infinite number of diagrams, the sum can be carried out indirectly, e.g., through the solution of the Beth-Goldstone equation. The idea is that the hard short-range core that makes the nucleon-nucleon interaction intractable is treated exactly through the sum over *all* two-particle states at high energy (short-range). The effective interaction $G[V]$ between low-energy states that results has a soft core because the effects of short-range physics have already been accounted for in the ladder sum (the argument V is meant to reinforce the fact that G depends on the "bare interaction" V).

After this nonperturbative construction of the G matrix, one can use perturbation theory in G to add the effects of states that are at low energy but still outside the model space. The set of diagrams in figure 3 includes all such effects up to second order in G, and some third-order effects. In this figure, the upward-going lines are low-lying "particle" states, that is, single-particle states in the six- or seven-shell gap between the shell-model levels and the high-energy levels (including the shell-model levels themselves). Downward going lines are "hole" states that correspond to the vacating of levels below the shell-model space.

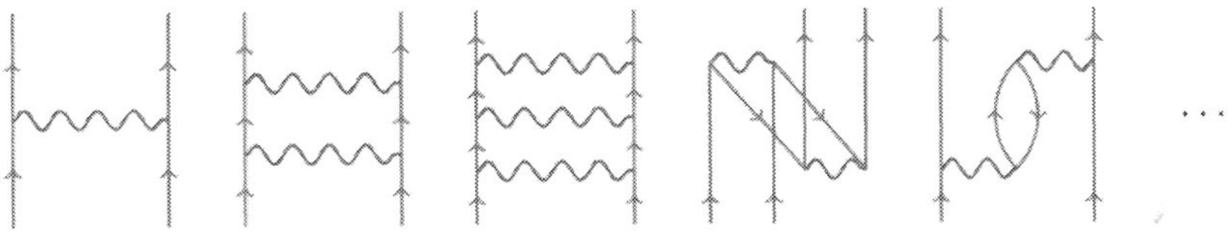

Figure 3. Low-order diagrams (in G) for the effective interaction H_{eff}.

Typically, more complicated graphs, including the folded ones, are added to the procedure. Even then, problems with convergence, three- and higher-body operators that are generally too complicated to include, etc., mean that the resulting interaction often must be tuned phenomenologically to reproduce spectra, especially in the monopole-monopole channel [7]. Sometimes, however, the perturbation theory by itself is enough to produce a pretty good interaction [6].

A similar construction can be carried out for any two-body operator $\mathcal{M}$. The expansion of equation (10) leads to a set of diagrams in which a horizontal line representing the bare operator appears once alongside an arbitrary number of interaction lines. The first step, in analogy with the construction of the G-matrix, is to construct an operator $\mathcal{M}_{\text{high}}$ by summing all diagrams with two particles excited to high energies in which one line in each diagram is the operator $\mathcal{M}$ rather than the interaction. The diagrams with different numbers of interaction lines before or after the operator line can be summed and replaced by G-matrices (or, more, precisely, the equivalent operator that acts on the entire space). If $\mathcal{M}$ is one of the double-beta operators appearing in equations (3), then two neutron lines become proton lines whenever it acts, and the ladder sum becomes that depicted in figure 4. Here all particle lines preceding the double-beta operator (red online) are neutrons and lines following the operator (blue online) are protons.

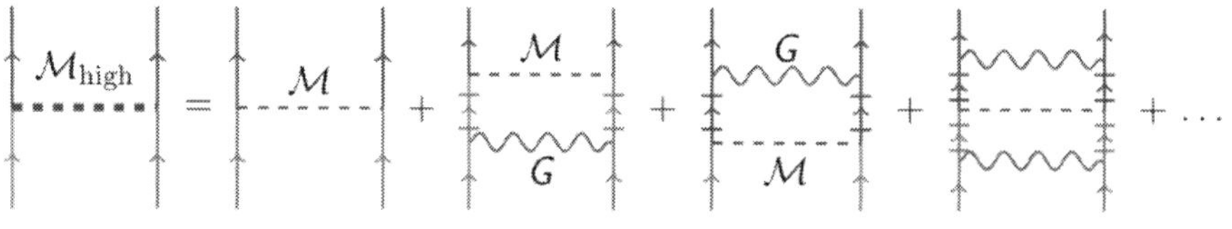

Figure 4. An effective double-beta operator that accounts for short-range ("high"-momentum) correlations in the nuclear wave function. This is the analog of the G matrix.

This sum of diagrams can be calculated simply from a G-matrix code through the trick

$$\mathcal{M} = \frac{d}{d\lambda} G[V + \lambda\mathcal{M}]|_{\lambda=0}\,, \tag{11}$$

that is by computing the G matrix corresponding to the interaction $V + \lambda\mathcal{M}$. The derivative filters out graphs that have one interaction replaced by a double-beta decay.

Having calculated this renormalized double-beta operator that includes the effects of short-range physics, one can use it, in parallel with the G-matrix, to calculate the additional renormalization from low-lying excitations. Here, we include the diagrams in figure 3, but with one G-matrix line replaced by $\mathcal{M}_{\text{high}}$. These diagrams, all first order in G, except for second-order ladders appear in figure 5. Short-range correlations are included at every vertex through the use of G and $\mathcal{M}_{\text{high}}$ in place of V and $\mathcal{M}$. Most of these diagrams can be calculated through the trick in equation (11). The core-polarization graphs (the last two in the figure) must be treated explicitly because the combination of neutron and proton lines that enters in those graphs for the effective operator in figure 5 are not the same as in the graphs for the effective interaction in figure 3.

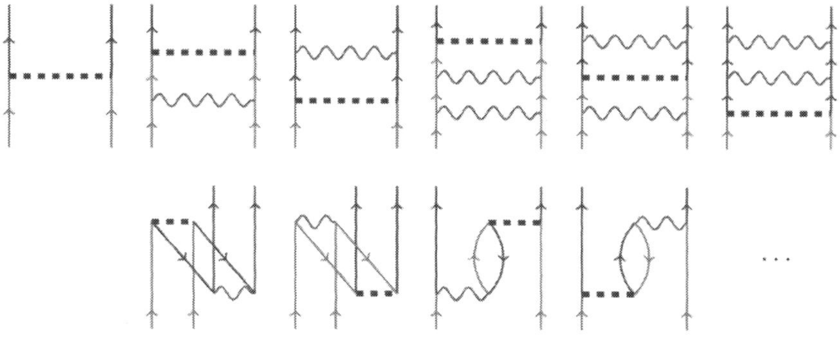

Figure 5. Low order diagrams (in G) contributing to the final effective double-beta operator. The thick dashed line is $\mathcal{M}_{\text{high}}$, expressed diagramatically in figure 3.

4. Results

The shell-model calculation of reference [2] used an interaction that was constructed largely through the effective-interaction theory discussed above (though it included higher-order and folded diagrams). The Bonn-C nucleon-nucleon interaction [8] was used as a starting point for the ladder-diagrams that enter the G-matrix. After cranking the theoretical machinery, the authors tweaked the interaction by fitting certain components to spectra.

In our calculation of the effective double-beta operator, to be as consistent as possible, we began from the same Bonn-C nucleon-nucleon interaction. After calculating the G-matrix and its extension to the Q space, we evaluated $\mathcal{M}_{\rm high}$ in figure 4 and the final effective-operator diagrams in figure 5. We then folded the matrix elements of this final effective operator with the two-body ground-state-to-ground-state transition densities from the shell-model calculation to obtain a transition matrix element for the decay ^{82}Se $\longrightarrow ^{82}$Kr. The preliminary results for the Gamow-Teller part of the matrix element (equation (5)) appear below in table 1.

Table 1. Renormalization of $M^{0\nu}_{GT}$ from equation (5) (preliminary results)

low-E border	bare	high-E ladders	all ladders	+ 4p-2h	+ 3p-1h
$5\hbar\omega$	3.33	3.06	4.17	5.38	—
$6\hbar\omega$	3.33	3.05	4.16	5.39	—
$7\hbar\omega$	3.33	3.06	4.17	5.39	—
$8\hbar\omega$	3.33	3.06	4.17	5.39	—

The table shows the contributions of various diagrams to the decay matrix element (except for the contributions of the core-polarization diagrams, which have not yet been correctly evaluated). Each row corresponds to a boundary, measured in $\hbar\omega$ from the valence fp shell, between the levels that are considered "high-energy" (the ruled lines in figures 2 and 4) and those that are not. The ladder and 4-particle-2-hole contributions increase the matrix element by some 60%. The core polarization (3-particle-1-hole) diagrams turn out to make it smaller again, though by how much we are not yet sure.

One thing that can be firmly deduced from these calculations is the size of the damping caused by short-range correlations (the high-E ladders in the third column), which are included in a self-consistent way for the first time here. As is apparent from the table, the short-range correlations reduce the matrix element by about 8%. This rigorous result is more in line with that of references [9, 10], which predicted a relatively small effect, than with that obtained from the phenomenological Jastrow function of reference [4].

Without the 3p-1h contributions, it is hard to draw a firm conclusion about the size of the overall correction to the matrix element[1]. Whatever the eventual result, however, it is clear that the individual corrections are significant. Since they are not all that much smaller than the matrix element itself, we shouldn't place too much trust in the perturbative treatment undertaken here. A more reliable nonperturbative evaluation of the corrections is very important, and we are working in that direction.

Acknowledgments

The author acknowledges Gaute Hagen, his collaborator in this work, and useful discussions with Morten Hjorth-Jensen. He thanks Alfredo Poves for supplying shell-model densities. This work was supported in part by the U.S. Department of Energy under grant DE-FG02-97ER41019.

References

[1] Avignone III F, Elliott S R and Engel J 2008 *Rev. Mod. Phys.* **80** 481
[2] Caurier E, Menendes J, Nowacki F and Poves A 2008 *Phys. Rev. Lett.* **100** 052503
[3] Šimkovic F, Faessler A, Rodin V, Vogel P and Engel J 2003 *Phys. Rev. C* **77** 045503
[4] Miller G and Spencer J 1976 *Ann. Phys.* **100** 562
[5] Ellis P J and Osnes E 1997 *Rev. Mod. Phys.* **1977** 777

[1] Complete results with the 3p-1h contributions, obtained after the writing of this paper, appear in reference [11]

[6] Hjorth-Jensen M, Kuo T T S and Osnes E 1995 *Phys. Rep.* **261** 125
[7] Dufour M and Zuker A P 1996 *Phys. Rev. C* **54** 1641
[8] Machleidt R 1989 *Adv. Nucl. Phys.* **19** 189
[9] Kortelainen M, Civitarese O, Suhonen J and Toivanen J 2007 *Phys. Lett. B* **647** 128
[10] Kortelainen M and Suhonen J 2007 *Phs. Rev. C* **75** 051303(R)
[11] Engel J and Hagen G 2009 (*Preprint* `arXiv:0904.1709`)

Nuclear Matrix Elements for Double Beta Decay in the QRPA approach: a critical review

Osvaldo Civitarese[1], **Jouni Suhonen**[2]

[1] Department of Physics, University of La Plata, c.c. 67 (1900) La Plata, Argentina
[2] Department of Physics, University of Jyväskylä, P. O. Box 35, FIN-40351, Jyväskylä, Finland

E-mail: osvaldo.civitarese@fisica.unlp.edu.ar

The excitement which goes with theoretical efforts towards the calculation of NME for DBD is, of course, due to the remarkable achievements of the active experimental groups. It is our pleasure and our honor to thank Profs. Ettore Fiorini and Frank Avignone for their life-long efforts devoted to the experimental exploration of DBD transitions. We wish for both of them many years of new contributions and inspirations.

Abstract. The calculation of nuclear matrix elements (NME) for double beta decay transitions (DBD) relies upon several approximations. The purpose of this note is to review some of these approximations, and their impact upon the NME. We shall present our results, which have been obtained in the framework of the proton-neutron quasiparticle random phase approximation (pnQRPA), and we shall focus on short range correlations, pairing, and symmetry effects.

PACS number(s): 21.60.Jz, 23.40.Bw, 23.40.Hc, 27.50.+e.

Keywords: Neutrinoless double beta decay, nuclear matrix elements, pairing correlations

1. Introduction

The calculation of NME for DBD transitions has been, and still it is, the target of choice for practitioners in the field of nuclear and nucleon structure. The body of literature is indeed huge and for the sake of brevity we shall only refer hereafter to few review articles [1, 2, 3, 4], and to some recently published papers [5, 6, 7, 8, 9]. The spreading of the theoretical results was so noticeable, even for physicists outside the field, that the credibility of the results was questioned severely, particularly with the claimed discovery of neutrinoless double beta decay [10, 11]. The situation was nearly catastrophic, with i) model predictions, obtained with the same models but varying within factors, and with ii) the proliferation of ad-hoc recipes, introduced to justify day by day changes of the theoretical estimations, sometimes also self-contradictory. To this we may add the almost recursive reference to back of the envelope calculations and hand waving arguments presented as illusive pieces of cakes. Fortunately enough the situation has improved recently, and the spread of the theoretical results is much smaller and therefore the reliability of the theoretical estimations of NME has improved considerably. In this note we would like to review some of the elements which, in our opinion, are crucial in determining the stability of the results, namely: a)the effect of short range correlations, b)the effect of pairing correlations, and c)symmetry violation effects introduced by the model assumptions. For details we shall refer the reader to the work of [5, 6, 7], for short range correlations, and to [12, 13] for pairing correlations. Concerning the issue of symmetry violation effects, which mostly reflect the failure of some of the extensions of the pnQRPA method, we shall refer the reader to refs. [14, 15]. The non-physical effects associated with attempts to go beyond the point of collapse of the pnQRPA approach are notorious, particularly in the case of some ad-hoc methods based on non-consistent corrections of the quasi-particle mean field. As a consequence, the results of some renormalized versions of the pnQRPA should be taken with extreme caution [14, 15].

2. Brief Review of Concepts

In the mass mode of the $0\nu\beta\beta$ decay a light virtual Majorana neutrino is exchanged by the two decaying neutrons of the initial nucleus. The measured $0\nu\beta\beta$-decay half-life is connected with the effective neutrino mass $\langle m_\nu \rangle$ through the relation

$$\left[t_{1/2}^{(0\nu)}(0_{\rm i}^+ \to 0_{\rm f}^+)\right]^{-1} = G_1^{(0\nu)} \left(\frac{\langle m_\nu \rangle}{m_{\rm e}}\right)^2 \left(M^{(0\nu)}\right)^2 , \tag{1}$$

where $G_1^{(0\nu)}$ is the leptonic phase-space factor [1, 2]. The $0\nu\beta\beta$ nuclear matrix element $M^{(0\nu)}$ consists of the Gamow–Teller, Fermi and tensor parts as

$$M^{(0\nu)} = M_{\rm GT}^{(0\nu)} - \left(\frac{g_{\rm V}}{g_{\rm A}}\right)^2 M_{\rm F}^{(0\nu)} + M_{\rm T}^{(0\nu)} . \tag{2}$$

Numerical calculations show that the tensor part in (2) is quite small and its contribution can be safely neglected. Most often people use the proton-neutron quasiparticle random-phase approximation (pnQRPA) [2] to treat the structure of double-beta decaying nuclei. Also some recent shell-model results are available [16, 17]. The pnQRPA contains a free parameter, the so-called particle-particle strength parameter, $g_{\rm pp}$, that controls the magnitude of the proton-neutron two-body interactions in the $T = 0$ pairing channel [18, 19]. The fixing of the value of this parameter has been done either by using the data on two-neutrino double beta ($2\nu\beta\beta$) decay or the data on single beta decay [20, 21].

3. Short Range Correlations

The average exchanged momentum is large, of the order of 100 MeV, so that the two neutrons tend to overlap. To prevent this a Jastrow type of correlation function has been introduced, in an ad-hoc manner, into the $0\nu\beta\beta$ calculations [22]. This way of using the Jastrow procedure is a very rudimentary way to introduce short-range correlations into many-nucleon systems. A more sophisticated microscopic approach for the inclusion of short-range correlations is the unitary correlation operator method (UCOM). In the UCOM one obtains the correlated many-particle state from the uncorrelated one by a unitary transformation and thus the norm of the correlated state is conserved and no amplitude is lost in the relative wave function. In [5, 6, 7] it was demonstrated that the Jastrow procedure leads to the excessive reduction of 25% – 40% in the magnitudes of the $0\nu\beta\beta$ nuclear matrix elements. At the same time the UCOM reduces the magnitudes of the matrix elements only by 4% – 16%. A fully consistent use of the UCOM method requires to treat not only the wave functions but the Hamiltonian, as well. Here we present our results of nuclear-structure calculations for the $0\nu\beta\beta$ ground-state-to-ground-state decays of ^{76}Ge, ^{82}Se, ^{96}Zr, ^{100}Mo, ^{116}Cd, ^{128}Te, ^{130}Te, and ^{136}Xe. In the following we shall talk

Table 1. Calculated $0\nu\beta\beta$ nuclear matrix elements, the used $g_{\rm pp}$ and $g_{\rm A}$ values and the resulting half-lives. The Jastrow (J) or UCOM (U) plus the other corrections are included. The half-lives $t_{1/2}^{(0\nu)}$ are expressed in units of $\mathrm{yr}/(\langle m_\nu\rangle[\mathrm{eV}])^2$. See [5, 6, 7] for further details.

Nucleus	$g_{\rm pp}$	$g_{\rm A}$	$M^{(0\nu)}$(J)	$t_{1/2}^{(0\nu)}$(J)	$M^{(0\nu)}$(U)	$t_{1/2}^{(0\nu)}$(U)
^{76}Ge	1.02	1.00	−5.077	4.0×10^{24}	−6.555	2.4×10^{24}
	1.06	1.25	−4.029	2.6×10^{24}	−5.355	1.4×10^{24}
^{82}Se	0.96	1.00	−3.535	1.9×10^{24}	−4.597	1.1×10^{24}
	1.00	1.25	−2.771	1.2×10^{24}	−3.722	6.9×10^{23}
^{96}Zr	1.06	1.00	−3.131	1.2×10^{24}	−4.319	6.1×10^{23}
	1.11	1.25	−2.065	1.1×10^{24}	−3.117	4.7×10^{23}
^{100}Mo	1.07	1.00	−3.526	1.2×10^{24}	−4.849	6.2×10^{23}
	1.09	1.25	−2.737	7.9×10^{23}	−3.931	3.8×10^{23}
^{116}Cd	0.82 (β^- decay)	1.25	−3.981	3.5×10^{23}	−4.928	2.3×10^{23}
	0.97	1.00	−3.681	1.0×10^{24}	−4.682	6.3×10^{23}
	1.01	1.25	−3.034	6.1×10^{23}	−3.935	3.6×10^{23}
^{128}Te	0.86 (β^- decay)	1.25	−4.068	9.5×10^{24}	−5.509	5.2×10^{24}
	0.89	1.00	−4.279	2.1×10^{25}	−5.841	1.1×10^{25}
	0.92	1.25	−3.383	1.4×10^{25}	−4.790	6.9×10^{24}
^{130}Te	0.84	1.00	−4.061	9.5×10^{23}	−5.442	5.3×10^{23}
	0.90	1.25	−2.993	7.0×10^{23}	−4.221	3.5×10^{23}
^{136}Xe	0.74	1.00	−2.864	1.8×10^{24}	−3.719	1.1×10^{24}
	0.83	1.25	−2.053	1.4×10^{24}	−2.802	7.6×10^{23}

about the comparison between Jyvaskylas results [5, 6, 7] and those of Tübingen [8, 9] and the recent shell-model calculations [16, 17]. To obtain physical values of the proton-neutron particle-particle interaction strength, $g_{\rm pp}$, of the pnQRPA, we fit it using the experimental rates of the $2\nu\beta\beta$ decay. The fitting included the experimental errors and the uncertainty $1.0 \le g_{\rm A} \le 1.25$ in the value of the axial-vector coupling constant $g_{\rm A}$. For ^{136}Xe only the experimental lower limit of the $2\nu\beta\beta$ half-life is available. The extracted values of $g_{\rm pp}$ and the corresponding values of $g_{\rm A}$ are listed in Table 1. We also used the available β^- decay data to fix the value of $g_{\rm pp}$. The lowest value of $g_{\rm pp}$, compatible with the β^- decay data, is also included in Table 1. For ^{100}Mo the interval determined by the β^- decay is contained in that of the $2\nu\beta\beta$ decay. From Table 1 one notices that for ^{116}Cd and ^{128}Te the β^- determined interval is in conflict with the one determined through the $2\nu\beta\beta$ data. This discrepancy is especially striking for ^{116}Cd, as was already pointed out in [20]. In Table 1 we also give the values of the nuclear matrix element of (1), calculated either by including the Jastrow or the UCOM short-range correlations on top of the other mentioned corrections. In addition, we include the corresponding half-lives $t_{1/2}^{(0\nu)}$, expressed in units of $\mathrm{yr}/(\langle m_\nu\rangle[\mathrm{eV}])^2$. As seen, the UCOM matrix elements are considerably larger in magnitude than those computed by using the traditional Jastrow approach. Such a big difference between the Jastrow and the UCOM results reflects on the estimated neutrino-mass sensitivities of the presently running and the future double-beta-decay experiments.

4. Pairing Effects

The recently measured neutron pairing correlations in the ^{76}Ge and ^{76}Se nuclei can be used as a guideline to adjust single-particle energies, so that the resulting BCS occupations reproduce the measured pairing data. The proton energies were inspected by using the odd-mass nuclei adjacent to ^{76}Ge and ^{76}Se as spectroscopic tools. The nuclear matrix element of the neutrinoless double beta ($0\nu\beta\beta$) decay of ^{76}Ge is calculated in this fitted single-particle basis by using the proton-neutron quasiparticle random-phase approximation (pnQRPA) in a realistic model space and by adopting effective microscopic two-nucleon interactions. We include the nucleon-nucleon short-range correlations and other relevant corrections at the nucleon level, as explained in the previous section. It is found that the resulting $0\nu\beta\beta$ matrix element is smaller than in the previous pnQRPA calculations, and closer to the recently reported shell-model results [16, 17]. The BCS method gives occupations of the single-particle orbitals and the occupation amplitudes are connected to the energy differences between orbitals. The customary way to determine the single-particle energy difference is to use the Woods–Saxon mean-field potential. Small adjustments of the resulting energies can be made based on the data on energy levels of odd-mass nuclei in the neighborhood of the nucleus where the pnQRPA calculation is performed. Recently a new measurement to determine the single-particle energy spectra has been reported[13]. In [13] the neutron pair correlations in the ^{76}Ge and ^{76}Se nuclei were measured by (p,t) reactions. As a result, the vacancies in the neutron subspace $28 \leq \mathrm{N} \leq 50$, hereafter called the pfg subspace, could be deduced. The adjusted basis exactly reproduces the measured spectroscopic factors in a BCS calculation [12] As a model space we have used the $N = 3$ and $N = 4$ oscillator

Table 2. Single-particle energies in the pfg subspace. Shown are the adjusted energies and their Woods–Saxon counterparts.

	^{76}Ge		^{76}Se	
Orbital	WS [MeV]	Adjusted [MeV]	WS [MeV]	Adjusted [MeV]
$\nu 1p_{3/2}$	−11.52		−12.77	
$\nu 0f_{5/2}$	−10.72	−10.30	−11.98	−11.40
$\nu 1p_{1/2}$	−9.797	−9.550	−10.98	−10.70
$\nu 0g_{9/2}$	−7.030	−9.500	−8.305	−11.00
$\pi 1p_{3/2}$	−9.015		−7.000	
$\pi 0f_{5/2}$	−8.212		−6.288	
$\pi 1p_{1/2}$	−6.789		−5.001	
$\pi 0g_{9/2}$	−5.311	−7.100	−3.371	−5.300

shells and the $0h_{11/2}$ single-particle orbital, both for protons and neutrons. The single-particle energies are obtained from the Coulomb-corrected Woods–Saxon potential. These energies are presented for the pfg subspace in Table 2. The measured neutron vacancies [13] in this sub-space have been summarized in Table 3. As can be seen from this table, the computed vacancies in the Woods–Saxon basis (WS) are far from the measured ones: the $0g_{9/2}$ orbital is much too thinly occupied and the other orbitals are too full. This can be cured by lowering the energy of the $0g_{9/2}$ orbital. Shifting also the $0f_{5/2}$ and $1p_{1/2}$ orbitals slightly, as indicated in Table 2, brings the computed vacancies right on the measured ones. This is indicated in Table 3 by the column corresponding to the adjusted basis (labelled Adj) of Table 2. It is harder to adjust the proton single-particle energies since no data on their spectroscopic factors exist yet. The computed proton vacancies in the Woods–Saxon basis are given in Table 4. One possible way to access the adjusted basis for protons is to compute the spectra of the proton-odd nuclei ^{77}As and ^{77}Br, adjacent to ^{76}Ge and ^{76}Se. For this one can use the quasiparticle-phonon coupling in diagonalizing the residual nuclear Hamiltonian. Comparing the energies of the computed one-quasiparticle states with the available data gives information on the quality of the underlying BCS calculation and thus on the mean-field single-particle energies. The critical proton orbit for the success of the quasiparticle-phonon coupling calculation is $\pi 0g_{9/2}$. The corresponding

Table 3. Measured and calculated neutron vacancies in the pfg subspace. The calculations were done in the Woods–Saxon and adjusted basis.

	^{76}Ge			^{76}Se		
Orbital	Exp.	Adj.	WS	Exp.	Adj.	WS
$\nu 1p_{1/2}$	-	0.717	0.195	-	1.016	0.282
$\nu 1p_{3/2}$	-	0.386	0.162	-	0.584	0.213
$\nu 1p_{1/2} + \nu 1p_{3/2}$	1.13	1.10	0.357	1.59	1.60	0.495
$\nu 0f_{5/2}$	1.44	1.43	0.500	2.17	2.16	0.618
$\nu 0g_{9/2}$	3.52	3.50	5.43	4.20	4.22	7.06

Table 4. Predicted proton vacancies in the pfg subspace. The calculations were done in the Woods–Saxon and adjusted basis.

	^{76}Ge		^{76}Se	
Orbital	Adj.	WS	Adj.	WS
$\pi 1p_{1/2}$	1.82	1.80	1.70	1.62
$\pi 1p_{3/2}$	2.19	1.92	1.75	1.44
$\pi 0f_{5/2}$	4.58	4.43	3.80	3.32
$\pi 0g_{9/2}$	9.28	9.73	8.61	9.49

adjusted energy is indicated in Table 2 and the calculated proton vacancies are given in Table 4. As can be seen from these tables, the proton and neutron energy shifts and the resulting vacancies behave qualitatively in the same way. However, the adjustments on the proton side do not have such drastic effects on the vacancies as on the neutron side. After settling the problem with the single-particle energies we are ready to compute the $2\nu\beta\beta$ and $0\nu\beta\beta$ NME's. As usual, we consider the two extreme values of the axial-vector coupling constant, namely the bare value $g_A = 1.25$ and the strongly quenched value $g_A = 1.00$. In Table 5 we list the corresponding g_{pp} values as extracted by comparing the measured $2\nu\beta\beta$ half-lives with the computed ones. By using these values of g_{pp} we have calculated the $0\nu\beta\beta$ NME's of (2) and we summarize their values in Table 5. Our computed results of Table 5 can be compared with the results of other

Table 5. Matrix elements of (2) computed in this work for different values of g_A and g_{pp}. For the short-range correlations both the Jastrow and UCOM prescriptions have been used.

		Jastrow			UCOM		
g_A	g_{pp}	$M_{GT}^{(0\nu)}$	$M_F^{(0\nu)}$	$M^{(0\nu)}$	$M_{GT}^{(0\nu)}$	$M_F^{(0\nu)}$	$M^{(0\nu)}$
1.25	1.12	2.288	−0.772	2.779	3.385	−1.143	4.112
1.00	1.10	1.700	−0.579	2.279	2.413	−0.818	3.231

recent works in the field. A selection of recent calculations including the Jastrow and/or the UCOM correlator is given in Table 6. The second and third columns of this table give the results of [7] where exactly the same methods as here were applied, the only difference being the use of a different set of single-particle energies, where the neutron $0g_{9/2}$ orbital was shifted a good one MeV to better reproduce the low-energy spectra of ^{77}Ge and ^{77}Se in a BCS calculation. By comparing the results of these two calculations, Tables 5 and 6, one notices a significant reduction in the value of the total $0\nu\beta\beta$ matrix element $M^{(0\nu)}$. The Tübingen results [8, 9] are smaller than the results of [7]. The Jastrow results of the shell model [17] are the smallest in

Table 6. Values of the matrix element $M^{(0\nu)}$ obtained in some other recent works. Here (J) stands for Jastrow and (U) for UCOM.

g_A	(J)[7]	(U)[7]	(J)[8]	(J)[9]	(U)[9]	(J)[17]
1.25	4.029	5.355	3.92 – 4.51	4.68	5.73	2.300
1.00	3.249	4.195	3.46 – 3.83	3.33	3.92	-

Table 6. Interestingly, our present results for the Jastrow correlator, $M^{(0\nu)} = 2.779$, are closer to the shell-model result than the previous values quoted in [7]. The reduction of the magnitude of the pnQRPA calculated NME, which yields a value close to the shell-model result, is significant. The reason for the reduction of the magnitude of the $0\nu\beta\beta$ NME can be summarized by looking at the multipole decomposition of the NME. For the Fermi matrix element the reduction stems from the 0^+ intermediate states. For the Gamow–Teller matrix element $M_{\rm GT}^{(0\nu)}$ the significant changes concentrate on the 1^+ and 2^- contributions. The wave function of the 2_1^- state plays a key role when seeking the reason for the reduction of the magnitude of the Gamow–Teller matrix element. The quality of the lowest 2^- state in the intermediate nucleus ^{76}As can be tested by

Table 7. Beta-minus decay $\log ft$ values for transitions from the $2^-_{\rm g.s.}$ of ^{76}As to the ground state and one- and two-phonon states in ^{76}Se.

	$0^+_{\rm g.s.}$	2^+_1	0^+_2	2^+_2	4^+_1
Exp.	9.7	8.1	10.3	8.2	11.1
Present calc.	9.7	7.4	9.0	8.4	10.7
[7]	9.0	7.7	9.2	8.7	10.9

computing the β^- decay $\log ft$ values for transitions from this state to the ground state and one- and two-phonon states in ^{76}Se. The results obtained are compared with the data and the calculations of [7] in Table 7. As can be seen there is a drastic improvement in the $\log ft$ value of the ground-state-to-ground-state transition. This transition tests exclusively the 2_1^- wave function whereas the rest of the transitions depend also on the final-state wave function, built from the 2_1^+ collective phonon in the ^{76}Se nucleus. It is worth pointing out that in the present calculation the quasiparticle spectrum is more compressed than in the calculation of Ref. [7]. This increases the collectivity of the 2_1^+ state in ^{76}Se and thus results in smaller effective charges when trying to reproduce the data on the E2 transition probability from this state. The single β^- decay is a non-trivial way to check the reduction in the $0\nu\beta\beta$ NME: The β^- NME is reduced by 58% from the old value [7]. This is in nice agreement with the 56% reduction of the $0\nu\beta\beta$ NME. The main component of the wave function driving both transitions is the proton $0f_{5/2}$ orbital coupled to the neutron $0g_{9/2}$ orbital. In the present calculation the wave function of the 2_1^- state is more fragmented and thus reduces the pnQRPA amplitude responsible for the transitions. On the other hand, the occupation of the neutron $0g_{9/2}$ orbital has increased which also reduces the decay amplitudes since they are proportional to the emptiness of $\nu 0g_{9/2}$. Similar considerations, though in a more complicated way, apply to the intermediate 1^+ contribution. Our calculations [12] show that the main contributions to the $0\nu\beta\beta$ NME come from inside the pfg subspace. The corrected occupations in the BCS calculation most likely reproduce better the shell-model occupations and thus the pnQRPA result comes closer to the shell-model result of [17]. The small contributions from outside the pfg subspace partly explain the deviations from the shell model result. Our presently computed variations $M^{(0\nu)} = 2.279 - 2.779$ (Jastrow) and $M^{(0\nu)} = 3.231 - 4.112$ (UCOM) in the $0\nu\beta\beta$ NME can be converted to the following half-life limits

$$t_{1/2}^{(0\nu)} = (5.36 - 8.04) \times 10^{24}\,\text{yr}/((\langle m_\nu \rangle[\text{eV}])^2 \text{ (Jastrow)}\,, \quad (3)$$

$$t_{1/2}^{(0\nu)} = (2.45 - 4.00) \times 10^{24}\,\text{yr}/((\langle m_\nu \rangle[\text{eV}])^2 \text{ (UCOM)}\,. \quad (4)$$

To conclude, we have performed a pnQRPA-based calculation of the nuclear matrix elements involved in the neutrinoless double beta decay of ^{76}Ge. The recently available data on the neutron pairing correlations in ^{76}Ge and ^{76}Se have been used to modify the basic Woods–Saxon mean field to reproduce the neutron occupations of the pfg subspace in a BCS calculation. The subsequently calculated $0\nu\beta\beta$ nuclear matrix elements are smaller in magnitude than the ones obtained in a standard calculation using the Woods–Saxon single-particle energies. This stems from the reduction in the contributions of the 0^+, 1^+ and 2^- intermediate states, with a special emphasis on the first 2^- state [12].

5. Acknowledgments

This work has been partially supported by the National Research Council (CONICET) of Argentina and by the Academy of Finland under the Finnish Centre of Excellence Programme 2006-2011 (Nuclear and Accelerator Based Programme at JYFL). We thank also the EU ILIAS project under the contract RII3-CT-2004-506222. One of the authors (O.C.) is grateful for the warm hospitality extended to him at the Department of Physics of the University of Jyväskylä, Finland. We acknowledge the very fruitful discussions with Prof. J.P. Schiffer. We also thank Dr. M. Kortelainen and Mr. M.T. Mustonen for making their computer codes available to us.

References

[1] M. Doi, T. Kotani, E. Takasugi, Prog. Theor. Phys. Suppl. 83 (1985) 1.
[2] J. Suhonen, O. Civitarese, Phys. Rep. 300 (1998) 123.
[3] A. Faessler, F. Šimkovic, J. Phys.G 24 (1998) 2139.
[4] F.T. Avignone III, S.R. Elliott, J. Engel, arXiv:0708.1033, and to be published in Rev. Mod. Phys. (2008).
[5] M. Kortelainen, O. Civitarese, J. Suhonen, J. Toivanen, Phys. Lett. B 647 (2007) 128.
[6] M. Kortelainen, J. Suhonen, Phys. Rev. C 76 (2007) 024315.
[7] M. Kortelainen, J. Suhonen, Phys. Rev. C 75 (2007) 051303(R).
[8] V.A. Rodin, A. Faessler, F. Šimkovic, P. Vogel, Nucl. Phys. A 793 (2007) 213 (Erratum).
[9] F. Šimkovic, A. Faessler, V. Rodin, P. Vogel, J. Engel, arXiv:0710.2055 [nucl-th].
[10] H.V. Klapdor-Kleingrothaus, et al.,Mod. Phys. Lett. A 16 (2001) 2409.
[11] H.V. Klapdor-Kleingrothaus, et al., Phys. Lett. B 586 (2004) 198.
[12] J. Suhonen, O. Civitarese, Phys.Lett. B 668 (2008), 277.
[13] S.J. Freeman, et. al., Phys. Rev. C 75 (2007) 051301(R); J.P. Schiffer, private communication.
[14] O. Civitarese, J. G. Hirsch, A. Mariano and M. Reboiro, Phys. Rev. C 76 (2007) 024303.
[15] D.R Bes and O. Civitarese Nuclear Physics **A 705**(2002)297.
[16] E. Caurier, et al., Rev. Mod. Phys. 77 (2005) 427.
[17] E. Caurier, F. Nowacki, A. Poves, Int. J. Mod. Phys. E 16 (2007) 552.
[18] P. Vogel, M.R. Zirnbauer, Phys. Rev. Lett. 57 (1986) 3148.
[19] O. Civitarese, A. Faessler, T. Tomoda, Phys. Lett. B 194 (1987) 11.
[20] J. Suhonen, Phys. Lett. B 607 (2005) 87.
[21] O. Civitarese, J. Suhonen, Phys. Lett. B 626 (2005)80; ibid Nucl. Phys. A 761 (2005) 313.
[22] T. Tomoda, Rep. Prog. Phys. 54 (1991) 53.

Journal of Physics: Conference Series **173** (2009) 012013 doi:10.1088/1742-6596/173/1/012013

Are Neutrinos Their Own Antiparticles?

Boris Kayser

Fermilab, MS 106, P.O. Box 500, Batavia, IL 60510, U.S.A.

Email: boris@fnal.gov

Abstract: We explain the relationship between Majorana neutrinos, which are their own antiparticles, and Majorana neutrino masses. We point out that Majorana masses would make the neutrinos very distinctive particles, and explain why many theorists strongly suspect that neutrinos do have Majorana masses. The promising approach to confirming this suspicion is to seek neutrinoless double beta decay. We introduce a toy model that illustrates why this decay requires nonzero neutrino masses, even when there are both right-handed and left-handed weak currents.

For given helicity h, is each neutrino mass eigenstate ν_i identical to its antiparticle, or different from it? Equivalently, do neutrinos have Majorana masses? If they do, then, as we shall explain, each ν_i is identical to its antiparticle: $\overline{\nu_i(h)} = \nu_i(h)$. Neutrinos of this nature are referred to as Majorana neutrinos, while ones for which $\overline{\nu_i(h)} \neq \nu_i(h)$ are called Dirac neutrinos.

Let us recall what a Majorana mass is. Out of, say, a left-handed neutrino field, ν_L, and its charge conjugate, ${\nu_L}^c$, one can build the "left-handed" (so called because it is constructed from ν_L) Majorana mass term

$$\mathcal{L}_L = m_L \overline{\nu_L} {\nu_L}^c , \qquad (1)$$

which absorbs a $(\overline{\nu})_R$ and creates a ν_L. As this illustrates, Majorana neutrino masses mix neutrinos and antineutrinos, so they do not conserve the lepton number L defined by $L(\nu) = L(\ell^-) = -L(\overline{\nu}) = -L(\ell^+)$, where ℓ is a charged lepton. Moreover, a Majorana mass for any fermion f mixes f with its antiparticle $\bar{f}$, or with a related antifermion having the same electric charge as $\bar{f}$. Thus, quark and charged-lepton Majorana masses are forbidden by electric charge conservation. However, neutrinos, being electrically neutral, are permitted to possess Majorana masses, and such masses would make them *very* distinctive. For, a Majorana neutrino mass cannot arise from the neutrino analogue of the Standard-Model (SM) coupling that gives quarks and charged leptons their masses. That analogue would be a Yukawa coupling of the form $H_{SM}\, \overline{\nu_R} \nu_L$, where H_{SM} is the SM Higgs field. Instead, Majorana masses must come from couplings such as $H_{SM} H_{SM} \overline{\nu_L{}^c} \nu_L$ or $H_{I_W=1} \overline{\nu_L{}^c} \nu_L$. The first of these leads to non-renormalizability, and so is outside the spirit of the SM. The second involves a Higgs boson with weak isospin $I_W = 1$, and there is no such boson in the SM. In addition, one can have the right-handed Majorana mass $m_R \overline{\nu_R{}^c} \nu_R$, which doesn't involve any Higgs field at all. One way or another, Majorana neutrino masses must have a different origin than the masses of quarks and charged leptons.

To see why the neutrino mass eigenstates will be Majorana neutrinos if neutrinos have Majorana masses, we note first that the objects ν_L and $\nu_L{}^c$ that appear in the Majorana mass of Eqn. (1) are not the mass eigenstates, but just the neutrinos in terms of which our neutrino model is constructed. As we have noted, the Majorana mass of Eqn. (1) induces mixing between $\nu_L{}^c$ (a right-handed antineutrino) and ν_L. Now, we recall that, as a result of $\overline{K^0} \leftrightarrow K^0$ mixing, the neutral K mass eigenstates are not K^0 and $\overline{K^0}$, but K_S and K_L. Neglecting CP violation, the latter particles are just the states

$$K_{S,L} = \left(K^0 \pm \overline{K^0}\right)/\sqrt{2}. \tag{2}$$

Clearly, each of these states is self-conjugate (apart from an irrelevant sign) under particle-antiparticle interchange. In a similar way, as a result of the $\nu_L{}^c \leftrightarrow \nu_L$ mixing induced by the Majorana mass of Eqn. (1), the neutrino mass eigenstate that results from this mass term is

$$\nu_i = \nu_L + \nu_L{}^c. \tag{3}$$

Clearly, this ν_i satisfies $\nu_i{}^c = \nu_i$. Thus, like K_S or K_L, ν_i is identical to its antiparticle.

Among theorists, there is a widespread belief that neutrinos do have Majorana masses. One reason for this prejudice is the following argument: The electroweak SM may be defined by a few principles that include $SU(2)_L \times U(1)_Y$ gauge invariance and renormalizability, and by its field or particle content. The original version of the SM did not include neutrino masses, and, leaving these masses aside, it is observed that anything allowed by the defining SM principles actually occurs in nature. Now, if we extend the SM to include neutrino masses, we note that right-handed Majorana masses are allowed by the defining SM principles. Therefore, it seems likely that Majorana neutrino masses occur in nature too.

To determine whether Majorana masses do occur in nature, the most promising approach is to seek neutrinoless double beta decay ($0\nu\beta\beta$).[1] This is the process $Nucl \rightarrow Nucl' + e^- e^-$, in which one nucleus decays into another plus two electrons. The observation of this decay at any nonzero level would imply the existence in nature of a neutrino Majorana mass term.[2] To see this, we note that at the quark level, $0\nu\beta\beta$ is the process $dd \rightarrow uu + e^- e^-$. If this process is observed, then, by crossing, the amplitude for the process $e^+ \bar{u} d \rightarrow e^- u \bar{d}$ must be nonzero. The SM tells us that the amplitudes for all the (virtual) processes $(\bar{\nu})_R \rightarrow e^+ W^-$, $W^- \rightarrow \bar{u} d$, $u \bar{d} \rightarrow W^+$, and $e^- W^+ \rightarrow \nu_L$ are nonzero as well. Thus, combining amplitudes, we conclude that the amplitude for the chain $(\bar{\nu})_R \rightarrow e^+ W^- \rightarrow e^+ (\bar{u} d) \rightarrow e^- (u \bar{d}) \rightarrow e^- W^+ \rightarrow \nu_L$ must be non-vanishing. But this chain results in $(\bar{\nu})_R \rightarrow \nu_L$, which is precisely the effect of the Majorana mass term of Eqn. (1). Hence, the observation of $0\nu\beta\beta$ would imply the existence of a non-vanishing amplitude that is equivalent to a Majorana mass term. Consequently, this observation would also imply that neutrinos are Majorana particles.

Although $0\nu\beta\beta$ can receive contributions from a variety of sources, one anticipates that it will be dominated by the diagram in Fig. 1. There, the vertices labeled "SM vertex" are SM charged-current vertices, U is the leptonic mixing matrix, and, as indicated, the amplitude is a coherent sum over the contributions of all the neutrino mass eigenstates ν_i.

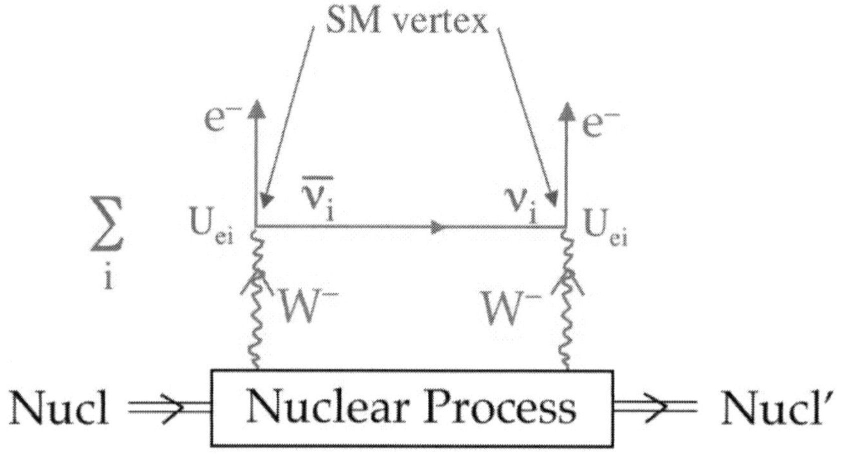

Figure 1. The diagram expected to dominate neutrinoless double beta decay.

The SM charged-current interaction that acts at each of the two leptonic vertices in Fig. 1 is lepton-number conserving. If the neutral particle that is absorbed at the vertex on the right is to create an electron, it must be a neutrino, not an antineutrino. However, at the vertex on the left, where this same particle is emitted, it is created by a W boson together with an electron, so at this vertex it must be an antineutrino, not a neutrino. Thus, the diagram in Fig. 1 does not exist unless $\overline{\nu_i} = \nu_i$. This is another way to see that the observation of $0\nu\beta\beta$ would imply that neutrinos are identical to their antiparticles.

Owing to the left-handed chiral character of the SM charged-current interaction, the "antineutrino" created by a W boson together with an e^- will be in a state that is dominantly of right-handed helicity, even if there is no difference between neutrinos and antineutrinos. Thus, the "$\overline{\nu_i}$" emitted at the leptonic vertex on the left in Fig. 1 is in a state that is mostly of right-handed helicity. However, if m_i is the mass of ν_i and E is its energy, this state does have a small component, of order m_i/E, with left-handed helicity. At the vertex on the right in Fig. 1, where the ν_i is absorbed to make an e^-, the SM left-handed charged current can absorb without suppression only its left-handed-helicity component. Thus, the contribution of ν_i exchange to the diagram of Fig. 1 is proportional to m_i. Summing over i and including the factors of U_{ei} that appear at the vertices, we see that if the diagram of Fig. 1 dominates, then the amplitude for $0\nu\beta\beta$, Amp[$0\nu\beta\beta$], is proportional to the quantity

$$\left|\sum_i m_i U_{ei}^2\right| \equiv m_{\beta\beta}. \tag{4}$$

This quantity is known as the effective Majorana neutrino mass for neutrinoless double beta decay.

We see that when the diagram of Fig. 1 dominates, Amp[$0\nu\beta\beta$] is proportional to neutrino mass. Our discussion of helicities can leave one with the misimpression that this proportionality to mass is due merely to a mismatch of helicities at the two leptonic vertices. One might be tempted to believe that if the current acting at the leptonic vertex on the right in Fig. 1 were a non-SM right-handed (RH) current, rather than the SM left-handed (LH) one, then the diagram could lead to $0\nu\beta\beta$ without the need for any neutrino mass. Indeed, some years ago, it was common for an experiment obtaining an upper bound on the rate for $0\nu\beta\beta$ to quote its result as a bound on $m_{\beta\beta}$, and, alternatively, as a bound on the strength of any RH current, as if a RH current could engender $0\nu\beta\beta$ all by itself, without there being any neutrino mass. However, we now understand that, even if the current acting at one of the leptonic vertices in Fig. 1 is a RH current, $0\nu\beta\beta$ still requires nonzero neutrino mass. We have already seen that, without any assumption about the underlying mechanism driving $0\nu\beta\beta$, the observation of this decay would imply a Majorana neutrino mass. To see the need for mass in a way that is more specific to the mechanism in Fig. 1, we note first that $0\nu\beta\beta$ does not conserve lepton number L. Its initial state, a nucleus, has $L=0$. Its final state, containing another nucleus, two leptons, and no antileptons, has $L=2$. Now, as already mentioned, the SM interactions in Fig. 1 *do* conserve L. Moreover, a SM-like leptonic interaction that merely has a RH current in place of the SM LH current would conserve L as well. Absent any non-SM L-violating interactions, the $\Delta L=2$ of $0\nu\beta\beta$ can only come from Majorana neutrino masses such as the one of Eqn. (1). By absorbing a $\overline{\nu}$ and creating a ν, this Majorana mass term causes the required $\Delta L=2$. If we were to turn off the neutrino masses, including their Majorana masses, no lepton number violation would remain, and $0\nu\beta\beta$ would not occur.

The essential role of neutrino mass, even when there are RH currents, is nicely illustrated by a parity-conserving toy model that contains both LH and RH currents. In this model, we assume that there is only one generation. We suppose that the W couples to the electron and neutrino fields via the parity-conserving interaction

$$-\mathcal{L}_W = \frac{g}{\sqrt{2}} W_{\lambda}^{-} \overline{e} \gamma^{\lambda} \nu + \text{h.c.} = \frac{g}{\sqrt{2}} W_{\lambda}^{-} \left(\overline{e_L} \gamma^{\lambda} \nu_L + \overline{e_R} \gamma^{\lambda} \nu_R \right) + \text{h.c.}, \tag{5}$$

where g is a coupling constant. Finally, we suppose that the neutrino mass term is the $L \Leftrightarrow R$ symmetric

$$-\mathcal{L}_M = \frac{1}{2} m_M \left(\overline{\nu_L^c} \nu_L + \overline{\nu_R^c} \nu_R \right) + m_D \overline{\nu_R} \nu_L + \text{h.c.}. \tag{6}$$

Here, the Majorana mass m_M, and the "Dirac mass" m_D (the neutrino analogue of the masses of other Dirac fermions such as the quarks), are both taken to be real and positive, and it is assumed that $m_M > m_D$.

In this model, there are two Majorana neutrino mass eigenstates:

$$\nu_2 = \frac{1}{\sqrt{2}} \left[(\nu_L + \nu_R) + (\nu_L + \nu_R)^c \right], \text{ with mass } m_2 = m_M + m_D, \tag{7}$$

and

$$\nu_1 = \frac{1}{\sqrt{2}}\left[(\nu_L - \nu_R) + (\nu_L - \nu_R)^c\right], \text{ with mass } m_1 = m_M - m_D. \qquad (8)$$

In terms of them,

$$-\mathcal{L}_W = \frac{g}{\sqrt{2}} W_\lambda^- \left[\overline{e_L}\gamma^\lambda \frac{1}{\sqrt{2}}(\nu_{2L} + \nu_{1L}) + \overline{e_R}\gamma^\lambda \frac{1}{\sqrt{2}}(\nu_{2R} - \nu_{1R}) \right] + \text{h.c.} . \qquad (9)$$

To see what this toy model predicts for $0\nu\beta\beta$, let us consider the particle-physics part of this process, pictured in Fig. 2. The current that acts at the vertices of the diagram in Fig. 2 can be the

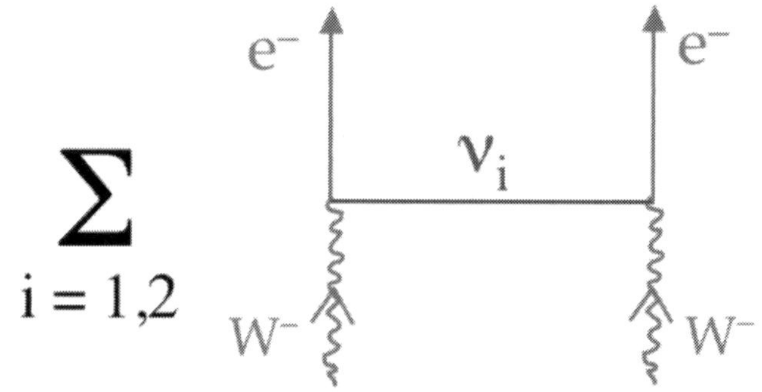

Figure 2. The particle-physics part of $0\nu\beta\beta$.

LH current in Eqn. (5) at both vertices, or the RH current in Eqn. (5) at both vertices, or the LH current at one vertex and the RH current at the other. We find that if the LH current acts at both vertices, the amplitude for the process involves the neutrino masses through the factor

$$A_{LL} = \frac{m_2}{q^2 - m_2^2} + \frac{m_1}{q^2 - m_1^2} \cong \frac{m_2 + m_1}{q^2} = \frac{2m_M}{q^2}. \qquad (10)$$

Here, q is the momentum transfer carried by the exchanged neutrino. In the second step in Eqn. (10) we have used the fact that in $0\nu\beta\beta$, $q \approx 50\,\text{MeV} >> m_{1,2}$, and in the last step we have used Eqns. (7) and (8).

We see that when the LH current acts at both vertices, the amplitude for $0\nu\beta\beta$ is proportional to the Majorana neutrino mass m_M. Fig. 3, in which neutrino mass is treated perturbatively,

Journal of Physics: Conference Series **173** (2009) 012013 doi:10.1088/1742-6596/173/1/012013

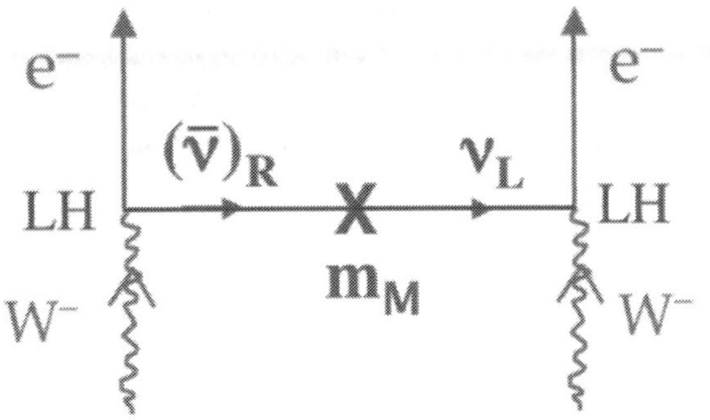

Figure 3. The particle-physics part of $0\nu\beta\beta$, with neutrino mass treated perturbatively, and LH currents acting at both vertices, as indicated.

makes clear the reason for this. A Majorana mass turns a $\bar{\nu}$ into a ν (and a ν into a $\bar{\nu}$), and, like any mass term, reverses the handedness of the particle. Thus, an insertion of the Majorana mass m_M is just what is needed to convert the *RH antineutrino* emitted together with an electron by the lepton-number-conserving LH current that acts at the vertex on the left in Fig. 3, into the *LH neutrino* that can be absorbed to make a second electron by the lepton-number-conserving LH current that acts at the vertex on the right.

When we go to higher order in neutrino mass, we can have multiple mass insertions in the neutrino line of Fig. 3. However, together these insertions must reverse the lepton number of the emitted antineutrino. Thus, there must be an odd number of insertions of the Majorana mass m_M. Together, the mass insertions must also reverse the handedness of the exchanged particle. Since both Majorana and Dirac masses reverse handedness, there must be an odd total number of mass insertions, including both m_M and m_D. The implication of this argument is that if the exact expression for A_{LL}, Eqn. (10), is expanded in powers of m_2 and m_1, and the expansion is rewritten in terms of m_M and m_D, every term must be of the form $m_M^{Odd\ Power} m_D^{Even\ Power}$. It is trivial to verify that this is indeed the case.

If the RH current in Eqn. (5) acts at both vertices in Fig. 2, the amplitude once again involves the neutrino masses through the same factor A_{LL} as before, so that it is again proportional, to lowest order in the mass, to m_M.

Now suppose the LH current in Eqn. (5) acts at the vertex on the left in Fig. 2, but the RH current acts at the one on the right. This is the combination that was once thought capable of producing $0\nu\beta\beta$ without the need for any neutrino mass. However, we find that in our illustrative toy model, this combination of currents leads to an amplitude for the process in Fig. 2 that involves the neutrino masses through the factor

$$A_{LR} = i\not{q}\left(\frac{1}{q^2 - m_2^2} - \frac{1}{q^2 - m_1^2}\right) \cong i\not{q}\left(\frac{m_2^2 - m_1^2}{\left(q^2\right)^2}\right) = \frac{i\not{q}}{\left(q^2\right)^2} 4 m_M m_D \quad . \tag{11}$$

In the last step in this relation, we have used Eqns. (7) and (8). We see that, despite the combination of a LH current and a RH one, the amplitude for $0\nu\beta\beta$ still requires neutrino mass. In fact, it is quadratic in neutrino mass,[3] being proportional to $m_M m_D$,[4] whereas the amplitude that results when a LH current acts at both leptonic vertices, as in the SM, is only linear in mass.

The reason that a LH-RH current combination leads to an amplitude quadratic in neutrino mass is made clear by Fig. 4. Since both currents are lepton-number conserving, the $\bar{\nu}$ created together with

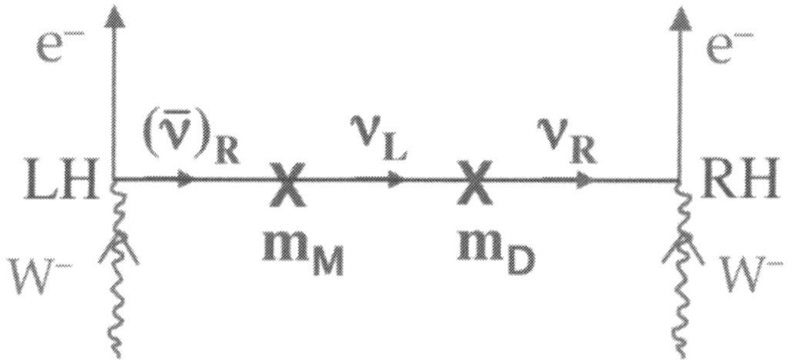

Figure 4. The particle-physics part of $0\nu\beta\beta$, with neutrino mass treated perturbatively, and a LH current acting at one vertex, but a RH one at the other, as indicated.

an electron by the current at the vertex on the left must be converted to a ν so that it can be absorbed to make another electron by the current at the vertex on the right. Hence, one Majorana mass m_M must be inserted along the neutrino line. But this mass will also flip the handedness of the exchanged particle, turning it into a LH neutrino, with the wrong handedness to be absorbed by a RH current. Thus, another mass insertion is needed to flip the handedness back again. Since this second mass insertion must not at the same time turn the exchanged particle back into a $\bar{\nu}$, it must be a Dirac mass m_D.[4]

From this discussion, it is clear that if we go to higher order in neutrino mass and have more than two mass insertions in Fig. 4, there must be an odd number of insertions of m_M, to flip the lepton number, and also an odd number of insertions of m_D, so that there is no net flip of handedness. Thus, if the expression for A_{LR} of Eqn. (11) is expanded in powers of m_2 and m_1, and the expansion is rewritten in terms of m_M and m_D, every term must be of the form $m_M^{Odd\ Power} m_D^{Odd\ Power}$. It is trivial to see that this is indeed the case.

This simple toy model nicely illustrates the fact that, absent any lepton-number-nonconserving interactions, $0\nu\beta\beta$ requires Majorana neutrino mass. This mass is needed to introduce the lepton-number violation without which $0\nu\beta\beta$ cannot occur.

At this Symposium, we paid special tribute to the work of Frank Avignone, Ettore Fiorini, and Peter Rosen. In collaboration with Henry Primakoff, Peter Rosen was a pioneer of the whole field of double beta decay. A half-century ago, Primakoff and Rosen contemplated $0\nu\beta\beta$ arising from non-SM lepton-number-nonconserving interactions. Then Majorana neutrino mass is not needed. However, we have not yet seen any evidence for lepton-number-nonconserving interactions, but we do expect Majorana neutrino masses.

Frank Avignone and Ettore Fiorini have helped lead the way to the present very exciting point in the experimental quest for $0\nu\beta\beta$. Of course, we cannot know in advance what the rate for $0\nu\beta\beta$ will prove to be, or even if the process occurs at all. But our present understanding of neutrino physics gives us good reasons to think that $0\nu\beta\beta$ most likely does occur. It also allows us to make the rather plausible estimate that the $m_{\beta\beta}$ of Eqn. (4) lies in the range $2\,\mathrm{meV} < m_{\beta\beta} < 40\,\mathrm{meV}$,[5] which leads to lifetimes τ for $0\nu\beta\beta$ in the rough range $10^{27}\,\mathrm{yr} < \tau < 10^{29}\,\mathrm{yr}$. Hopefully, the next one or two generations of $0\nu\beta\beta$ experiments will achieve the required sensitivity, and we will see neutrinoless double beta decay at last.

References

[1] The existing neutrino data from processes other than $0\nu\beta\beta$ do not tell us anything about whether neutrino Majorana masses occur in nature. In particular, neutrino oscillation probabilities have no dependence at all on the answer to this question.

[2] Schechter J, and Valle J 1982 *Phys. Rev.* D **25**, 2951

[3] Work of Kayser B, Petcov S and Rosen P, described in, for example, Kayser B 1992 *Nucl. Phys.* A **546**, 399c and in Kayser B 1986 in the *Proceedings of the International Symposium on Nuclear Beta Decays and Neutrino*, eds. Kotani T, Ejiri H and Takasugi E (World Scientific, Singapore) p. 473

[4] Enqvist K, Maalampi J and Mursula K 1984 *Phys. Lett.* B **149**, 217

[5] See, for example, Pascoli S and Petcov S 2008 *Phys. Rev.* D **77**, 113003

Carolina International Symposium on Neutrino Physics IOP Publishing
Journal of Physics: Conference Series **173** (2009) 012014 doi:10.1088/1742-6596/173/1/012014

CN Neutrinos and the Sun's Primordial Core Metalicity

Wick Haxton
Institute for Nuclear Theory and Department of Physics
Box 351550, University of Washington, Seattle WA 98195
E-mail: haxton@phys.washington.edu

Abstract. I discuss the use of neutrinos from the CN cycle and pp chain to constrain the primordial solar core abundances of C and N at an interesting level of precision. A comparison of the Sun's deep interior and surface compositions would test a key assumption of the standard solar model (SSM), a homogeneous zero-age Sun. It would also provide a cross-check on recent photospheric abundance determinations that have altered the once excellent agreement between the SSM and helioseismology. Motivated by the discrepancy between convective-zone abundances and helioseismology, I discuss the possibility that a two-zone Sun could emerge from late-stage metal differentiation in the solar nebula connected with formation of the gaseous giant planets.

1. Introduction

One of the initial motivations for pursuing solar neutrino physics was to test our understanding of main-sequence stellar evolution and the Standard Solar Model (SSM). In the past two decades this goal was put aside, as difficulties in understanding the pattern of solar neutrino fluxes led to the discovery of solar neutrino oscillations. But because of the precision with which the relevant flavor physics is now known – and because the solar neutrino problem also spurred progress in the nuclear physics of the Sun and the development of high-statistics detectors such as Super-Kamiokande (SK) and SNO – the use of neutrinos as quantitative solar probe is now a practical possibility. Specifically, this talk summarizes recent arguments by Aldo Serenelli and me [1] that a measurement of the CN neutrino flux could test a key assumption of the SSM, the homogeneity of the zero-age main sequence Sun. This assumption, the basis for equating the SSM's primordial core metal abundances to today's surface metal abundances, may now be in some degree of conflict with observation: recent 3D modeling of photospheric absorption lines has led to a downward revision in the metal content of the solar convective zone [2], altering helioseismology in the upper radiative zone, where the temperature $\sim$ 2-5 $\times$ 10^6 K [3, 4, 5, 6]. In this region C, N, O, Ne, and Ar are partially ionized and particularly O and Ne have a significant influence on the radiative opacity.

A quantitative comparison between the Sun's surface and core abundances could prove useful in understanding the chemical evolution of other gaseous bodies in our solar system, whose interiors are not as readily probed. The Galileo and Cassini missions found significant metal enrichments in the H/He atmospheres of Jupiter and Saturn, e.g., abundances of C and N of $\sim$ four times solar for Jupiter and $\sim$ 4-8 for Saturn [7]. Planetary models that take account of these

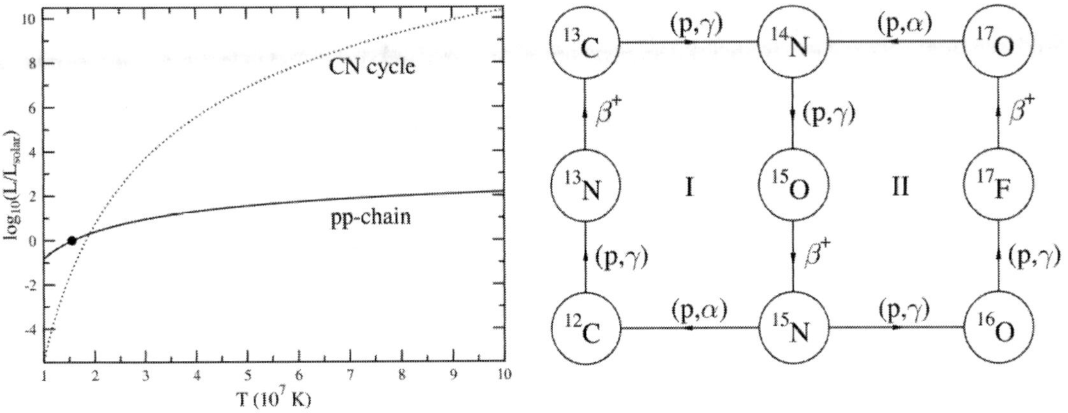

Figure 1. The right panel shows the CNO bi-cycle for hydrogen burning. The left compares the energy produced in the CN cycle with that produced in the pp-chain, as a function of temperature T_7, measured in units of 10^7 K. The results are normalized to the pp-chain energy production at the Sun's center and to solar metalicity, assuming equilibrium burning. The sharper CN-cycle dependence on temperature is apparent. If approximated as a power law T^x, x ranges between $\sim$ 19 and $\sim$ 22 over the range of temperatures typical of the Sun's hydrogen-burning core. The dot marks the point corresponding to the Sun's center, $T_7 \sim 1.57$.

data show that the gaseous giants are very significant solar-system metal reservoirs. The metal excess in these planets, estimated to be $\sim 40-90M_\oplus$, is very similar to the depletion that would be needed in the convection zone to bring the photospheric abundances and helioseismology into agreement. Serenelli and I raised the possibility that a two-zone Sun resulted from a process in which the last $\sim$ 1% of nebular gas was scoured of its metals by the process of planetary formation, then deposited on the early Sun to dilute the convective zone.

2. The CNO Bi-Cycle and its Neutrinos

The need for two mechanisms to burn hydrogen was recognized in the pioneering work of Bethe and collaborators. The first, the pp-chain, dominates energy production in our Sun and other low-mass main-sequence stars and can be considered a primary process in which the chain's "catalysts," deuterium, ^{3}He, and ^{7}Be/^{7}Li (the elements participating in the steps leading to ^{4}He), are synthesized as the chain burns to equilibrium.

But the sharper temperature dependence of a second mechanism, the CNO bi-cycle, is needed to account for the structure of more massive main-sequence stars. Unlike the pp-chain, the CNO bi-cycle (Fig. 1) is a secondary process: the catalysts for H burning are the pre-existing metals. Thus the CNO contribution to energy generation is proportional to stellar metalicity. The CN-cycle, denoted by I in Fig. 1, operates in the Sun. The cycle conserves the number abundance, but alters the distribution of metals as it burns into equilibrium, eventually achieving equilibrium abundances proportional to the inverses of the respective rates.

While responsible for only $\sim$ 1% of solar energy generation, the CN cycle produces measurable neutrino fluxes. The slowest, rate-controlling reactions in the CN cycle are ^{14}N(p,γ) and ^{12}C(p,γ). The latter has reached equilibrium over most of the solar core. In fact, the rapid burning of ^{12}C, as the cycle seeks equilibrium at the start of the main sequence, likely drives convective mixing of the early solar core for $\sim 10^8$ y. In contrast, the ^{14}N lifetime is shorter than the age of the Sun only in regions where $T_7 \gtrsim 1.33$, which corresponds to $R \lesssim 0.1R_\odot$ in today's Sun, or equivalently the central 7% of the Sun by mass. Consequently, over a significant portion of the outer core, ^{12}C has been converted to ^{14}N, but further reactions are inhibited by the ^{14}N(p,γ) bottleneck.

The BSP08(GS) SSM [8] – which employs recent updates of metal abundances and of the ^{14}N(p,γ) S-factor – finds a modest CN-cycle contribution to solar energy generation of 0.8% and corresponding neutrino fluxes

$$^{13}\mathrm{N}(\beta^+)^{13}\mathrm{C} \quad E_\nu \lesssim 1.199 \text{ MeV} \quad \phi = (2.93^{+0.91}_{-0.82}) \times 10^8/\mathrm{cm}^2\mathrm{s}$$

$$^{15}\mathrm{O}(\beta^+)^{15}\mathrm{N} \quad E_\nu \lesssim 1.732 \text{ MeV} \quad \phi = (2.20^{+0.73}_{-0.63}) \times 10^8/\mathrm{cm}^2\mathrm{s}.$$

The ranges reflect conservative abundance uncertainties as defined empirically in [9]. The first reaction is part of the path from ^{12}C to ^{14}N, while the latter follows ^{14}N(p,γ). Thus neutrinos from ^{15}O β decay are produced in the central core: 95% of the flux comes from the CN-equilibrium region, described above. About 30% of the ^{13}N neutrinos come from outside this region, primarily because of the continued burning of primordial ^{12}C: this accounts for the somewhat higher flux of these neutrinos.

The SSM makes several reasonable assumptions, including local hydrostatic equilibrium (the balancing of the gravitational force against the gas pressure gradient), energy generation by proton burning, a homogeneous zero-age Sun, and boundary conditions imposed by the known mass, radius, and luminosity of the present Sun. It assumes no significant mass loss or accretion. The homogeneity assumption allows the primordial core metalicity to be fixed to today's surface abundances. Corrections for the effects of diffusion of He and the heavy elements over 4.57 b.y. of solar evolution are included, and generally been helpful in improving the agreement between SSM predictions and parameters probed in helioseismology.

The SSM postulate of a homogeneous zero-age Sun is based on the assumptions 1) that the early pre-main-sequence Sun passed through a fully convective, highly luminous Hayashi phase, mixing the Sun; and 2) that no chemical differentiation in the gas accreted onto the Sun occurred after this phase, when the Sun develops a radiative core distinct from the convective surface zone.

Solar surface abundances are known, determined from analyses of photospheric atomic and molecular spectral lines. Traditionally the associated atmospheric modeling has been done in one dimension, in a time-independent hydrostatic analysis that incorporates convection via mixing-length theory. But much improved 3D models of the solar atmosphere have been developed recently to treat the radiation-hydrodynamics and time dependence of this problem. This 3D analysis led to a revision in solar metalicity from the previous standard, Z=0.0169 [10], to Z=0.0122 [2], thus altering SSM predictions. Hereafter we denote these as the GS and and AGS abundances, respectively.

The predictions of solar models that use the GS solar composition, the most up to date of which is the BPS08(GS) [8] but including also the BP00 [11], BP04 [12] and BS05(OP) [4] models, are in excellent agreement with helioseismology. But those computed with the revised abundances are in much poorer agreement, with discrepancies exceeding 1% in the region just below the convective zone (R $\sim$ 0.65 $-$ 0.70R$_\odot$). Associated properties of the SSM, such as the depth of the convective zone and the surface He abundance, are also now in conflict with helioseismology. As discussed in [13], the discrepancies are significantly above measurement and solar model uncertainties.

The reduced core opacity also lowers the SSM prediction of the temperature-dependent ^{8}B neutrino flux by about 20%: the predicted ^{8}B flux using the GS abundances and Opacity Project [14] opacities (model BPS08(GS)) is $5.95 \times 10^6/\mathrm{cm}^2\mathrm{s}$, which drops to $4.72 \times 10^6/\mathrm{cm}^2\mathrm{s}$ when AGS abundances are used (model BPS08(AGS)). These results can be compared to the ^{8}B neutrino flux deduced from the NCD-phase SNO data set of $[5.54\ ^{+0.33}_{-0.31}\ \text{(stat)}\ ^{+0.36}_{-0.34}\ \text{(sys)}] \times 10^6/\text{cm}^2\text{s}$ [15].

3. The Sun as a Calibrated Laboratory

It has been recognized for many years that a measurement of the CN-cycle solar neutrino flux would, in principle, determine the metalicity of this core zone, allowing a comparison with abundance determined from the solar atmosphere. In the past several years new developments have occurred that may make such a measurement practical:

- Accurate calibrations of the solar core temperature by SNO and SK;
- Tight constraints on the oscillation parameters and matter effects that determine the flavor content of the CN and ^{8}B neutrino fluxes;
- Recent measurements of the controlling reaction of the CN cycle, ^{14}N(p,γ), that have significantly reduced the nuclear physics uncertainties affecting SSM predictions of CN-cycle fluxes; and
- New ideas for high-counting rate experiments that would be sensitive to CN-cycle neutrinos, and from which reliable terrestrial fluxes could be extracted.

The analysis of [1], which examined whether CN neutrino experiments could place a significant constraint on the solar core's primordial metalicity, used previous SSM work in which the logarithmic partial derivatives $\alpha(i,j)$ for each neutrino flux ϕ_i are evaluated for the SSM input parameters β_j,

$$\alpha(i,j) \equiv \frac{\partial \ln [\phi_i/\phi_i(0)]}{\partial \ln [\beta_j/\beta_j(0)]}, \tag{1}$$

where $\phi_i(0)$ and $\beta_j(0)$ denote the SSM best values. This information, in combination with the assigned uncertainties in the 19 β_j of the SSM, then provides an estimate of the uncertainty in the SSM prediction of ϕ_i. In particular, crucial to the current analysis is the dependence [9] on the mass fractions (measured relative to hydrogen) of different heavy elements,

$$\beta_j = \frac{\text{mass fraction of element j}}{\text{mass fraction of hydrogen}} \equiv X_j. \tag{2}$$

Having this information not as a function of the overall metalicity Z, but as a function of the individual abundances, allows one to separate the "environmental" effects of the metals in the solar core from the special role of primordial C and N as catalysts for the CN cycle. By environmental effects we mean the influence of the metals on the opacity and thus the ambient core temperature, which controls the rates of neutrino-producing reactions of both the pp-chain and CN cycle. In [1] the temperature-dependent ^{8}B neutrino flux was used to calibrate the environmental effects of the metals and of other SSM parameters, thus isolating the special CN-cycle dependence on primordial C+N. This primordial abundance can be expressed, with very little residual solar model uncertainty, in terms of the measured ^{8}B neutrino flux and nuclear cross sections that have been determined in the laboratory.

The partial derivatives allow one to define the power-law dependencies of neutrino fluxes, relative to the SSM best-value prediction $\phi_i(0)$

$$\phi_i = \phi_i(0) \prod_{j=1}^{N} \left[\frac{\beta_j}{\beta_j(0)}\right]^{\alpha(i,j)} \tag{3}$$

where the product extends over $N = 19$ SSM input parameters. This expression can be used to evaluate how SSM flux predictions will vary, relative to $\phi_i(0)$, as the β_j are varied. Alternatively, the process can be inverted: a flux measurement could in principle be used to constrain an uncertain input parameter.

The baseline SSM calculation used in [1], BPS08(AGS) [8], employed the recently determined AGS abundances for the volatile elements C, N, O, Ne, and Ar, rather than the previous GS standard composition. It should be noted that AGS includes a downward revision by 0.05 dex of the Si photospheric abundance compared to GS and, accordingly, a similar reduction in the meteoritic abundances. The partial derivatives needed in the present calculation are summarized in Tables 1 (solar model parameters and nuclear cross sections) and 2 (abundances).

The SSM estimate of uncertainties in the various solar neutrino fluxes ϕ_i can be obtained by folding the partial derivatives with the uncertainties in the underlying β_j. In particular, it is convenient to decompose Eq. (3) into its dependence on solar parameters and non-CN metals, nuclear S-factors, and the primordial C and N abundances,

$$\phi_i = \phi_i^{SSM} \prod_{j\in\{\text{Solar and Metals}\neq\text{C,N}\}} \left[\frac{\beta_j}{\beta_j(0)}\right]^{\alpha(i,j)} \prod_{j\in\{\text{Nuclear}\}} \left[\frac{\beta_j}{\beta_j(0)}\right]^{\alpha(i,j)} \prod_{j\in\{\text{C,N}\}} \left[\frac{\beta_j}{\beta_j(0)}\right]^{\alpha(i,j)} \tag{4}$$

where the first term will be designated the "environmental" uncertainty – SSM solar parameters and metal abundances that primarily influence neutrino flux predictions through changes they induce in the core temperature. These are, respectively, the uncertainties in the photon luminosity $L_\odot$, the mean radiative opacity, the solar age, and calculated He and metal duffusion; and the fractional abundances of O, Ne, Mg, Si, S, Ar, and Fe. The estimated 1σ fractional uncertainties for the these parameters are given in [1]. The abundances of Mg, Si, S, and Fe are meteoritic, while those of the volatile elements C, N, O, Ne, and Ar are photospheric.

The next term contains the effects of nuclear cross section uncertainties on flux predictions. The β_j are the S-factors for p+p (S_{11}), ^{3}He + ^{3}He (S_{33}), ^{3}He+^{4}He (S_{34}), p + ^{7}Be (S_{17}), e + ^{7}Be (S_{e7}), and p + ^{14}N (S_{114}). The estimated 1σ fractional uncertainties are also given in [1].

The last term is the contribution of the primordial C and N abundances. As Table 2 shows, pp-chain neutrino fluxes are relatively insensitive to variations in these abundances, as the heavier nuclei like Fe have a more important influence on the core opacity. But the expected, nearly linear response of the ^{13}N and ^{15}O neutrino fluxes to these abundances is apparent. These are the abundances we would like to constrain by a future measurement of the ^{13}N and ^{15}O solar neutrino fluxes. Such a measurement begins to be of interest if these abundances could be determined with an accuracy significantly better than 30%.

Were one to vary the 11 SSM parameters designated as "environmental" according to their assigned uncertainties (taking them to be uncorrelated), a 7.5% SSM net uncertainty in $\phi(^{13}\text{N})$ would be obtained. This uncertainty would be one of dominant ones in the analysis of present interest. But, as discussed in [1] in more detail, these environmental uncertainties influence neutrino fluxes through their impact on the core temperature, regardless of the details of which parameters are varied. Consequently, there are strong environmental correlations between different neutrino fluxes, allowing one to form ratios of fluxes that are much less sensitive to environmental effects. Furthermore these correlations remain valid for parameter variations far outside accepted SSM uncertainties.

This is illustrated in the SSM Monte Carlo tests represented in Fig. 3. One finds that a flux ratio such as

$$\frac{\phi(^{15}\text{O})}{\phi(^{8}\text{B})^K}, \tag{5}$$

where the exponent $K \sim 0.828$ is taken from the fit shown in Fig. 3, is much less sensitive to environmental uncertainties than either the numerator or denominator separately. This is apparent from the entries in Tables 1 and 2. As $\phi(^8\text{B})$ can be taken from SK and SNO measurements, most of the environmental uncertainty in predicting $\phi(^{15}\text{O})$ can be eliminated.

Journal of Physics: Conference Series **173** (2009) 012014 doi:10.1088/1742-6596/173/1/012014

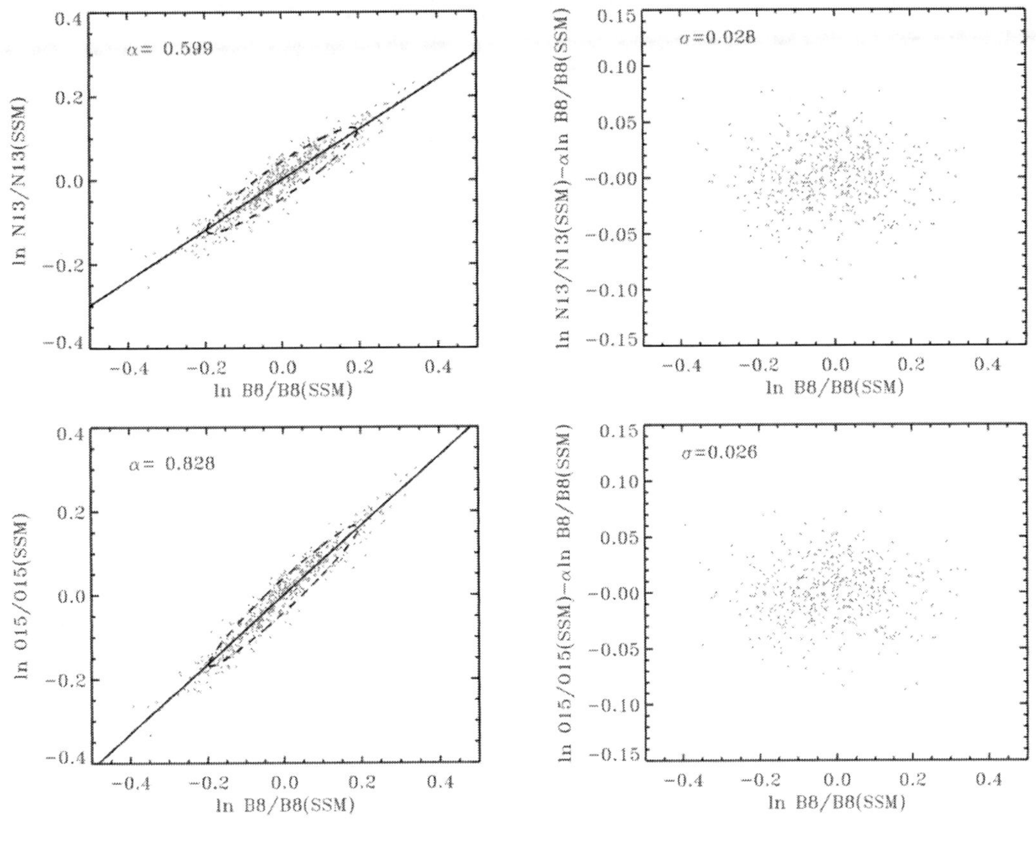

Figure 2. Results from SSM Monte Carlo simulations in which the 11 environmental parameters (see text) have been varied. The two left panels show the correlations between the ^{8}B flux and the two CN-cycle neutrino fluxes ^{13}N and ^{15}O. The slopes of the correlations are given in the plots, together with the 68.3% confidence level contours. The panels on the right show the residuals from the fits, 2.8% and 2.6% for the ^{13}N and ^{15}O fluxes respectively – the residual environmental uncertainty that remains after making use of the ^{8}B flux constraint.

Effectively, $\phi(^8\mathrm{B})$ becomes a "thermometer" constraining core temperature changes induced by varying the environmental β_j. In this way one obtains a more precise relationship between the CN flux, a quantity that should be measured quite accurately in next-generation experiments like SNO+, and the core abundances of C and N.

4. The Analysis for Elastic Scattering and Neutrino Oscillations

The analysis requires a number of steps that will be summarized here, as more detail can be found in Ref. [1]. The total ^{8}B flux (the instantaneous solar flux), normalized to the SSM best value, can be related to rates that would be measured in terrestrial detectors by

$$\frac{\phi(^8\mathrm{B})}{\phi^{SSM}(^8\mathrm{B})} = \frac{\phi(^8\mathrm{B})\langle\sigma^{SK}(^8\mathrm{B},\delta m^2_{12},\theta_{12})\rangle}{\phi^{SSM}(^8\mathrm{B})\langle\sigma^{SK}(^8\mathrm{B},\delta m^2_{12},\theta_{12})\rangle} \equiv \frac{R^{SK}_{exp}(^8\mathrm{B})}{R^{SK}_{cal}(^8\mathrm{B},\delta m^2_{12},\theta_{12})} \tag{6}$$

Table 1. Partial derivatives $\alpha(i,j)$ of neutrino fluxes with respect to solar environmental and nuclear cross section parameters.

	Environmental β_j				Nuclear β_j					
Source	$L_\odot$	Opacity	Age	Diff.	S_{11}	S_{33}	S_{34}	S_{17}	S_{e7}	S_{114}
$\phi(^8B)$	7.16	2.70	1.38	0.28	-2.73	-0.43	0.85	1.0	-1.0	-0.020
$\phi(^{13}N)$	4.40	1.43	0.86	0.34	-2.09	0.025	-0.053	0.0	0.0	0.71
$\phi(^{13}N)/\phi(^8B)^{0.599}$	0.11	-0.19	0.03	0.17	-0.45	0.28	-0.56	-0.60	0.60	0.72
$\phi(^{15}O)$	6.00	2.06	1.34	0.39	-2.95	0.018	-0.041	0.0	0.0	1.00
$\phi(^{15}O)/\phi(^8B)^{0.828}$	0.07	-0.18	0.20	0.16	-0.69	0.37	-0.74	-0.83	0.83	1.02

Table 2. Partial derivatives $\alpha(i,j)$ of neutrino fluxes with respect to fractional abundances of the primordial heavy elements.

	C, N β_j		Environment Abundance β_j						
Source	C	N	O	Ne	Mg	Si	S	Ar	Fe
$\phi(^8B)$	0.027	0.001	0.107	0.071	0.112	0.210	0.145	0.017	0.520
$\phi(^{13}N)$	0.874	0.142	0.044	0.030	0.054	0.110	0.080	0.010	0.268
$\phi(^{13}N)/\phi(^8B)^{0.599}$	0.858	0.141	-0.020	-0.013	-0.013	-0.016	-0.007	0.000	-0.043
$\phi(^{15}O)$	0.827	0.200	0.071	0.047	0.080	0.158	0.113	0.013	0.393
$\phi(^{15}O)/\phi(^8B)^{0.828}$	0.805	0.199	-0.018	-0.012	-0.013	-0.016	-0.007	-0.001	-0.038

Here $\langle\sigma^{SK}\rangle$ is an effective cross section that takes into account all of the neutrino flavor and detector response issues (trigger efficiencies, resolution, cross section uncertainties, oscillations, etc.) that determine the relationship between a measured detector rate and the instantaneous solar flux. The numerator of the ratio on the right is a directly measured experimental quantity: the SK elastic scattering rate for producing recoil electrons with apparent energies between 5.0 and 20 MeV, per target electron per second. The denominator is a theoretical quantity, computed by folding the SSM best-value ^{8}B flux produced in the Sun with the cross section for (ν_x, e) elastic scattering, averaged over a normalized ^{8}B spectrum, defined for the specific experimental conditions of SK, and including the effects of flavor mixing that alter the flux during transit from the Sun to the detector. This cross section is calculated from laboratory measurements of detector properties, the β decay spectrum, the underlying neutrino-electron cross sections, and most critically, the parameters governing oscillations. We describe these factors below.

The experimental rate comes from the 1496 days of measurements of SK I [16]. From the SK I rate/kiloton/year

$$520.1 \pm 5.3(\mathrm{stat})\ ^{+18.2}_{-16.6}(\mathrm{sys})\ \mathrm{kton}^{-1}\ \mathrm{y}^{-1}. \tag{7}$$

we find $R^{SK}_{exp}(^8\mathrm{B})$,

$$4.935 \pm 0.05(\mathrm{stat})\ ^{+0.17}_{-0.16}(\mathrm{sys})\ \times 10^{-38}\ \mathrm{electron}^{-1}\mathrm{s}^{-1} \tag{8}$$

(or $\sim$ 0.049 Solar Neutrino Units, or SNUs). The dominant systematic error includes estimates for the energy scale and resolution, trigger efficiency, reduction, spallation dead time, the gamma ray cut, vertex shift, background shape for signal reduction, angular resolution, and lifetime uncertainties. The combined statistical and systematic error is $\sim \pm$ 3.6%.

To evaluate the denominator in Eq. (6) we need the suitably averaged cross section, defined for the window used by the SK I collaboration,

$$\langle\sigma^{SK}(^8\mathrm{B}, \delta m^2_{12}, \theta_{12})\rangle = \int dE_\nu \phi^{^8\mathrm{B}}_{norm}(E_\nu) \left[P_{\nu_e}(E_\nu, \delta m^2_{12}, \sin^2 2\theta_{12}) \int_{T=0}^{T^{max}(E_\nu)} dT\ \sigma^{es}_{\nu_e}(T) \right.$$

$$+ P_{\nu_\mu}(E_\nu, \delta m^2_{12}, \sin^2 2\theta_{12}) \int_{T=0}^{T^{max}(E_\nu)} dT\ \sigma^{es}_{\nu_\mu}(T) \Bigg] \int_{5\ \mathrm{MeV}}^{20\ \mathrm{MeV}} d\epsilon_a f_{\mathrm{trig}}(\epsilon_a) \rho(\epsilon_a, \epsilon_t = T + m_e) \quad (9)$$

where $\phi^{^8\mathrm{B}}_{norm}(E_\nu)$ is the normalized ^{8}B neutrino spectrum. Equation (9) involves an integral over the product of this spectrum and the energy-dependent oscillation probabilities. ($P_{\nu_e} + P_{\nu_\mu}$=1, assuming oscillations into active flavors. P_{ν_μ} can be defined as the oscillation probability to heavy flavors, if the effects of three flavors are considered.) A given E_ν fixes the range of kinetic energies T of the scattered electron, over which an integration is done; in the laboratory frame $T^{max} = 2E^2_\nu/(m_e + 2E_\nu)$. The integrand includes the elastic scattering cross sections $\sigma^{es}(T)$ for electron and heavy-flavor neutrinos and the SK resolution function $\rho(\epsilon_a, \epsilon_t)$, where $\epsilon_t = T + m_e$ is the true total electron energy while ϵ_a is apparent energy, as deduced from the number of phototube hits in the detector. Finally, an integral must be done over the window used by the experimentalists, apparent electron energies ϵ_a between 5 and 20 MeV. The deduced counting rate includes the triggering probability that an event of apparent energy ϵ_a will be recorded in the detector. Resolution and triggering functions for SK are given in [1].

Similarly, the CN-cycle neutrino response for a detector like SNO+ is

$$\frac{\phi(^{15}\mathrm{O})}{\phi^{SSM}(^{15}\mathrm{O})} \equiv \frac{R^{B/S}_{exp}(^{15}\mathrm{O})}{R^{B/S}_{cal}(^{15}\mathrm{O}, \delta m^2_{12}, \theta_{12})} = \frac{R^{B/S}_{exp}(\mathrm{CN})/(1 + \alpha(0.8, 1.3))}{R^{B/S}_{cal}(^{15}\mathrm{O}, \delta m^2_{12}, \theta_{12}))} \quad (10)$$

where the experimental rate for ^{15}O neutrinos has been written in terms of the total CN-neutrino rate $R^{B/S}_{exp}$(CN) by introducing a correction factor $\alpha \sim 0.12$ that accounts for the ^{13}N neutrino contribution. As discussed in [1], α can be measured in principle, but can also be evaluated from theory, with negligible uncertainty. No measurement of $R^{B/S}_{exp}$(CN) currently exists, of course. But such measurements could be made in Borexino or SNO+, existing or planned detectors using large volumes of organic scintillator and placed quite deep underground. A window for the apparent kinetic energy T of the scattered electron of 0.8-1.3 MeV has been discussed by the Borexino group. As the ^{7}Be 0.866 MeV line corresponds to $T^{max} \sim 0.668$ MeV, this window would limit contamination from ^{7}Be neutrino recoil electrons.

As discussed in [1], one can use Eqs. (6) and (10) in expressions based on Eq. (4), then on dividing obtain

$$\frac{R^{B/S}_{exp}(\mathrm{CN})}{R^{B/S}_{cal}(^{15}\mathrm{O}, \delta m^2_{12}, \theta_{12}))} = (1.120 \pm 0.003) \left[\frac{R^{SK}_{exp}(^8\mathrm{B})}{R^{SK}_{cal}(^8\mathrm{B}, \delta m^2_{12}, \theta_{12})} \right]^{0.828}$$
$$\times [1 \pm 2.6\%(\text{resid. envir.}) \pm 7.6\%(\text{nuclear})] \left(\frac{X(^{12}\mathrm{C})}{X(^{12}\mathrm{C})_{SSM}} \right)^{0.805} \left(\frac{X(^{14}\mathrm{N})}{X(^{14}\mathrm{N})_{SSM}} \right)^{0.199} . \quad (11)$$

Effectively the SK rate has been used to limit SSM "environmental" uncertainties, leaving an error budget dominated by the nuclear physics. But this source of error is under laboratory control and will be reduced as nuclear reaction measurements continue. The last two terms are the primordial abundances one would like to constrain. The role of the SSM in this equation is to define a set of parameters and thus a set of reference rates, about which we then explore possible variations. Those variations generate the environmental and nuclear uncertainties discussed above.

The R_{cal} factors in Eq. (11) contain additional uncertainties discussed in [1], including one important one, that associated with neutrino oscillations. Apart from the dependence on the solar density profile, this should be considered a laboratory uncertainty, analogous to nuclear cross sections. Uncertainties in oscillation parameters will continue to be refined by astrophysical, accelerator and reactor measurements.

The LMA parameter uncertainties in SK and Borexino/SNO+ are anti-correlated. Most of the low-energy ^{15}N neutrinos do not experience a level crossing, residing instead in a portion of the MSW plane where the oscillations are close to the vacuum oscillation limit,

$$P_{\nu_e}(E_\nu) \to 1 - \frac{1}{2}\sin 2\theta_{12}, \tag{12}$$

so that an increase in the vacuum mixing angle θ_{12} decreases the ν_e survival probability. The higher energy ^{8}B neutrinos are largely within the MSW triangle, described by an adiabatic level crossing. The limiting behavior for an adiabatic crossing is

$$P_{\nu_e}(E_\nu) \to \frac{1}{2}(1 - \cos 2\theta_{12}) \tag{13}$$

so that an increase in θ_{12} increases the survival probability. This anti-correlation thus leads to larger effects in the ratio.

The impact of this uncertainty on Eq. (11) was analyzed in [1], using the allowed regions for θ_{12} and δm^2_{12} obtained in KamLAND's combined analysis, yielding

$$\frac{R^{B/S}_{cal}(^{15}\mathrm{O}, \delta m^2_{12}, \theta_{12}))}{R^{SK}_{cal}(^8\mathrm{B}, \delta m^2_{12}, \theta_{12})^{0.825}} = (1 \pm 0.049)\left[\frac{R^{B/S}_{cal}(^{15}\mathrm{O}, \delta m^2_{12}, \theta_{12}))}{R^{SK}_{cal}(^8\mathrm{B}, \delta m^2_{12}, \theta_{12})^{0.825}}\right]^{BV} \tag{14}$$

where BV denotes the SSM best-value ratio.

Thus the overall uncertainty budget in Eq. (11) includes the experimental uncertainty of the SK "thermometer" of 3%, residual solar environmental uncertainties at 2.6%, LMA parameter uncertainties at 4.9%, and nuclear S-factor uncertainties of 7.6%. The overall uncertainty in the "theoretical" relationship between a future SNO+ or Borexino CN-neutrino flux and core C/N metals is thus about 9.6%. As the nuclear physics uncertainty dominates, one would expect this relationship to become more precise when ongoing analyses of data obtained by the LUNA collaboration and others for ^{14}N(p,γ) are completed. An appropriate goal would be 3.5% in this S-factor, a 30% improvement. The uncertainty in ^{14}N(p,γ) would no longer dominate the nuclear physics error budget, but instead would be comparable to the contributions from S_{33} and S_{34}. However, the current 9.6% uncertainty is not a bad starting point, as first-generation CN-cycle neutrino experiments are expected to measure this flux to an accuracy of about 10%. That is, the theoretical uncertainty will not dominate the experimental uncertainty, even without anticipated nuclear physics improvements.

5. Future Experiments and Summary

The work reported here was motivated in part by new detectors that might enable a high-statistics measurement of the CN-cycle neutrinos. Three possibilities are Borexino, SNO+, and Hanohano, detectors based on ultra-clean organic scintillation liquids. Borexino, which operates within Gran Sasso, must deal with a serious background, cosmogenic ^{11}C (a β^+ source). The collaboration has discussed a possible triple-coincidence veto [17] to limit this background.

Because of SNOLab's 6.0 km.w.e. depth, ^{11}C will be much less troublesome in SNO+, an experiment that will use the SNO cavity and about three times more scintillator than Borexino. Figure 3 shows a simulation of the expected SNO+ response, performed by the experimenters (Chen, private communication). (Note that the simulation is based on the BS05(OP) SSM and the best-fit LMA solution to the solar neutrino problem, rather than the updated BPS08(AGS) used in this paper.) The CN-neutrino event rate for an energy window above 0.8 MeV was found to be 2300 counts/year. The experimenters concluded that SNO+ could determine the CN-neutrino rate to an accuracy of approximately 10%, after three years of running [18]. This

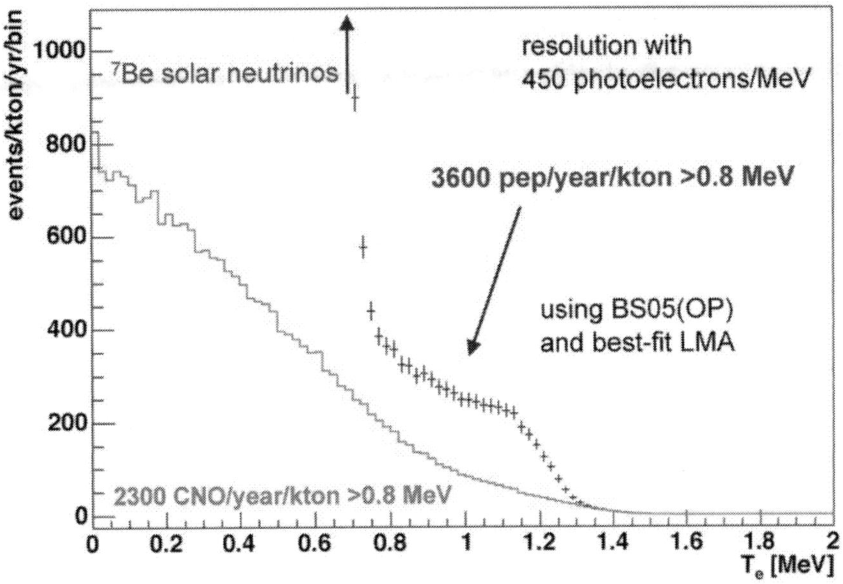

Figure 3. A simulation of the ^{7}Be, pep, and CNO electron recoil spectrum expected in SNO+. This figure is due to M. C. Chen [18].

is an appropriate goal for such a first-generation CN-cycle neutrino measurement, as it would approach the accuracy with which that flux could be related theoretically to the Sun's primordial core C and N abundances, as argued in this paper.

A still larger experiment, Hanohano [19], is being explored by a collaboration from the universities of Hawaii, Maryland, and Alabama, together with Makai Ocean Engineering. This experiment would be deployed deep underwater and utilized 10 kilotons of scintillator.

One main point of this talk is that if a 10% measurement can be made, an analysis could be done that limits theoretical uncertainties to below this level, provided one uses SK as a thermometer to eliminate environmental uncertainties. Indeed, both of the limiting errors in such an analysis can be addressed in future laboratory measurements, namely nuclear cross sections ($\sim$ 7.6%) and LMA oscillation parameters ($\sim$ 4.9%). One goal should be the reduction of these error bars to the level of uncertainty of the SK thermometer, $\sim$ 3%.

In my view, the primary motivation for such a measurement is to test the SSM assumption of a homogeneous zero-age-main-sequence Sun, given recent revisions of metal abundances derived from analyses of photospheric absorption lines. One could try to reconcile helioseismology with lower photospheric metal abundances by violating this SSM assumption, adopting a two-zone Sun with a low-Z convective zone.

It is intriguing that a possible mechanism for diluting the convective zone exists. The solar system's primary reservoir for metals, the gaseous giant planets, formed late in the evolution of the solar nebula, incorporating an excess of metal estimated at 40-90 $M_\oplus$. This mass is similar to the deficit of metals in the convective zone, were one to interpret the helioseismology/photospheric abundance discrepancy in the most naive way. This raises a provocative question: is it possible that the process that concentrated metals in the gaseous giants also produced a large volume of metal-depleted gas that subsequently was accreted onto the Sun's surface? If so, late-stage accretion of depleted gas onto the Sun would have not only diluted the convective zone, but generated a transition zone in the modern Sun's upper radiative zone – one that might alter helioseismology in that region. One would also expect abundance anticorrelations between the atmospheres of the Sun and gaseous giants. While the suggestion of a common chemical mechanism linking the convection zone and the gaseous giants is speculative, it provides additional motivation for exploiting the CN neutrinos as a quantitative probe of solar core metalicity.

6. Acknowledgments

This work was supported in part by the Office of Nuclear Physics, US Department of Energy, under grant DE-FG02-00ER-41132.

References

[1] Haxton W C and Serenelli A M 2008, to appear in *Ap. J.*
[2] Asplund M, Grevesse N, and Sauval A J 2005 in *Cosmic Abundances as Records of Stellar Evolution and Nucleosynthesis* vol 336, ed T G Barnes III and F N Bash (Astronomical Society of the Pacific) p 25
[3] Bahcall J N, Basu S, Pinsonneault M, and Serenelli A M 2005 *Ap. J.* **618** 1049
[4] Bahcall J N, Serenelli A M, and Basu S 2005 *Ap. J.* **621** L85
[5] Antia H M and Basu S 2005 *Ap. J.* **620** L129
[6] Montalbán J, Miglio A, Noels A, Grevesse N, and di Mauro, M P 2004 in *SOHO 14 Helio- and Asteroseismology: Towards a Golden Future* vol 559, ed D Danesy (ESA Special Publication) p 574
[7] Guillot, T 2005 *Annual Review of Earth and Planetary Sciences* **33** 493
[8] Peña-Garay C and Serenelli A M 2008, in preparation
[9] Bahcall J N and Serenelli A M 2005 *Ap J* **626** 530
[10] Grevesse N and Sauval A J 1998 *Space Science Reviews* **85** 161
[11] Bahcall J N, Pinsonneault M H, and Basu S 2001 *Ap J* **555** 990
[12] Bahcall J N and Pinsonneault M H 2004 *Phys. Rev. Lett.* **92** 121301
[13] Bahcall J N, Serenelli A M, and Basu S 2006, *Ap. J. Suppl.* **165** 400
[14] Badnell N R et al. 2005 *MNRAS* **360** 458
[15] SNO Collaboration 2008 to appear in *Phys. Rev. Lett.* (arXiv:0806.0989)
[16] Hosaka J, *et al.* 2006 *Phys. Rev. D* **73** 112001
[17] The BOREXINO Collaboration 2005 *Nucl. Phys. B Proc. Suppl.* **145** 29
[18] Chen, M C 2007 in *Next Generation Nucleon Decay and Neutrino Dectectors*, vol 944, ed J R Wilkes (AIP) p 25
[19] Learned J, Dye S, Pakvasa S, and Svoboda R 2006 http://www.phys.hawaii.edu/ sdye/hanohano.html and arXiv hep-ex/0612002/

Unconventional applications of the Ge detector and the axion

F.T. Avignone III

Department of Physics and Astronomy, University of South Carolina,
Columbia, South Carolina 29208,
E-mail: avignone@physics.sc.edu

Abstract: A brief discussion of the early history of unconventional uses of Ge detectors is given, followed by a more detailed discussion focusing on their uses for axion searches. The main purpose of this discussion is to explore the possibility of pushing the envelope of sensitivity of solar axion searches with future large Ge detector arrays applied to searches employing coherent Bragg-Primakoff conversion, as well as the axio-electric effect.

1. Introduction

The first unconventional application of Ge detectors of interest to particle physicists was the early search for neutrino-less double-beta decay ($0\nu\beta\beta$-decay) by Fiorini et al. [1]. This led to a long series of experiments [2], finally resulting in searches with detector arrays involving kg quantities of germanium enriched to ~86% in ^{76}Ge [3,4].
The Ge detector was also employed in the first terrestrial search for Cold Dark Matter in the galactic halo, or Weakly Interacting Massive Particles (WIMPS) [5]. This experiment was inspired by A.K. Drukier and L. Stodolsky [6]. Although very rudimentary, it placed constraints on the mass and coupling of WIMPS to nuclei that eliminated heavy Dirac neutrinos with masses from ~10 GeV to almost 1 TeV as the dominant component of the Cold Dark Matter. There were several follow-up experiments using the same technique [7,8,9]. Later, the Ge technology was expanded to include collecting both phonon and photon signals to distinguish between ionization, caused by photons, and recoil signals that would be caused by neutrons or by WIMPS [10]. This technique led to the CDMS [11] and EDELWEIS [12] experiments. At this time, there are two large Ge detector arrays in various stages of development and construction, GERDA [13] and MAJORANA [14]. These arrays are primarily designed to search for $0\nu\beta\beta$−decay; however, under certain circumstances they could be very effective for searches for CDM and solar axions. It is their application to axion searches that is of main interest here.

2. Axions

The axion, the Goldstone boson associated with the spontaneous symmetry breaking of the Peccei-Quinn Symmetry [15,16], could be generated in the core of the sun via Primakoff interactions with nuclear electromagnetic fields. The axions could similarly be reconverted to photons in magnetic fields perpendicular to their velocities and detected with photon detectors at the end of a magnetic helioscope, or by exchanging a virtual photon with nuclei in a detector. The magnetic helioscope, CAST, briefly discussed later, has placed the most stringent constraints on the coupling of solar axions to the electromagnetic field. Both the coupling to nuclear electric fields and helioscope magnetic fields are driven by the Primakoff diagram via the Hamiltonian,

$$\mathcal{L}_{a\gamma\gamma} = a\vec{B} \cdot \vec{E} / M \quad (1)$$

In (1), $\vec{B}$ and $\vec{E}$ are magnetic and electric fields, respectively, a is the axion field, and $M \equiv 2\pi f_a / \alpha$, where f_a is the scale of the spontaneous symmetry breaking. For a value of $f_a \approx 10^7 GeV$ ($M \approx 10^{10} GeV$), and $m_a << 1keV$, for example, one expects a solar axion flux from the Primakoff process of $\Phi_a \approx 2.1\times10^{11} / cm^2 \cdot s \cdot keV$ with a maximum in the spectrum at about $4keV$ [17].

3. The axio-electric effect

The first search for solar axions with Ge detectors was with the Battelle-Carolina ultra-low background Ge prototype of a $\beta\beta-$decay detector in the Homestake goldmine [18]. It utilized the axio-electric effect, which is simply related to the photo-electric effect as follows [19]:

$$\sigma_{ae} = \frac{\alpha_{axion}}{\alpha_{EM}}\left(\frac{\hbar\omega}{m_e c^2}\right)^2 \sigma_{photo-electric} \text{ and }, \quad \alpha_{axion} = \frac{1}{4\pi}\left(\frac{2x'_e m_e c^2}{f_a}\right)^2 \tag{2a}$$

$$\text{where } x'_e \approx 1, \text{ and } \alpha_{axion} = \frac{8.312\times10^{-8} GeV^2}{f_a^2} \tag{2b}$$

$$\sigma_{axioelectric} = \frac{8.312\times10^{-8} GeV^2 (E_a / keV)^2}{f_a^2} \times \sigma_{photoelectric} \tag{2c}$$

In equations (2), f_a is the Peccei-Quinn scale in GeV. This search was only a small pilot experiment, and application of this technique to a large array of low-background Ge detectors could be a very effective search for solar axions as discussed later.

4. Axions in M1 nuclear transitions

A completely different search technique involves the search for axions generated in competition with M1 gamma rays in nuclear transitions, and detecting them via the Primakoff conversion to photons in a Ge detector [20]. The Lagrangian describing the coupling of axions to hadrons can be written:

$$L = a\overline{\Psi} i\gamma_5 (g_0\beta - g_3)\Psi, \tag{3}$$

where g_0 and g_3 are the isoscalar and isovector coupling constants respectively. The axion to photon branching ratio for $M1$ transitions was derived by Haxton [20,21], and is written as follows:

$$\frac{\Gamma_a}{\Gamma_\gamma} = \frac{1}{2\pi\alpha(1-\delta^2)}\left[\frac{g_0\beta - g_3}{(\mu_0 - 1/2)\beta + \mu_3 - \eta}\right]^2, \tag{4}$$

where μ_0 is the isoscalar magnetic moment $(\mu_0 - 1/2) \approx 0.38$, while $\mu_3 \approx 4.71$, is the isovector magnetic moment. The parameters, η and β, are nuclear structure dependent and written as follows:

$$\eta \equiv -\frac{\left\langle J_f \left\| \sum_i^A l(i)\tau(i) \right\| J_i \right\rangle}{\left\langle J_f \left\| \sum_i^A \sigma(i)\tau_3(i) \right\| J_i \right\rangle}, \quad \text{and} \quad \beta \equiv \frac{\left\langle J_f \left\| \sum_{i=1}^A \sigma(i) \right\| J_i \right\rangle}{\left\langle J_f \left\| \sum_{i=1}^A \sigma(i)\tau_3(i) \right\| J_i \right\rangle}. \tag{5}$$

The parameter $\beta \approx +1$ for an unpaired proton, and $\beta \approx -1$ for an unpaired neutron, while $\eta \approx +0.80$ in the case of ^{57}Fe, for example, which has an unpaired neutron, and $\eta \approx -3.74$ for ^{55}M and -1.20 for ^{23}Na, both of which have unpaired protons [21].

The axion-nucleon coupling constants depend on several parameters and were given by Haxton and Lee [21]. The isoscalar and isovector couping constants are:

$$g_0 = -1.61\times10^{-7}\left\{\frac{3F-D+2S}{f_a}\right\}, \quad \text{and} \quad g_3 = -4.84\times10^{-7}\left\{\frac{(D+F)(1-z)}{f_a(1+z)}\right\}, \tag{6}$$

where F and D are invariant matrix elements of the axial current with values: $F \approx 0.48; D \approx 0.77$. In (6), $z = m_u/m_d \approx 0.56$ in the quark model, and the Peccei-Quinn mass scale, f_a, is given in units of $10^6 GeV$. The quantity, S, is the flavor singlet, axial vector matrix element and plays a crucially important role in the cross section.

5. Moriyama's ^{57}Fe Solar Axion Search Proposal

In 1995, Shigetaka Moriyama [22] used the calculations of Haxton and Lee [21] to design a technique to search for the $14.4keV$ axion line from the $14.4keV$ $M1$ transition from the first excited $7/2^+$nuclear level to the $9/2^+$ground state in the ^{57}Fe in the sun. This level is excited by the bath of thermal photons in the solar core. While this transition lies in the high-energy tail of the predicted axion spectrum [23], there is enough ^{57}Fe in the sun to produce a substantial flux of axions. Haxton and Lee calculated the solar abundance of this isotope to be 3.26×10^{-5} of the solar mass. They calculated: $\beta = -1.19$, and $\eta = 0.80$. Even though ^{57}Fe has an unpaired neutron, and $g_0\beta$ and g_3 have opposite signs, the relatively small excitation energy, $14.4keV$, yields a favorable Boltzman factor, $e^{-14.4keV/kT}$, which also compensates for the slow transition rate $(\tau = 140ns; \Gamma_\gamma = 4.7\times10^{-9}eV)$. The temperature of the core of the sun would Doppler broaden the $14.4keV$ axion line to a $FWHM \cong 5eV$. From this, one concludes that the portion of line in the axion flux, effective in the resonant excitation of cold ^{57}Fe in a detector on earth, would only be $\Phi_{eff} \cong 10^{-9}\Phi_{14.4keV}$.

Moriyama calculated the axion flux based on [21], but with a slight modification in the solar model given by Turck-Chie`ze et al. [24]. The resulting expression is:

$$\frac{d\Phi_a}{dE_a} = 2.0\times10^{13} cm^{-1}s^{-1}keV^{-1}\left(\frac{10^6 GeV}{f_a}\right)^2 C^2 \tag{7}$$

where,

$$C = -1.19\left(\frac{3F-D+2S}{3}\right) + (D+F)\left(\frac{1-z}{1+z}\right). \tag{8}$$

We choose the value $S = 0.35$ [25] for the following sample calculation. Accordingly $C^2 = 0.0365$. In reference [22], the nuclear recoil and red-shift energies were evaluated as $1.9\times10^{-3}eV$ and $1.5\times10^{-1}eV$ respectively, and were neglected.

Assembling these pieces, the expected detection rate in a detector of ^{57}Fe is written:

$$R_{\det} = \frac{d\Phi_a}{dE_a}\pi\sigma_{0\gamma}\frac{\Gamma_a}{\Gamma_\gamma}\Gamma_{tot}, \tag{9}$$

where $\sigma_{0\gamma} = 2.6\times10^{-18}cm^2$, is the resonant $\gamma - ray$ excitation cross section for this process [26], and $\Gamma_{tot} = 4.7\times10^{-12}keV$ is the level width obtained from a measurement of the mean life of the $14.4 - keV$ level. The quantity Γ_a/Γ_γ is the axion to gamma branching ratio of the $14.4 - keV$ axion line computed using equation (4). In the following analysis we will use the top of the open hadronic-axion window given by Raffelt [27], for which the Peccei-Quinn scale is $f_a \le 6\times10^5 GeV$. Substituting these values into equation (9), the differential flux is: $d\Phi_a/dE_a = 2.0\times10^{12}cm^{-2}s^{-1}keV^{-1}$, and the rate can be expressed as:

$$R_{\det} = (7.67\times10^{-17}s^{-1})\frac{\Gamma_a}{\Gamma\gamma} \tag{10}$$

per ^{57}Fe atom in the detector fiducial volume. Substituting into equation (4), we see $\Gamma_a/\Gamma_\gamma = 4.2\times10^{-14}$, and the detection rate per atom in the detector becomes $R_{\det} = 3.2\times10^{-30}s^{-1}$. If the iron is isotopically enriched to 85% in ^{57}Fe, there would be 8.98×10^{24} ^{57}Fe atoms per kg. The detection rate then becomes: $3.2\times10^{-30}/(\sec\cdot{}^{57}Fe)\times(8.98\times10^{24}\,{}^{57}Fe/kg) = 2.9\times10^{-5}\sec^{-1}kg^{-1} \approx 2.5d^{-1}kg^{-1}$. While there are uncertainties in the various input parameters to this calculation, it is clear that there exist reasonable scenarios in which several events per day could be detected in a detector of reasonable mass.

This experiment has only been attempted with a very small ^{57}Fe foil on top of a Ge detector [28]. For a detector with kg quantities of ^{57}Fe, and high efficiency, the only choice would be a bolometer of a compound with a significant quantity of ^{57}Fe. No such compound has yet been found. Later in this article, another way will be discussed to search for this solar-axion line using the material discussed with the multi kg detector arrays currently under development or construction.

6. The magnetic helioscope

In 1983, Sikivie proposed a technique to search for the invisible "axion" [29] that depended on the coupling of the axion to the electromagnetic field via the Primakoff vertex. At first glance, it appears that for a given magnetic induction, B, $P_{a\gamma}(L) \propto (BL)^2$ would continually increase as the square of the length. This is true in the special case of massless axions. For axions with mass there exists a coherence length, L_{coh}, beyond which the conversion probability vanishes.

The wave function of an axion entering the region of a magnetic field perpendicular to its velocity will be in a state, which is a coherent linear superposition of axion and photon wave functions. If the axion has a finite mass, its group velocity in the vacuum will be less than that of the speed of light, and after a distance the wave function will loose coherence, unless it would have already collapsed into a pure photon state. If the wave function is still in the superposition after traveling this maximum length of coherence, L_{coh}, the probability of conversion to a photon vanishes. This condition is met when the following relation is satisfied: $(c - v_a)t = \lambda/2 = (1-\beta_a)ct \Rightarrow ct \equiv L_{coh}$. In this case we see, $(1-\beta_a^2) = m_a^2c^4/\hbar^2\omega^2$ and since $(1+\beta_a) \cong 2$, we can write:

$$\frac{1}{(1+\beta_a)} \cong \frac{2\hbar^2\omega^2}{m_a^2c^4} \Rightarrow L_{coh} = \frac{\lambda}{2(1-\beta_a)} = \frac{2\pi\hbar c}{\hbar\omega}\cdot\frac{1}{2}\cdot\frac{2\hbar^2\omega^2}{m_a^2c^4} \text{ and } L_{coh} \cong \frac{2\pi\hbar c E_a}{m_a^2c^4} \qquad (11)$$

From equation (11) we see that for an axion of $m_ac^2 = 1eV$, $L_{coh} = 0.5cm$ for an axion with $4keV$ of energy. In fact we see that a 10-m magnet begins to loose coherence for axions with masses of about $0.023eV$.

There is a technique that makes the physical length of a given magnet the coherence length, i.e., $L_{coh} = L_{magnet}$. The technique uses a filler gas in the magnet bore to provide an index of refraction for the photon such that $v_\gamma < c$. The more massive the axion, the higher the required gas density needed for an increase in the index of refraction that will make $v_\gamma \approx v_a$. There is an upper limit of this density beyond which the increase in gas density reaches the point of diminishing returns.

The CAST magnetic helioscope experiment is the most sensitive experiment to date [30]. It is a 9.26 m, 9.0T magnet mounted on a rotating frame that tracks the sun for about 1.5 hours at both sunrise and sunset. When searching for mass-less axions, a sensitivity of $g_{a\gamma\gamma} \le 10^{-10} GeV^{-1}$ was reached. When cold ^{3}He was added, CAST rapidly lost sensitivity for an axion mass of about 1eV [31]. A recent analysis by Creswick, Nussinov and Avignone concluded [32]: "--- CAST helioscope, the most sensitive experiment to date, is near the limit of sensitivity in axion mass. Increasing the length, gas density, or tilt angle all have negative influences, and will not improve the sensitivity." The question then is, have axion searches themselves reached their point of diminishing returns? Could Ge detectors in large arrays (GERDA, MAJORANA) play important roles? If so, how?

7. Bragg Coherent Primakoff Conversion in Single Crystal Detectors

The original ideas for this technique were taken from Buchmuller and Hoogeveen [33], from Pascos and Zioutas [34], and were brought to our group by Zioutas [35]. They were developed into a concise formulation by our group (see Creswick et al., [36]). Three experiments were conducted using this technique [37,38] which placed the best laboratory bounds for axion masses greater than $\approx 0.2eV$. The proposed large $0\nu\beta\beta$-decay experiments utilizing single crystals might be able to improve these bounds by another order of magnitude. The cross section was derived by Buchmuller and Hoogeveen [33], and independently verified by a different method by Nussinov [39], and is written as:

$$\frac{\partial\sigma}{\partial\Omega}=\frac{g_{a\gamma\gamma}^2}{16\pi^2}F_a^2(2\theta)\sin^2\theta\text{ , where }\quad F_a(2\theta)=F_a(qk)=\frac{Zek^2}{r_0^{-2}+q^2}\text{ ,}\tag{12}$$

where $q=2k\sin\theta, r_0$ is the atomic-screening length, and $g_{a\gamma\gamma}$ is the axion coupling constant in GeV^{-1}. Integrating over all angles:

$$\bar{\sigma}(\eta)=\frac{1}{4\pi}\int\frac{d\sigma}{d\Omega}\Omega=\frac{Z^2\alpha\hbar^2g_{a\gamma\gamma}^2}{8\pi c^2}\left[\frac{2\eta^2+1}{4\eta^2}\ln(1+4\eta^2-1\right].\tag{13}$$

In (13), $\eta=r_0k$, and $Z^2\alpha\hbar^2g_{a\gamma\gamma}^2/8\pi c^2=1.15\times10^{-44}cm^2\equiv\sigma_0$ when $g_{a\gamma\gamma}=10^{-8}GeV^{-1}$.

For light axions, the Primakoff process of conversion from axion to photon will be coherent, similar to the Bragg reflection of x-rays, when the Bragg condition is met. That is to say, when the momentum transfer to the crystal is the reciprocal lattice vector, $\vec{G}\equiv2\pi/a_0(h,k,l)$, where a_0 is the dimension of a conventional cubic cell, and h,k,l are integers that define a given crystal plane.

The conversion rate in the detector is written in the following compact form [36]:

$$\frac{d\acute{N}}{d\Phi}=2\hbar c\frac{V}{v_c^2}\sum_G|S(G)|^2\frac{d\sigma}{d\Omega}(G)\frac{1}{|\vec{G}|^2}\delta\left|E_a-\frac{\hbar c|\vec{G}|^2}{2\vec{k}\cdot\vec{G}}\right|\tag{14}$$

In equation (14), $S(\vec{G})$ is the lattice structure function, written in equation below, V is the total volume of the crystal, and, v_c is the volume of a unit cell. The structure function is written as follows:

$$S(G)=\left[1+e^{i\pi/2(h+k+l)}\right]\left\{1+e^{i\pi(h+k)}+e^{i\pi(h+l)}+e^{i\pi(k+l)}\right\}.\tag{15}$$

Integrating over the solar axion flux, equation (14) can be written in the following very convenient form:

Equation (14) is the predicted rate of the conversion of axions for a given axion energy given the position of the sun, $\hat{k}$, and the reciprocal lattice vector, $\vec{G}$. The detector will have a finite energy resolution and to account for this, the rate is smoothed with a Gaussian with a finite width ΔE. The smoothed version of (14) is:

$$R(\hat{k},E)=\hbar c\frac{V}{v_c}\sum_{G}|S(G)|^2\frac{d\sigma}{d\Omega}\frac{1}{|\vec{G}|^2}\frac{d\Phi}{dE}\left[erf\left(\frac{E-E_a(\hat{k},\vec{G})}{\sigma\sqrt{2}}\right)-erf\left(\frac{E-E_a(\hat{k},\vec{G})-\Delta E}{\sigma\sqrt{2}}\right)\right] \quad (16)$$

where $E_a(\hat{k},\vec{G})=\hbar c|\vec{G}|^2/2\hat{k}\cdot\vec{G}$ and $erf(x)=2/\sqrt{\pi}\times\int e^{-t^2}dt$ is the error function.

The following theorem applies to the structure function, $S(G)$: " the individial indices h, k, and l, must all be either even or odd if $S(G)\neq 0$. If they are all even, they must all add up to a multiple of 4". Further, it can be shown that if (h, k, l) are all even $|S(G)|^2=64$; if they are all odd, $|S(G)|^2=32$, or $|S(G)|^2=\{3+(-1)^{h+k+l}\}\times 16$.

It is of course necessary to know when the vector, $\vec{s}$, from the solar core to the detector is parallel to the crystal's inverse lattice unit vectors, $\vec{g}$. To this end we recall that Ge detector crystals are pulled by the Czrachralski method along either the 100 or 011 axis. In the SOLAX experiment discussed in reference [36,37], the z-axis, which was the vertical axis, was the 100 axis. For an experiment to have real discovery potential, the angles between the other two axes and true north, for example, must also be known. The determination of the instantaneous vector, $\vec{s}$, between the solar core and the detector can be made using the U.S. Naval Observatory Vector Astronomy Subroutines, NOVAS [40]. Fig.1 shows an example of the detection rate on a given day, at a given location for $g_{a\gamma\gamma}=10^{-8}GeV^{-1}$, with zero background.

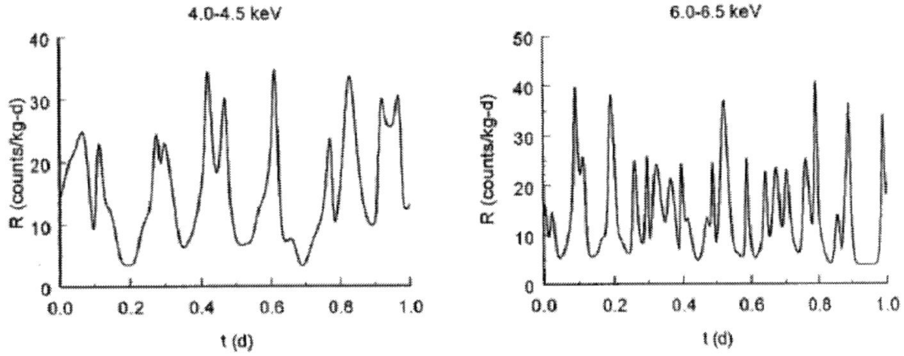

Figure 1. A sample predicted modulation of the signal in a 1-kg Ge detector computed with equation (16). The location was in southern Argentina at the SOLAX site [36,37].

8. Possible searches with large Ge arrays

The SOLAX experiment used the above technique in Sierra Granda Argentina [37]. It had a 1-kg detector, 4-keV energy threshold and a background of 3.4 keV^{-1}kg^{-1}d^{-1}. It operated for 708 kg.d, and achieved a limit, $g_{a\gamma\gamma}\leq 2.7\times 10^{-9}GeV^{-1}$. The COSME experiment [38], run later, used a 0.25 kg detector with a 2.5-keV threshold, a background of 0.7 keV^{-1}kg^{-1}d^{-1}. It operated for 72.7 kg.d, and achieved an upper bound of $g_{a\gamma\gamma}\leq 2.75\times 10^{-9}GeV^{-1}$. It is tempting to attempt to scale these parameters to determine what a large array could achieve by $g_{a\gamma\lambda}(sensitivity)\propto[b/Mt]^{1/8}$. This would be adequate for a counting experiment, but the sensitivity of a Bragg-Primakoff, SOLAX type experiment depends very sensitively on the analysis of the time-modulation structure in the data. Just scaling the experiment to a 1-ton

Ge experiment for a 10-y exposure would result in a sensitivity of $g_{a\gamma\gamma} \approx 6\times10^{-10} GeV$. This is an oversimplification; what is needed is a completely new look at the analysis technique to minimize the effect of background. The goal should be to achieve a sensitivity of $g_{a\gamma\gamma} \leq 10^{-10} GeV$, to cover the axion-model space in axion mass, which cannot be done with a helioscope.

Another approach depends involves the search for the axions from the 14.4 keV M1 transition in ^{57}Fe in the sun. Following the reasoning of Moriyama discussed above, $\Phi_a(14.4keV) \approx 2.8\times10^{10} cm^{-2}s^{-1}$. The photoelectric cross section from standard tables is: $\sigma_{photo} = 8.8\times10^{-21} cm^2$. Applying equations (2a,b,c), $\sigma_{axioelelectric} = 1.2\times10^{-42} cm^2$, using the Peccei-Quinn scale corresponding to $g_{a\gamma\gamma} = 10^{-10} GeV^{-1}$. Accordingly, the detection rate would be: R= $N\sigma_a\Phi_a \approx 2.4\times10^{-2} kg^{-1}d^{-1}$, or about 525 detections in a 60 kg Ge detector array per year.

The rate of detection of solar axions produced by the Primakoff process is also interesting. In this case we use the flux suggested by Raffelt [27] and integrate the product of flux and cross section over axion energy $E_a = \hbar\omega$, where $d\Phi_a(E_a)/dE_a = 6\times10^{10}\cdot E_a^{2.481}(e^{-E_a/1.205})$. The result is:

$$R = \int_{E_1}^{E_2} \Phi_a(E_a)\sigma(E_a)dE_a = 0.134 kg^{-1}d^{-1}\,. \tag{17}$$

This is a healthy rate of ~8 per day in a 60 kg array, like the Majorana Demonstrator, or 134 per day in a 1-ton array. This rate will fall rapidly as $g_{a\gamma\gamma}$ decreases.

9.Conclusions

While the search for axions has almost run into a stone wall, there is still some hope for discovery with large germanium detector arrays, for coupling constants not too much smaller that with a Peccei-Quinn scale corresponding to $M \approx 10^{10} GeV$. Since large arrays are being proposed by the GERDA and Majorana collaborations, it is well worth including these searches in their scientific programs. It is also still worth- while continuing the search for a compound containing ^{57}Fe that would make a good bolometer, although this appears to be a long shot at this time.

References,

[1] E.Fiorini et al., Il Nuovo Cim. **13**, 747 (1973).
[2] F.T. Avignone III et al., Phys. Rev. Lett. **50**, 721 (1983); **54**, 2309 (1985); Phys. Rev. C **34**, 666 (1986); H.S. Miley et al., Phys. Rev. Lett. **65**, 3092 (1990);
F.T. Avignone III et al., Phys. Lett. B **256**, 559 (1991).
[3] C.E. Aalseth et al.,(IGEX) Phys. Rev. D **65**, 092007 (2002); D **70**, 078302 (2004).
[4] L. Baudis et al., (Heidelberg-Moscow) phys. Rev. Lett. **83**, 41 (1999).
[5] S.P. Ahlen et al., Phys. Lett. B **195**, 603 (1987).

[6] A.K. Drukier and L. Stodolsky, Phys. Rev. D **30**, 2295 (1984).
[7] D.O. Caldwell et al., Phys. Rev. Lett. **61**, 510 (1988).
[8] D. Abriola et al., Astroparticle Physics **6**, 63 (1996).
[9] A. Morales et al., Astroparticle Physics **16**, 325 (2002); B **32**, 532 (2002).
[10] N.Wang et al., Physica B **3**, 165 (1990).
[11] Z. Ahmed et al., (CDMS Collaboration) arXiv: astro-ph/0802.3530.
[12] V. Sanglard et al., (EDELWEISS Collaboration) Phys. Rev. **71**, 122002 (2005).
[13] I. Abt et al., Nucl. Instrum. Methods Phys. Res. A **570**, 470 (2007).
[14] F.T. Avignone III (for the Majorana Collaboration) arXiv: nucl-ex/ 0711.4808; R.Gaitskill et al., arXiv: nucl-ex/0311013.
[15] R. D. Peccei and Helen Quinn, Phys. Rev. Lett. **38**, 1440 (1977);Phys Rev. D **16**, 1791 (1977).
[16] S. Weinberg, Phys. Rev. Lett. **40**, 223 (1978); F. Wilczek. Phys. Rev. Lett. **40**, 279 (1978).
[17] K. van Bibber et al., Phys. Rev. D **39**, 2089 (1989).
[18] F.T. Avignone III et al., Phys. Rev. D **35**, 2752 (1987).
[19] S. Dimopoulos et al., Mod. Phys. Lett. A **1**, 491 (1986).
[20] F.T. Avignone III et al., Phys. Rev. D **37**, 618 (1988).
[21] W.C. Haxton and K.Y. Lee, Phys. Rev. Lett. **66**, 2557 (1991).
[22] S. Moriyama, Phys. Rev. Lett. **75**, 3222 (1995).
[23] G.G. Raffelt, "Stars as Laboratories for Fundamental Physics" (University of Chicago Press , Chicago 1996).
[24] S. Turck- *Chie'ze* et al., Phys. Rep., **230**, 57 (1993).
[25] G. Altarelli et al., Phys. Lett. B **496**, 337 (1997).
[26] A. *DeRu'jula* and K. Zioutas, Phys. Lett. B **217**, 354 (1989).
[27] G. Raffelt, arXiv: hep-ph/0611118.
[28] *Krčmar* et al., Phys. Lett B **442**,38 (1998).
[29] P. Sikivie, Phys. Rev. Lett. **51**, 1415 (1983).
[30] K. Zioutas et al., Phys. Rev. Lett. 94, 121301 (2005).
[31] C. Eleftheriadis (for the CAST Collaboration) Journal of Cosmology and Astroparticle Physics, **4**, 10 (2007).
[32] S. Nussinov et al., arXiv: astro-ph/0805.2142
[33] W. Buchmuller and F. Hoogeeven, Phys. Rev. Lett. B **237**, 278 (1994).
[34] E.A. Pascos and K. Zioutas, Phys. Lett. B **323**, 367 (1994).
[35] K. Zioutas (private communication August, 1996).
[36] R.J. Creswick et al., Phys. Lett. B **427**, 235 (1998).
[37] F.T. Avignone III et al., (SOLAX Collab.) Phys. Rev. Lett. **81**, 5068 (1998).
[38] A. Morales et al., (COSME Collab.) Astropart. Phys. **16**, 325 (2002).
[39] S. Nussinov (private communication).
[40] U.S. Naval Observatory Vector Astronomy Subroutines, (supplied by J.A. Baggert)

Journal of Physics: Conference Series **173** (2009) 012016 doi:10.1088/1742-6596/173/1/012016

Direct Determination of Neutrino Mass

R.G. Hamish Robertson

Center for Experimental Nuclear Physics and Astrophysics, University of Washington, Seattle, WA, 98195, USA

E-mail: rghr@u.washington.edu

Abstract. The sum of the masses of the three neutrino mass eigenstates is now constrained both from above and below, and lies between 55 and 6900 meV. The lower limit is set by neutrino oscillations and the fact that masses are non-negative. The upper limit is set by laboratory measurements on the beta decay of tritium. These determinations share a common characteristic of being essentially model independent, or "direct". The clustering on large scales in the universe depends on this quantity, and, within the framework of ΛCDM cosmology, favors a mass sum below about 600 meV. In this article, the laboratory approach to neutrino mass via beta decay is emphasized, particularly an experiment now under construction, KATRIN, on the beta decay of tritium. Another candidate beta-active nuclide, ^{187}Re, offers an advantage in phase space but calls for a very different experimental approach.

1. Introduction

While it is now certain from neutrino oscillation experiments that there are three neutrinos with different masses, the absolute masses remain unknown. The masses are of importance to cosmology, because neutrino mass influences the formation of large-scale structure in the universe, tending to damp clustering until an epoch when the neutrinos have cooled sufficiently to become non-relativistic. The scale-dependence of clustering observed in the universe, interpreted in the framework of a ΛCDM cosmology, favors a mass sum below about 600 meV [1]. A laboratory determination of the mass would be valuable in constraining cosmological models, because there are significant correlations between neutrino mass and both the running of the spectral index and the equation of state of dark energy [2].

Neutrino mass is also an important issue in the development of the next standard model. In this case, it is the *patterns* of mass rather than the numerical details that can be expected to shed light on the mechanisms of mass generation. A quasi-degenerate, an inverted hierarchical, and a normal hierarchical spectrum would each send a unique message about the role of mass and possible new symmetries beyond the standard model.

2. Oscillations

Neutrino oscillations yield the differences between the squares of the eigenmasses. Two such differences ('solar' and 'atmospheric') have been measured in a variety of experiments on solar, atmospheric, reactor, and accelerator neutrinos. Given that the differences must close cyclically, this is sufficient to determine all three such quantities,

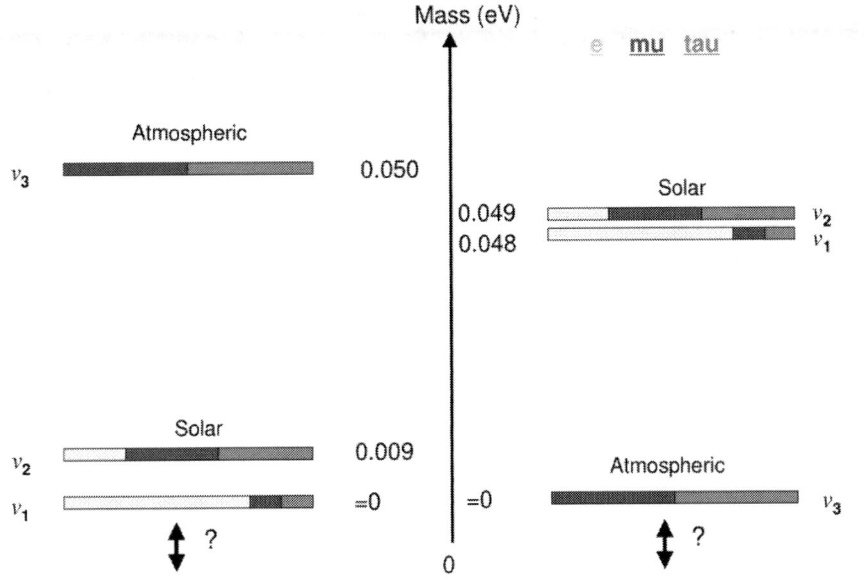

Figure 1. Schematic illustration (mass intervals not to scale, mass values given for lightest mass = 0) of the decomposition of the neutrino mass eigenstates ν_i in terms of their flavor intensities. There may also be a small admixture of ν_e into ν_3. Which of the two hierarchies is correct is not at present known.

except that the differences are signed quantities. The sign is known for the solar mass gap δm^2 from the effects of matter enhancement, but it is not known for the atmospheric gap Δm^2. The two possibilities lead to the 'normal' and 'inverted' hierarchy, as illustrated in Fig. 1. The most recent determinations yield [1] the following values (95% CL) for the mass-squared differences:

$$\delta m^2 = (7.66 \pm 0.35) \times 10^{-5}\ \text{eV}^2 \quad (1)$$
$$\Delta m^2 = (2.38 \pm 0.27) \times 10^{-3}\ \text{eV}^2. \quad (2)$$

The smallest possible sum (with 1-σ uncertainties) of the three masses is found in the normal hierarchy, with $m_1 = 0$:

$$\sum m_i \geq (58.3^{+1.4}_{-1.5})\ \text{meV} \quad (3)$$

If the hierarchy is inverted, the smallest possible sum would be,

$$\sum m_i \geq (97.8^{+2.8}_{-2.8})\ \text{meV} \quad (4)$$

The precision of the lower bound on the sum of the masses is remarkable.

3. Beta Decay

In beta decay the phase-space available to the electron near the endpoint is modified by neutrino mass, a fact realized immediately by Fermi [3] as he formulated the theory of beta decay.

In the intervening seven decades of experimental searches for neutrino mass, tritium has been the beta-active nucleus of choice because it has a low endpoint energy, which makes the modification caused by neutrino mass a larger fraction of the total spectrum:

$$^3\mathrm{H} \rightarrow {}^3\mathrm{He}^+ + \mathrm{e}^- + \overline{\nu}_\mathrm{e} \quad + 18580 \text{ eV}.$$

It is also a very simple atom, and atomic or molecular effects are important at the eV level. The decay is superallowed, with a short half-life, which reduces the amount of source material needed for a given counting rate and sensitivity.

A recent complete treatment of the spectrum shape in beta decay is given by Masood *et al.* [4]. In the standard analysis, the electron energy spectrum for neutrinos with eigenmasses m_i is given to a good approximation by

$$\frac{dN}{dT} = CF(Z,T)p(T+m_e)(T_0-T)\sum_{i=1,3}|U_{ei}|^2[(T_0-T)^2-m_i^2]^{\frac{1}{2}}\Theta(T_0-T-m_i), \quad (5)$$

where T denotes the kinetic energy of the electron, U_{ei} is the electron-flavor amplitude in the ith mass eigenstate, m_e is the mass of the electron, p is the electron momentum, T_0 corresponds to the maximum electron energy for $m_i = 0$ (endpoint energy), $F(Z,T)$ is the Fermi function, and the step function $\Theta(T_0-T-m_i)$ ensures energy conservation. The constant C is given by

$$C = \frac{G_F^2}{2\pi^3}\cos^2\theta_C|M|^2 \ . \quad (6)$$

Here G_F is the Fermi constant, θ_C is the Cabibbo angle and M is the nuclear matrix element. As both M and $F(Z,T)$ are independent of m_i, the dependence of the spectral shape on m_i is given by the phase space factor only. Moreover, the bound on the neutrino mass from beta decay is independent of whether the electron neutrino is a Majorana or a Dirac particle.

Above 200 meV the masses m_i are essentially equal and degenerate, the unitarity of $\mathbf{U}$ gives $\sum_i |U_{ei}|^2 = 1$, and the resulting beta spectrum is indistinguishable from what a single massive 'electron neutrino' would produce if mass eigenstates were also flavor eigenstates. The current best limit is that reported by the Mainz group, $m_i \leq 2300$ meV (95% CL) [5]. The corresponding limit on the sum of the masses is 6900 meV, and the sum is therefore known from model-independent constraints to lie, at the 95% CL, in the range,

$$55.3 \leq \sum m_i \leq 6900 \text{ meV}.$$

4. Future Beta Decay Experiments

Two beta-active nuclides are attractive for the next phase of experimental work. Tritium will be the basis for the next-generation KATRIN experiment now under construction in Germany [6], and ^{187}Re is the basis of the MARE project in Italy [7]. These nuclides have the lowest endpoint energies. Table 1 summarizes some of the basic quantities relevant to each.

Because the spectrum near the endpoint is quadratic in both cases, ^{187}Re enjoys a factor of about 300 advantage over tritium in the branching ratio to the last eV because of the low endpoint energy of the Re. On the other hand, the forbidden nature of the decay and the correspondingly low specific activity means that to obtain one event per day in the last 200 meV of the spectrum requires 20 μg of tritium, but 13 kg of ^{187}Re. If nothing is seen in the current round of experiments down to 200 meV, it will be necessary to develop a program with 20 meV sensitivity, for which the corresponding source masses are 20 mg of tritium and 13 tonnes of ^{187}Re.

Table 1. Parameters of ^{3}H and ^{187}Re (taken to be isotopically pure for all listings except isotopic abundance). Each nuclide is treated as though having only a single final state.

Parameter	^{3}H	^{187}Re	Unit
Spin sequence	$1/2^+ \to 1/2^+$	$5/2^+ \to 1/2^-$	
Half-life	12.32	4.32×10^{10}	y
Isotopic abundance	–	62.6	%
Endpoint energy	18.58	2.47	keV
Specific activity	3.58×10^{14}	1.64×10^3	Bq/g
Branch to last eV	2×10^{-13}	6×10^{-11}	
Specific activity, last eV	72	1.1×10^{-7}	Bq/g

The experimental approaches in the two cases differ fundamentally. For tritium experiments the source and detector are separate, which permits filtering out the bulk of uninteresting decays before they reach the detector, but which also sets a limit on how large the source can be while still getting the electrons out of it. For ^{187}Re, on the other hand, the experiments are done with microcalorimeters in order to capture all the energy, independent of the complexities of atomic final state excitations. In that case the source and detector are the same, obviating the fundamental limit for separated source and detector, but raising the problem that the detector must detect every event no matter whether it is interesting or not. Avoidance of spectral distortions and background caused by pileup then becomes vitally important.

5. The MARE Experiment

MARE, Microcalorimeter Arrays for a Rhenium Experiment, brings together the technical advances of earlier programs, MANU [8] and MIBETA [9, 10], with a view to optimizing the approach for the next phases [7]. MARE-1 aims for 2-eV sensitivity, and MARE-2 200 meV. The MANU approach involves superconducting Re crystals and transition-edge sensors made of an Ir-Au alloy, while the MIBETA approach has used crystals of $AgReO_4$ and resistance thermometers made of doped Si. MIBETA has measured the endpoint energy of the ^{187}Re beta decay to be

$$T_0 = (2465.3 \pm 0.5 \pm 1.6)\ \mathrm{eV},$$

and the the neutrino mass limit to be

$$m_\nu^2 = (-112 \pm 207 \pm 90)\ \mathrm{eV}^2,$$

corresponding to an upper limit of 15 eV on the mass.

6. The KATRIN Experiment[1]

The KArlsruhe TRItium Neutrino experiment [6] is a very large scale tritium-beta-decay experiment to determine the mass of the neutrino. It is presently under construction at the Forschungszentrum Karlsruhe, and makes use of the Tritium Laboratory built there

[1] These sections are an updated version of an earlier presentation on the KATRIN experiment [11].

for the ITER project. The combination of a very large retarding-potential electrostatic-magnetic spectrometer and an intense gaseous molecular tritium source makes possible a sensitivity to neutrino mass of 0.2 eV, about an order of magnitude below present laboratory limits.

KATRIN consists of seven major subsystems (see Fig. 2), a gaseous tritium source, the tritium processing and recirculation system, a differential pumping section, a cryogenic Ar frost pumping section, a pre-spectrometer, a main spectrometer, and the detector system.

Figure 2. Layout of the KATRIN Experiment

7. Source and Tritium Recirculation

The gaseous molecular source consists of a tube 9 cm in diameter and 10 m in length maintained at the desired temperature (27 K) by circulation of two-phase Ne. An axial magnetic field of 3.6 T guides electrons toward the spectrometers. Four turbomolecular pumps at the ends of the source tube collect T_2 for recirculation in an 'inner loop' with Pd filters to remove contaminants, particularly ^{3}He. A 'demonstrator' for the source is being constructed by ACCEL Instruments GmbH to verify the thermal performance.

8. Differential Pumping Section

Further reduction of tritium pressure is achieved in this section, which consists of a tube with a chicane, superconducting solenoids to guide the electrons, and four 2800 l/s magnetic-bearing turbomolecular pumps to scavenge the tritium. The subsystem is being built by Ansaldo Superconduttori for delivery in 2008.

9. Cryogenic Pumping Section

Any remaining tritium that escapes through the differential pumping section is trapped in Ar frost, which forms a highly efficient, large-area, chemically inert, and radiation-immune surface. The contract for this system is in place.

10. Pre-spectrometer

There are two spectrometers in tandem in KATRIN, both of the retarding-potential type. The pre-spectrometer operates at a cutoff potential typically 100 eV below the endpoint, preventing most electrons from reaching the main spectrometer, which is expected to improve the background near the endpoint. The pre-spectrometer was the first KATRIN subassembly to be completed, in order to permit some key concepts to be tested before other design elements were frozen. An important discovery is the existence of parasitic Penning-trap configurations, which cause increased backgrounds or breakdown in the presence of high voltage and a magnetic field. Electrode redesign is in progress. A new feature in this type of spectrometer is a grid inside the shell to suppress the emission of low-energy electrons ejected by cosmic rays and radioactivity from the shell.

11. Main Spectrometer

The main spectrometer is a large stainless-steel vessel 10 m in diameter and 24 m in length. The 200-tonne chamber was placed in its building November 29, 2006. Thermal insulation and heating tubes were installed, and 6 turbopumps. The tank was baked out to 350 C from July 16, 2007 to July 25. The base pressure 2 months later was 0.9×10^{-9} mbar and the outgassing rate (predominantly H_2) 1.2×10^{-12} mbar l s^{-1} cm^{-2}. When the interior grid structure and 1000 m of non-evaporable getter strips are installed later, the predicted base pressure will be 3×10^{-11} mbar.

12. Detector

Electrons surmounting the potential barriers in the spectrometers enter a high axial magnetic field region and are detected in a monolithic 148-pixel Si PIN diode array 10 cm in diameter. Two superconducting solenoids can produce up to 6 T, defining the electron beam diameter and the maximum pitch angle accepted from the source. The magnets are being made by Cryomagnetics, Inc. and the detector by Canberra. Between the solenoids is a region in which various calibration devices – gamma sources and a photoemissive electron gun – can be inserted. Selection of materials, shielding, and an active veto are being adopted to keep non-beam-associated backgrounds below 1 mHz.

13. Summary

The KATRIN experiment is scheduled for initial data-taking in 2012. Subsystems are in an advanced state of construction and commissioning. Funding for KATRIN is being provided by the Helmholtz Gemeinschaft, the Bundesministerium für Bildung und Forschung, and the US Department of Energy. MARE is in the R&D phase, with support from INFN, Italy.

This paper was prepared for the Carolina International Symposium on Neutrino Physics, Columbia, SC, May 15 - 17, 2008, and the author acknowledges with great respect the advances in neutrino physics springing from the work of Frank T. Avignone, Ettore Fiorini, and the late S. Peter Rosen. This research was supported by the US DOE under Grant DE-FG02-97ER41020.

References

[1] Fogli G L *et al.* 2008 (*Preprint* 0805.2517)
[2] Fogli G L *et al.* 2007 *Phys. Rev.* **D75** 053001 (*Preprint* hep-ph/0608060)
[3] Fermi E 1934 *Nuovo Cim.* **11** 1–19
[4] Masood S S *et al.* 2007 *Phys. Rev.* **C76** 045501 (*Preprint* 0706.0897)
[5] Kraus C *et al.* 2005 *Eur. Phys. J.* **C40** 447–468 (*Preprint* hep-ex/0412056)
[6] Angrik J *et al.* (KATRIN) FZKA-7090, http://www-ik.fzk.de/katrin/
[7] Andreotti E *et al.* 2007 *Nucl. Instrum. Meth.* **A572** 208–210
[8] Pergolesi D *et al.* 2006 *Nucl. Instrum. Meth.* **A559** 349–351
[9] Sisti M *et al.* 2004 *Nucl. Instrum. Meth.* **A520** 125–131
[10] Arnaboldi C *et al.* 2006 *Phys. Rev. Lett.* **96** 042503
[11] Robertson R G H (KATRIN) 2007 (*Preprint* 0712.3893)

Searching for Physics Beyond the Standard Model with Accelerator Neutrino Experiments

William C. Louis
LANL, Physics Division, Los Alamos, NM 87545
E-mail: louis@lanl.gov

Abstract. The MiniBooNE experiment at Fermilab was designed to test the LSND evidence for $\bar{\nu}_\mu \rightarrow \bar{\nu}_e$ oscillations [1]. The first MiniBooNE oscillation result in neutrino mode [2] shows no significant excess of events at higher energies ($E_\nu > 475$ MeV), although a sizeable excess is observed at lower energies ($E_\nu < 475$ MeV). The lack of a significant excess at higher energies allows MiniBooNE to rule out simple $2-\nu$ oscillations as an explanation of the LSND signal. However, the low-energy excess is presently unexplained. Additional antineutrino data and NuMI data may allow the collaboration to determine whether the excess is due, for example, to a neutrino neutral-current radiative interaction [3] or to neutrino oscillations involving sterile neutrinos [4, 5, 6]. If the excess is consistent with being due to sterile neutrinos, then future experiments at FNAL (MicroBooNE & BooNE) or ORNL (OscSNS) could confirm their existence.

1. Introduction

The MiniBooNE experiment was designed to test the $\bar{\nu}_\mu \rightarrow \bar{\nu}_e$ oscillation interpretation of the LSND signal in both neutrino and antineutrino modes. MiniBooNE has approximately the same L/E_ν as LSND but with an order of magnitude higher baseline and energy. Due to the higher energy and dissimilar event signature, MiniBooNE systematic errors are completely different from LSND. The MiniBooNE detector design has worked very well, and follow-up experiments at FNAL (BooNE) or ORNL (OscSNS) could take advantage of this detector design. BooNE would employ a second detector similar to MiniBooNE at a different distance, while OscSNS would involve a MiniBooNE-like detector at a distance of ~ 60 m from the SNS beam dump. Another follow-up experiment, MicroBooNE, which was recently approved at FNAL, will consist of a ~ 70 ton fiducial volume Liquid Argon Time Projection Chamber (LArTPC).

2. MiniBooNE Experiment

A schematic of the MiniBooNE experiment is shown in Fig. 1. The experiment is fed by 8-GeV protons from the Booster that interact in a 71-cm long Be target located at the upstream end of a magnetic focusing horn. The horn pulses with a current of 170 kA and, depending on the polarity, either focuses π^+ and K^+ and defocuses π^- and K^- to form a very pure neutrino beam or focuses π^- and K^- and defocuses π^+ and K^+ to form a less pure antineutrino beam. The produced pions and kaons then decay in a 50-m decay pipe, and the resulting neutrinos and antineutrinos [7] can then interact in the MiniBooNE detector, which is located 541 m downstream of the Be target. For the MiniBooNE results presented here, a total of 5.58×10^{20} protons on target (POT) were collected in neutrino mode.

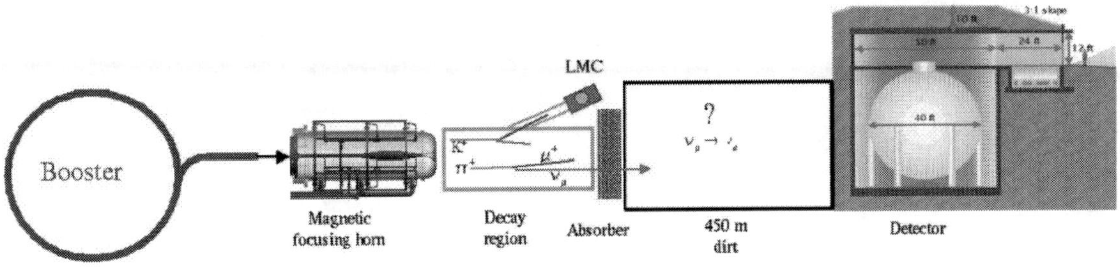

Figure 1. A schematic of the MiniBooNE experiment.

The MiniBooNE detector [8] consists of a 12.2-m diameter spherical tank filled with approximately 800 tons of mineral oil (CH_2). There are a total of 1280 8-inch detector phototubes (covering 10% of the surface area) and 240 veto phototubes. The fiducial volume is a 5-m radius that corresponds to approximately 450 tons. The fraction of bad phototube channels was $< 2\%$ over the course of the run.

3. MiniBooNE Results

MiniBooNE has published several interesting results. First, MiniBooNE has made a precision measurement of ν_μ charged-current quasi-elastic (CCQE) scattering events [9]. Fig. 2 shows the ν_μ CCQE Q^2 distribution for data (points with error bars) compared to the Monte Carlo simulation (histograms). A strong disagreement between the data and the original simulation (dashed histogram) was first observed. However, by increasing the axial mass, M_A, to 1.23 ± 0.20 GeV and by introducing a new variable, $E_{lo} = 1.019 \pm 0.011$, where E_{lo} is the increase in the incident proton threshold, the agreement between data and the simulation (solid histogram) is greatly improved. It is impressive that such good agreement is obtained by adjusting these two variables.

MiniBooNE has also collected the world's largest sample of neutral-current π^0 events [10], as shown in Fig. 3. By fitting the $\gamma\gamma$ mass and $E_\pi(1 - \cos\theta_\pi)$ distributions, the fraction of π^0 produced coherently is determined to be $19.5 \pm 1.1 \pm 2.5\%$. Excellent agreement is obtained between the data and the Monte Carlo simulation.

Fig. 4 shows the reconstructed neutrino energy distribution for candidate ν_e data events (points with error bars) compared to the Monte Carlo simulation (histogram) [2]. Good agreement between the data and the Monte Carlo simulation is obtained for $E_\nu > 475$ MeV; however, an unexplained excess of electromagnetic events is observed for $E_\nu < 475$ MeV. Several improvements have been made to the data analysis since this data was published, including an improved background estimate, an additional fiducial volume cut that greatly reduces the background from events produced outside the tank, and an increase in the data sample from 5.579×10^{20} POT to 6.462×10^{20} POT. The results of this improved analysis will be available soon.

An excess of ν_e candidate events is also observed in MiniBooNE from the NuMI beam. The NuMI beam, as shown in Fig. 5, differs from the Booster neutrino beam (BNB) in several respects. First, the NuMI beam is off axis by 110 mrad, whereas the BNB is on axis. Second, neutrinos from NuMI travel $\sim$ 700 m, compared to $\sim$ 500 m for neutrinos from the BNB. Also, the NuMI beam has a 6% contribution from electron-neutrinos and a 14% contribution from antineutrinos, while the BNB percentages are 0.5% and 2%, respectively. Fig. 6 shows the comparison between data events (points with error bars) and the Monte Carlo simulation (histogram) for ν_μ CCQE candidate events (bottom) and ν_e CCQE candidate events (top). Although the systematic errors are presently very large, the data is observed to be systematically low for ν_μ CCQE candidate events and systematically high for ν_e CCQE candidate events.

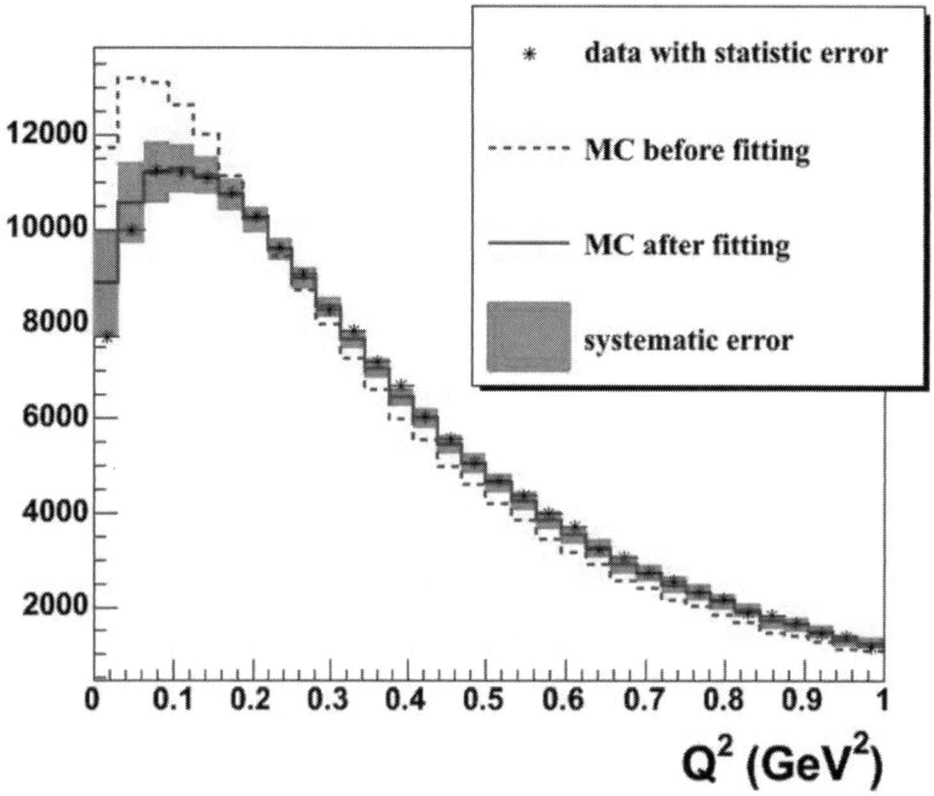

Figure 2. The ν_μ CCQE Q^2 distribution for data (points with error bars) compared to the Monte Carlo simulation (histograms).

Updated results should be available later this year with three times the data sample and with reduced systematic errors by constraining the normalization to the ν_μ sample.

4. MiniBooNE Future

MiniBooNE is approved to run for another year in antineutrino mode. So far MiniBooNE has collected $\sim 10^{21}$ POT, corresponding to $\sim 7 \times 10^{20}$ POT in neutrino mode and $\sim 3 \times 10^{20}$ POT in antineutrino mode. For the future, it will be imperative to understand the MiniBooNE low-energy excess. This excess is very interesting in its own right and crucial for future long-baseline experiments such as T2K. T2K will have a very similar neutrino energy distribution to MiniBooNE and will, therefore, be affected by the same low-energy excess. By analysing the MiniBooNE antineutrino data, NuMI data, and SciBooNE data, it may be possible to determine whether any of the published models [3, 4, 5, 6] can provide an explanation for the excess. If the low-energy excess continues to be consistent with a signal, then new experiments at FNAL (BooNE) or ORNL (OscSNS) should be proposed to explore physics beyond the Standard Model.

5. MicroBooNE

The MicroBooNE experiment, which was recently approved at Fermilab, exploits the precise differentiation of photons versus electrons in a detector, as is uniquely available from a Liquid Argon Time Projection Chamber (LArTPC). The detector consists of a ~ 70 ton fiducial volume

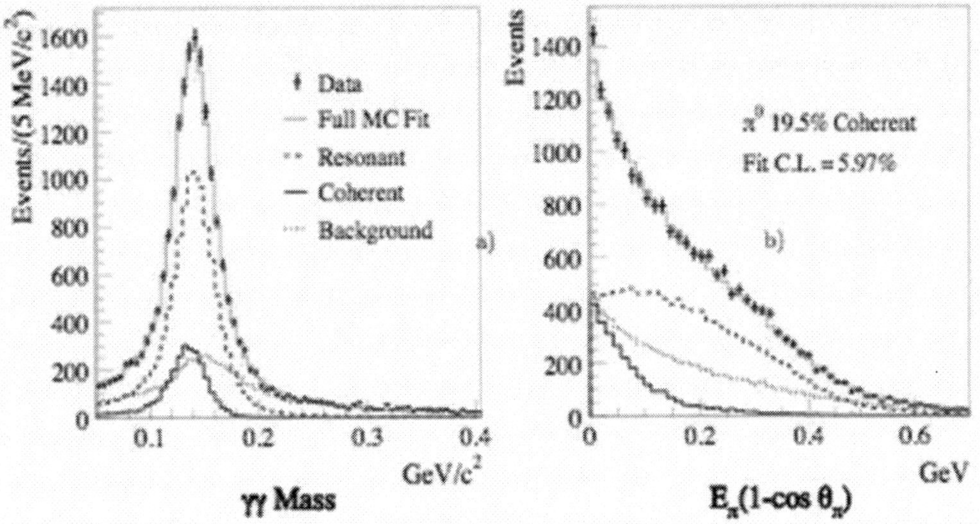

Figure 3. The neutral-current π^0 $\gamma\gamma$ mass and $E_\pi(1 - \cos\theta_\pi)$ distributions for data (points with error bars) compared to the Monte Carlo simulation (histograms).

LArTPC. It will run near the MiniBooNE enclosure on the BNB with an expected exposure of 6×10^{20} protons on target. The high spatial resolution and energy measurement down to the MeV scale substantially improves on information available from the MiniBooNE detector. The experiment will run in 2011.

The MicroBooNE detector can separate electron showers from photon showers using the energy deposited in the first 2.4 cm of the track. For an electron efficiency of 80%, γ contamination is expected to be $< 5\%$, based on Monte Carlo studies. Given the excellent e/γ separation, MicroBooNE can identify the source of the low energy events observed in MiniBooNE. MicroBooNE's sensitivity to the low energy excess is $\sim 5\sigma$ if the signal is electron-like and $\sim 3\sigma$ if the signal is photon-like, in a strictly counting-based experiment. Fits to shape-signatures may increase the significance.

6. BooNE

The BooNE experiment would involve building a second detector at a cost of $\sim$ \$8M along the BNB at FNAL at a different distance. With two detectors, many of the systematic errors would cancel, as the neutrino flux varies as $1/r^2$ to good approximation, so that a simple ratio of events in the two detectors would provide a sensitive search for ν_e appearance and ν_μ disappearance. Furthermore, by comparing the rates for a neutral-current reaction, such as neutral-current π^0 scattering, a sensitive search for sterile neutrinos can be made.

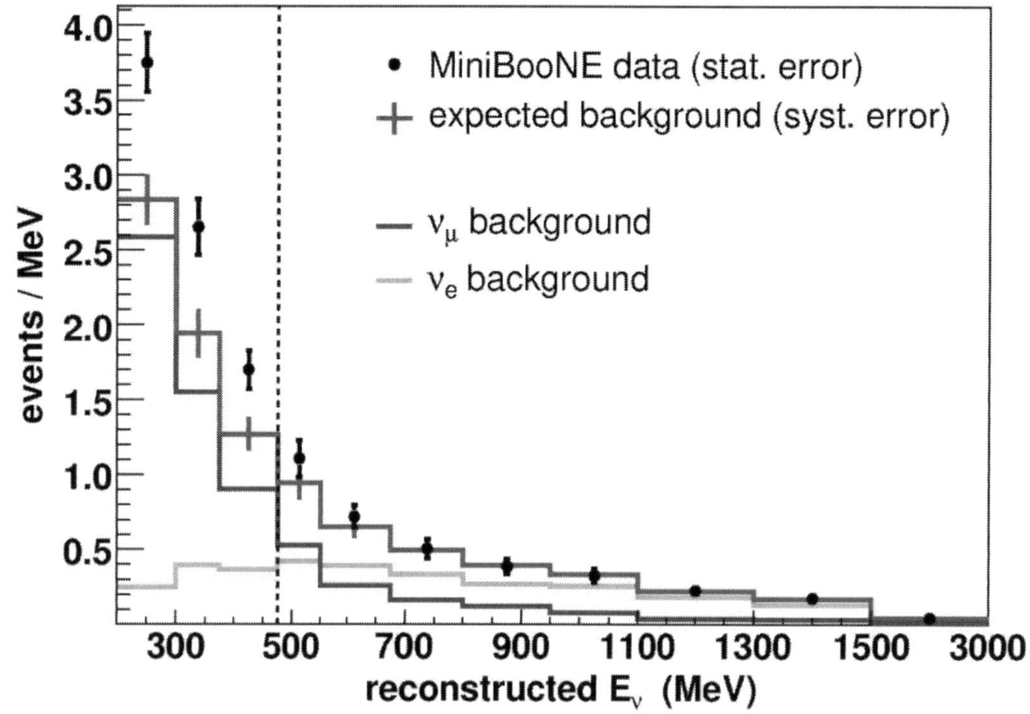

Figure 4. The reconstructed neutrino energy distribution for candidate ν_e data events (points with error bars) compared to the Monte Carlo simulation (histogram) [2].

7. OscSNS

The OscSNS experiment would involve building a MiniBooNE-like detector at a distance of ~ 60 m from the SNS beam dump at ORNL. The detector would be the same as MiniBooNE except with a higher phototube coverage of 25% and the addition of ~ 0.031 g/l of b-PBD scintillator. Due to the higher phototube coverage, the estimated cost is $\sim$ \$12M. Fig. 7 shows the layout of the SNS, which is running with a proton energy of 1 GeV and will eventually reach a beam intensity of ~ 1.4 MW. The great advantage of the SNS is that the neutrino flux is extremely intense and known almost perfectly, and the neutrino cross sections are known very well. The right plot of Fig. 8 shows the neutrino flux energy distribution, which includes a monoenergetic 30 MeV ν_μ from π^+ decay at rest and ν_e and $\bar{\nu}_\mu$ from μ^+ decay at rest. Furthermore, as shown in the left plot of Fig. 8, the monoenergetic ν_μ can be identified by timing.

With the SNS neutrino flux, OscSNS would be capable of making precision measurements of ν_e and $\bar{\nu}_e$ appearance and ν_μ disappearance and proving the existence of sterile neutrinos via the neutral-current reaction $\nu_\mu C \rightarrow \nu_\mu C^*(15.11)$. Any observed reduction of this cross section would be evidence for active-sterile neutrino oscillations. Fig. 9 shows the expected active-sterile neutrino oscillation sensitivity as a function of Δm^2 and $\sin^2 2\theta$ [11]. Other physics goals include precision measurements of $\nu e \rightarrow \nu e$ elastic scattering (and the world's best sensitivity for the ν_μ magnetic moment) and $\nu_e C \rightarrow e^- N$ charged-current scattering.

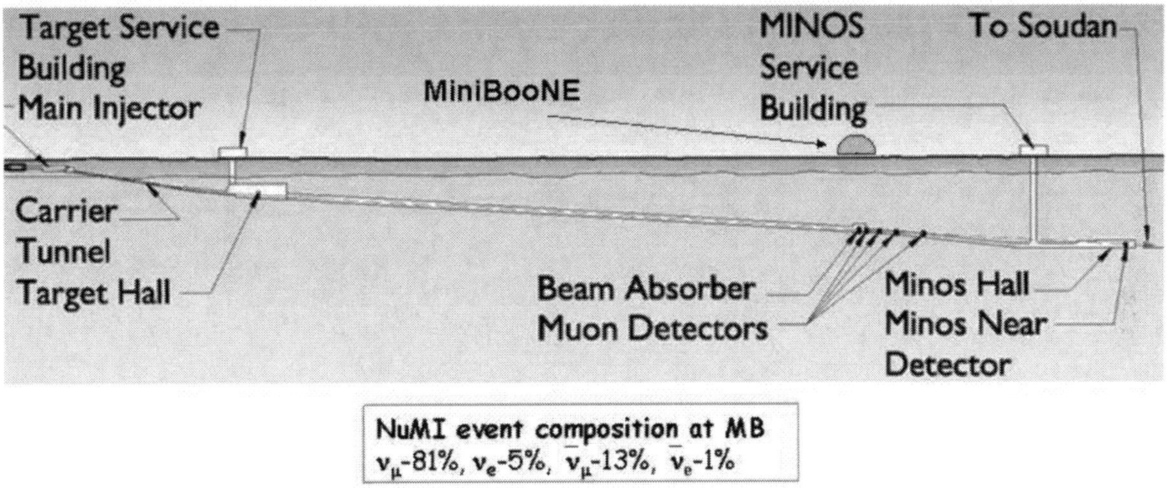

Figure 5. The NuMI beam.

8. Conclusions

The MiniBooNE experiment observes an unexplained excess of electromagnetic events at low energies, which may be due, for example, to either a neutral current radiative interaction [3] or to neutrino oscillations involving sterile neutrinos [4, 5, 6]. If the excess is consistent with being due to sterile neutrinos, then future experiments at FNAL (MicroBooNE & BooNE) or ORNL (OscSNS) would confirm their existence.

Acknowledgments

We wish Frank Avignone and Ettore Fiorini happy 75th birthdays and thank them for their numerous contributions to neutrino physics! We also honor the memory of Peter Rosen, who helped make our MiniBooNE experiment a reality.

References

[1] C. Athanassopoulos *et al.*, Phys. Rev. Lett. 75, 2650 (1995); 77, 3082 (1996); 81, 1774 (1998); A. Aguilar *et al.*, Phys. Rev. D 64, 112007 (2001).
[2] A. Aguilar-Arevalo *et al.*, Phys. Rev. Lett. 98, 231801 (2007).
[3] Jeffrey A. Harvey, Christopher T. Hill, Richard J. Hill, [arXiv:0708.1281]; [arXiv:0712.1230].
[4] Heinrich Paes, Sandip Pakvasa, Thomas J. Weiler, Phys. Rev. D 72, 095017 (2005); [arXiv:hep-ph/0504096].
[5] Michele Maltoni and Thomas Schwetz, Phys. Rev. D 76 093005 (2007); [arXiv:0705.0197].

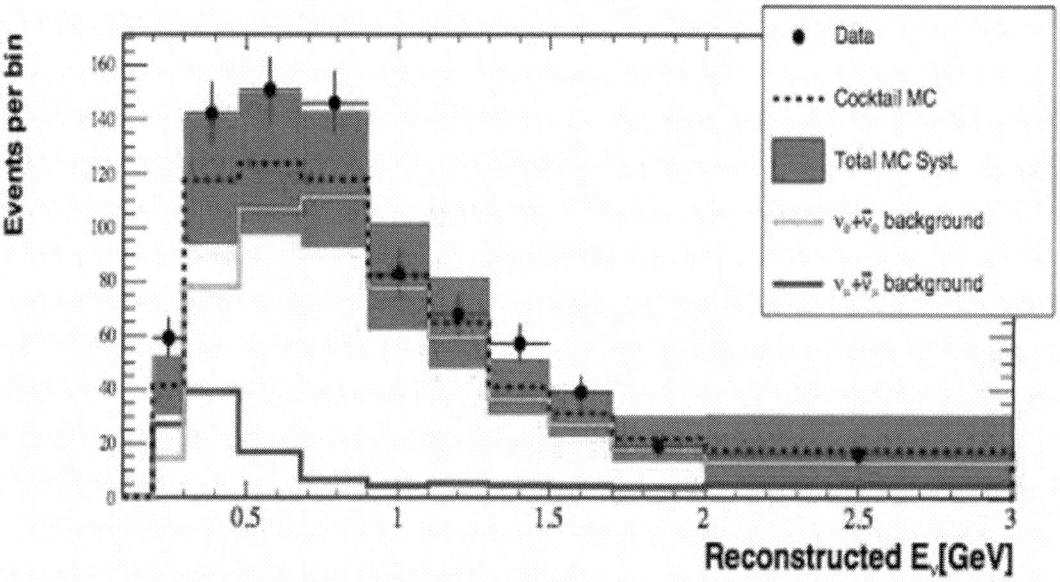

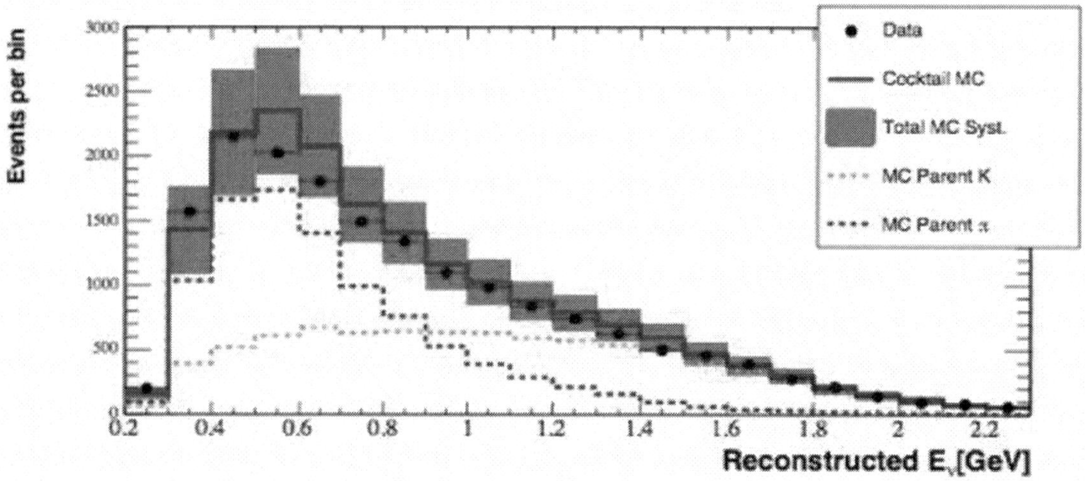

Figure 6. The comparison between data events (points with error bars) and the Monte Carlo simulation (histogram) for ν_μ CCQE candidate events (bottom) and ν_e CCQE candidate events (top).

[6] Ann E. Nelson and Jonathan Walsh, [arXiv:0711.1363].
[7] A. Aguilar-Arevalo *et al.*, [arXiv:0806.1449].
[8] A. Aguilar-Arevalo *et al.*, [arXiv:0806.4201].
[9] A. Aguilar-Arevalo *et al.*, Phys. Rev. Lett. 100, 032301 (2008).
[10] A. Aguilar-Arevalo *et al.*, Phys. Lett. 664B, 41 (2008).
[11] G. T. Garvey *et al.*, Phys. Rev. D 72, 092001 (2005).
[12] Michel Sorel, Janet Conrad, and Michael Shaevitz, Phys. Rev. D 70, 073004 (2004).

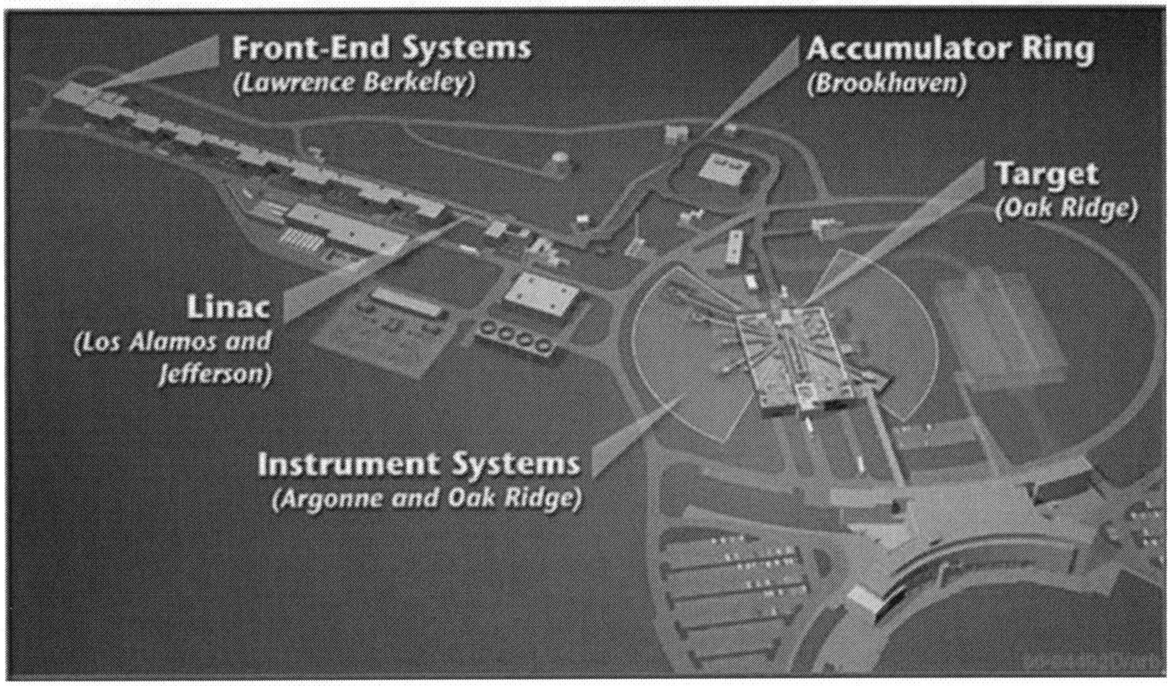

Figure 7. The layout of the SNS.

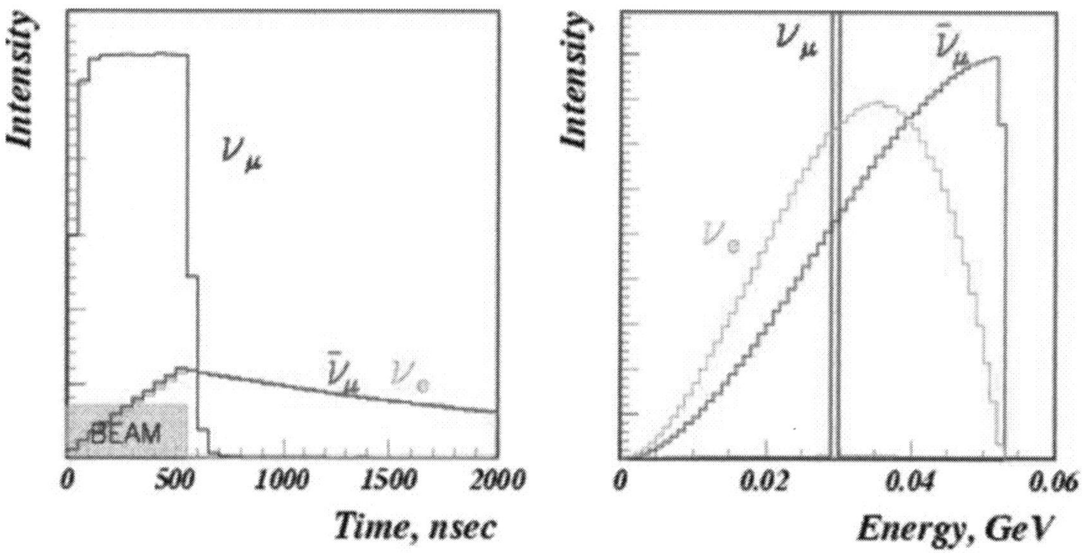

Figure 8. The right plot shows the neutrino flux energy distribution, which includes a monoenergetic 30 MeV ν_μ from π^+ decay at rest and ν_e and $\bar{\nu}_\mu$ from μ^+ decay at rest. The left plot shows the neutrino time distribution. The monoenergetic ν_μ can be identified by timing.

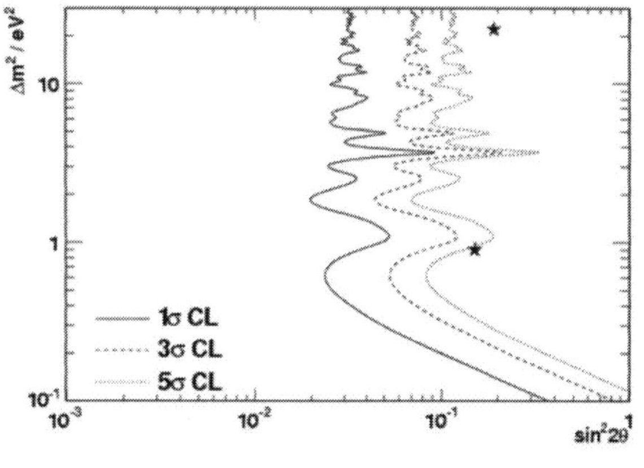

Figure 9. The expected OscSNS active-sterile neutrino oscillation sensitivity as a function of Δm^2 and $\sin^2 2\theta$ [11]. The stars represent the best fit of a 3+2 model [12].

Journal of Physics: Conference Series **173** (2009) 012018 doi:10.1088/1742-6596/173/1/012018

Neutrinos and Cosmology: a Lifetime relationship

Pasquale D. Serpico

Center for Particle Astrophysics, Fermi National Accelerator Laboratory, Batavia, IL 60510-0500 USA

E-mail: serpico@fnal.gov

Abstract. We consider the example of neutrino decays to illustrate the profound relation between laboratory neutrino physics and cosmology. Two case studies are presented: In the first one, we show how the high precision cosmic microwave background spectral data collected by the FIRAS instrument on board of COBE, when combined with Lab data, have greatly changed bounds on the radiative neutrino lifetime. In the second case, we speculate on the consequence for neutrino physics of the cosmological detection of neutrino masses even as small as $\sim$0.06 eV, the lower limit guaranteed by neutrino oscillation experiments. We show that a detection at that level would improve by many orders of magnitude the existing limits on neutrino lifetime, and as a consequence on some models of neutrino secret interactions.

FERMILAB-CONF-08-144-A

1. Introduction

Historically, there has been a deep relation between neutrino physics and cosmology: the bounds on the number of generations from primordial nucleosynthesis, the neutrinos as dark matter candidates (and their effect on structure formation), or their masses as related to the origin of baryon asymmetry represent well known examples of this link. Here we take the example of neutrino decays to illustrate the interplay of the Lab and Cosmology in the neutrino sector with two case studies: In the first case, we show how the high precision cosmic microwave background spectral data collected by the FIRAS instrument on board of COBE, when combined with data from neutrino oscillation experiments and direct bounds on absolute masses, have greatly improved the bounds on the radiative neutrino lifetime. The relevant formalism is summarized in Sec. 2, while in Sec. 3 we present the bounds obtained. In the second case, presented in Sec. 4, we speculate on the consequence for neutrino physics of the cosmological detection of neutrino masses even as small as $\sim$0.06 eV, the lower limit guaranteed by neutrino oscillation experiments. We show that a detection at that level would improve by many orders of magnitude the existing limits on neutrino lifetime, and as a consequence on neutrino secret interactions with (quasi-)massless particles as in majoron models. In Sec. 5 we present our conclusions. For details on the two cases see respectively [1] and [2], on which this article is mostly based.

2. Radiative neutrino decays

Traditionally, constraints on neutrino radiative lifetime coming from cosmology were based on the diffuse Cosmic Infrared Background (CIB) and assumed strongly hierarchical masses in the eV range [3, 4]. These constraints are now outdated and strictly speaking inapplicable. The neutrino mass splittings squared provided by oscillation experiments and present upper bounds on the neutrino mass scale constrain neutrino mass differences to fall in the microwave energy range ($E \sim 10^{-3}$ eV), in most of the allowed parameter space. This implies that current bounds can take advantage of the high precision cosmic microwave background (CMB) data collected by the FIRAS instrument on board of COBE, which tested the blackbody nature of the spectrum at better than 1 part in 10^4 [5, 6].

Let us denote by ν_i the neutrino fields respectively of masses m_i, where $i = 1, 2, 3$. The radiative decay $\nu_i \to \nu_j + \gamma$ can be thought of as arising from an *effective* interaction Lagrangian of the form

$$\mathcal{L}_{\rm int} = \frac{1}{2}\bar{\nu}^i \sigma_{\alpha\beta}(\mu_{ij} + \epsilon_{ij}\gamma_5)\nu^j F^{\alpha\beta} + \text{h.c.} \tag{1}$$

where $F^{\alpha\beta}$ is the electromagnetic field tensor, $\sigma_{\alpha\beta} = [\gamma_\alpha, \gamma_\beta]$ where γ_μ are the Dirac-matrices and $[.,.]$ is the commutator and μ_{ij} and ϵ_{ij} are the magnetic and electric transition moments usually expressed in units of the Bohr magneton μ_B. The convention to sum over repeated indices is used. In general, μ_{ij} and ϵ_{ij} are functions of the transferred momentum squared q^2, so that constraints obtained at a different q^2 are independent. The radiative decay rate for a transition $i \to j$ is written

$$\Gamma^\gamma_{ij} = \frac{|\mu_{ij}|^2 + |\epsilon_{ij}|^2}{8\pi}\left(\frac{m_i^2 - m_j^2}{m_i}\right)^3 \equiv \frac{\kappa_{ij}^2}{8\pi}\left(\frac{m_i^2 - m_j^2}{m_i}\right)^3 .$$

In the following, we shall assume that the radiative decay rate is very low compared with the expansion rate of the universe; neither the cosmological evolution or the primordial neutrino spectrum is affected by the additional coupling we are going to introduce. A posteriori, this is known to be an excellent approximation. For the same reason, we shall also neglect "multiple decays". We shall take our input data for neutrino mass eigenstate densities from the calculation performed in [7] without any extra parameter, as non-vanishing chemical potentials. With present data, the latter are anyway constrained to be well below $\mathcal{O}(1)$ [8], so dropping this assumption would not change much our conclusions.

¿From simple kinematical considerations it follows that in a decay $\nu_i \to \nu_j + \gamma$ from a state of mass m_i into one of mass $m_j < m_i$, the photon in the rest frame of the decaying neutrinos is thus monochromatic (two-body decay), with an energy

$$\varepsilon_{ij} = \frac{m_i^2 - m_j^2}{2\,m_i} . \tag{2}$$

At present, the neutrino mass spectrum is constrained by the well-known values of the two squared mass splittings for the atmospheric (Δm_H^2) and the solar (Δm_L^2) neutrino problems. We take their best-fit values and 2σ ranges from [9], $\Delta m_L^2 = 7.92\,(1 \pm 0.09) \times 10^{-5}$ eV^2, $\Delta m_H^2 = 2.6\,(1^{+0.14}_{-0.15}) \times 10^{-3}$ eV^2 . The remaining unknowns in the neutrino spectrum are the absolute mass scale (equivalently, the mass of the lightest eigenstate m_1) and the mass hierarchy. Namely, in normal hierarchy (NH) the mass pattern would be $\{m_1\,, m_2 = \sqrt{m_1^2 + \Delta m_L^2}\,, m_3 = \sqrt{m_1^2 + \Delta m_L^2 + \Delta m_H^2}\,\}$ while in inverted hierarchy (IH) one would have $\{m_1, m_2 = \sqrt{m_1^2 + \Delta m_H^2}\,, m_3 = \sqrt{m_1^2 + \Delta m_L^2 + \Delta m_H^2}\}$.

In the limiting case of normal hierarchy and $m_1 = 0$, the lightest neutrino for which a decay is possible has a mass $m_2 \simeq 9 \times 10^{-3}$ eV and is non-relativistic for most of the universe lifetime, namely in the redshift range $z \lesssim 50$. We can thus safely work in the approximation of all neutrinos decaying effectively at rest. In this limit, we can also neglect the momentum distribution of the neutrino spectra. We shall vary the mass scale in $0 \lesssim m_1 \lesssim 2$ eV as allowed by the Mainz experiment on the ^{3}H beta decay endpoint [10].

Let F_E be the present energy flux of photons with present energy E produced in the decay. The differential energy flux φ_E (energy flux F_E per unit energy and solid angle) is related to the differential number flux φ_n (the particle flux F_n per unit energy and solid angle) at present by

$$\varphi_E \equiv \frac{\mathrm{d}^2 F_E}{\mathrm{d}E\,\mathrm{d}\Omega} = E\,\frac{\mathrm{d}^2 F_n}{\mathrm{d}E\,\mathrm{d}\Omega} \equiv E\,\varphi_n, \tag{3}$$

and it can be shown that, if the lifetime τ_i of the neutrino of mass m_i is much greater than the universe lifetime it holds [11]

$$\varphi_E = \frac{\Gamma^\gamma_{32}}{4\pi}\frac{n_3}{H(z_{32})} + \frac{\Gamma^\gamma_{31}}{4\pi}\frac{n_3}{H(z_{31})} + \frac{\Gamma^\gamma_{21}}{4\pi}\frac{n_2}{H(z_{21})}, \tag{4}$$

where $n_i \simeq 113$ cm^{-3} is the present number density of the i–th neutrino *in absence of decay*, the Hubble function is (assuming, for simplicity, a flat cosmology) $H(z) = H_0\sqrt{\Omega_M(1+z)^3 + \Omega_\Lambda}$, $H_0 \simeq 73$ km s^{-1} Mpc^{-1} being the present Hubble expansion rate, and $\Omega_M \simeq 0.26$ and $\Omega_\Lambda \simeq 0.74$ respectively the matter and the cosmological constant energy density relative to the critical one. The dependence on energy enters implicitly via the quantities $0 \leq z_{ij} = \varepsilon_{ij}/E - 1$.

In practice, to a very good approximation one can write a general equation of the kind

$$\varphi_E = \frac{\Gamma^\gamma_H}{4\pi}\frac{n_H}{H(z_H)} + \frac{\Gamma^\gamma_L}{4\pi}\frac{n_L}{H(z_L)}, \tag{5}$$

where the meaning of the factors however depends on the hierarchy. In NH, in the first two terms of the sum in Eq. (4) it holds $z_{32} \simeq z_{31} \equiv z_H$, and one can identify $z_L = z_{21}$, $\Gamma^\gamma_L = \Gamma^\gamma_{21}$, $\Gamma^\gamma_H = \Gamma^\gamma_{31} + \Gamma^\gamma_{32}$. In IH, it is the last two terms of the sum in Eq. (4) which have $z_{31} \simeq z_{21} \equiv z_H$, and using $n_2 \simeq n_3$ one can identify $z_L = z_{32}$, $\Gamma^\gamma_L = \Gamma^\gamma_{32}$, $\Gamma^\gamma_H \equiv \Gamma^\gamma_{31} + \Gamma^\gamma_{21}$.

In Fig. 1 we represent the unredshifted photon energy ε_{ij} from decaying neutrinos [Eq. (2)] as a function of the lightest neutrino mass eigenstate m_1 in the case of normal and inverted hierarchy, where the meaning of $\varepsilon_{L,H}$ is clear from the previous discussion. We also indicate by an horizontal band the energy range of the CMB spectrum (2.84×10^{-4} eV $\leq E \leq 2.65 \times 10^{-3}$ eV) measured by FIRAS [5]. We also show the CIB range in the energy band above the FIRAS range up to (conventionally) 0.15 eV [12]. For $m_1 \lesssim 0.5$ eV, ε_H falls in the CIB range.

For photons emitted at $z = 0$ in the FIRAS range, the effect of radiative decays is most prominent and results in a feature on the CMB spectrum. Actually even if photons are emitted at higher energy the effect is still strong, since photons emitted at a redshift $z \sim 10$ enter the FIRAS spectrum because of cosmological redshift; it is easy to check that one has thus some sensitivity to κ_H in the whole range for m_1. On the other hand, for $m_1 \gtrsim 0.14$ eV the photons corresponding to the smaller splitting are falling in the radio band, below the frequency range probed by COBE, where measurements are more uncertain and thus one has no sensitivity to κ_L and the corresponding bound disappears.

Journal of Physics: Conference Series **173** (2009) 012018 doi:10.1088/1742-6596/173/1/012018

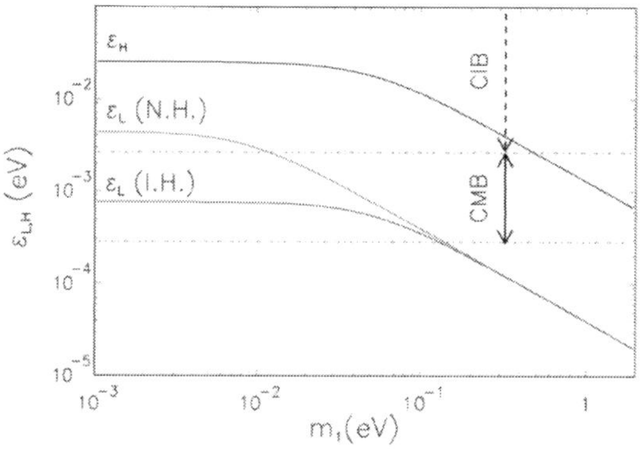

Figure 1. Unredshifted photon energy ε from decaying neutrinos [Eq. (2)] as a function of the lightest neutrino mass eigenstate m_1, for the two neutrino mass splittings (L,H) in normal and inverted hierarchy (see text for details). The horizontal band represents the energy range of the CMB spectrum measured by FIRAS [5]. The CIB energy range is also shown.

3. The radiative decay bound

To constrain the neutrino electromagnetic decay we use the COBE/FIRAS data for the experimentally measured CMB spectrum, corrected for foregrounds [5, 6]. The $N = 43$ data points $\Phi_i^{\rm exp}$ at different energies E_i are obtained by summing the best-fit blackbody spectrum to the residuals reported in Ref. [5]. The experimental errors σ_i and the correlation indices ρ_{ij} between different energies are also available. In the presence of neutrino decay, the original radiance of the "theoretical blackbody" at temperature T

$$\Phi^0(E,T) = \frac{E^3}{4\pi^3}[\exp(E/T) - 1]^{-1} \tag{6}$$

would gain an additional term so that the intensity becomes

$$\Phi^0(E,T) \quad \to \quad \Phi(E,T,\kappa^2_{L,H},m_1) = \Phi^0(E,T) + \varphi_E(\kappa^2_{L,H},m_1).$$

The bound are derived via a χ^2 technique, taking into account the covariance matrix of the data points and leaving the parameter T entering initially in $\Phi^0(E,T)$ free. Indeed, it needs not to be fixed at the standard value $T_0 = 2.725 \pm 0.002$ K [6], which is the best fit of the "distorted" spectrum eventually observed now, but is left to be determined in the minimization procedure.

In Fig. 2 we report the exclusion plot in the plane $\tau_{L,H} \equiv (\Gamma^\gamma_{H,L})^{-1}$ vs. m_1, where the regions below the solid curves are excluded at 95 % C.L. For small values of m_1 the most stringent limit is $\tau_L \gtrsim 4 \times 10^{20}$ s in IH (slightly better than in NH case), while the bound on τ_H is about an order of magnitude smaller, say $\tau_H \gtrsim 2 \times 10^{19}$ s, since for low m_1 only photons produced by H decays at high redshift are in FIRAS range. On the contrary, for $m_1 \gtrsim 0.14$ eV, the bound on τ_L disappears, while the bound on τ_H becomes more stringent, being $\tau_H \gtrsim 5 \times 10^{20}$ s. The "fuzzy" behaviour of the bounds is due to the sharp edge of the photon spectrum at $E = \varepsilon_{H,L}$: when the photon energy embeds a new FIRAS bin, the χ^2 function has a sharp discontinuity. If one translates the bounds of Fig. 2 into bounds on $\kappa_{L,H}$, the factor $(\delta m^2_{ij}/m_i)^3$ plays a significant role. For the NH case, $\kappa_L \lesssim 3 \times 10^{-8}\,\mu_B$, while in the IH case, $\kappa_L \lesssim 3 \times 10^{-7}\,\mu_B$. In agreement with our

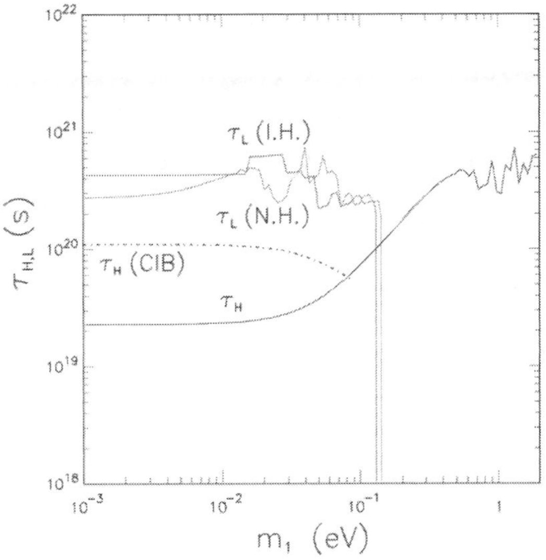

Figure 2. Bounds on τ_H and τ_L vs. m_1, for the two cases of NH and IH. The regions below the solid curves are excluded at 95 % C.L. The NH and IH curves for τ_H coincide, although the definition of τ_H is different (see text). The dot-dashed line is the limit obtained from the CIB.

previous considerations, the bound on κ_L disappears for $m_1 \gtrsim 0.14\,\mathrm{eV}$. On the contrary, the bound for κ_H is always present, and it corresponds to $\kappa_H \lesssim 8 \times 10^{-9}\,\mu_B$ apart for the degenerate region, where it degrades down to $10^{-7}\,\mu_B$ or even more.

It is worth noting that an improved bound on τ_H for low m_1 can be obtained from observations of the CIB, which differently from the CMB does not origin in the early universe, being rather the relic emission of all the galaxies at wavelengths larger than a few microns. Recently, a new estimate of the CIB flux has been established using the Spitzer Observatory data [12]. Using this determination one can obtain a rough bound on τ_H simply requiring that the total energy flux of the photons coming from ν decay does not exceed the CIB flux:

$$\int_{E_{\rm min}}^{\varepsilon_H} \varphi_E \, dE < \Phi_{\rm CIB} \sim 24\,\mathrm{nW\,m^{-2}\,sr^{-1}} \quad , \tag{7}$$

where we consider as lower limit of the CIB range the upper value of the FIRAS range, i.e. $E_{\rm min} = 2.65 \times 10^{-3}$ eV. The bounds on τ_H from Eq. (7) is shown in Fig. 2 by the dot-dashed line. Although this bound is stronger than those obtained by the FIRAS data in the same range of m_1, we emphasize that it should be considered only as indicative, due to the larger uncertainties in the CIB normalization and spectral shape. Interestingly, if it turns out that $m_1 \lesssim 0.1\,\mathrm{eV}$, an improvement on the bound on $\tau_{\rm H}$ will clearly take advantage of a better measurement of the CIB flux and a more detailed knowledge of the astrophysical sources contributing to it.

4. Invisible decays and secret neutrino interactions

Recent years have seen an impressive improvement on the cosmological constraints to the sum of neutrino masses $\Sigma = \sum m_i$ (for reviews see [13, 14]), with current limit from CMB only at the level of 1.3 eV [15], while combinations of different datasets produce bounds below 1 eV, with the most aggressive bounds (but also the most fragile ones with respect to unaccounted systematics) already at $\Sigma \lesssim 0.2$ eV, 95% C.L. [16, 17]. Several forecast analyses suggest that cosmological probes will reach in the future an incredible sensitivity to the effects of even a tiny mass of the cosmic background neutrinos. In particular, cosmic microwave background (CMB) lensing extraction may be sensitive to $\Sigma \simeq 0.035\,\mathrm{eV}$ [18]; CMB plus weak galaxy lensing with tomography may also push the sensitivity to Σ below the level of $\sim 0.05\,\mathrm{eV}$ [19, 20], with an error as low as $\sim 0.013\,\mathrm{eV}$ [19]. Also galaxy cluster surveys may probe Σ down to $\sim 0.03\,\mathrm{eV}$ [21], and a sensitivity down to $0.05 \pm 0.015\,\mathrm{eV}$ may be reached combining CMB with the data from the Square Kilometre Array survey of large scale structures [22]. These forecasts show that cosmology has a potential sensitivity to neutrino masses well below the 0.1 eV level. Of course, the ultimate level of the systematics to beat has yet to be reliably established. On the other hand, the synergy between different strategies and probes may help to identify the systematics, and also to improve over the above-mentioned figures of merit.

The interest of these expectations relies on the fact that neutrino oscillation data imply $\Sigma \gtrsim 0.06\,\mathrm{eV}$, where the minimum $\Sigma \simeq 0.061 \pm 0.004\,\mathrm{eV}$ is attained for a normal hierarchy (NH; values quoted at 2 σ, see [17]). For the case of an inverted mass hierarchy (IH), the oscillation data imply $\Sigma \simeq 0.1\,\mathrm{eV}$. The following arguments assume that neutrinos have a hierarchical spectrum of either inverted or normal sign, as favored by many theoretical models, including the simplest seesaw ones. We argue that, if a positive cosmological mass detection is achieved as expected, one will be able to put a remarkably strong constraint on the neutrino lifetime. Note that previous attention has been paid to the cosmological signatures of decaying neutrinos [23]. Yet the mass range explored in those papers is very large compared to present bounds, and the main signature considered was the impact on the integrated Sachs-Wolfe effect on the CMB. In our considerations, the bound comes from the impact that massive neutrinos have in the background evolution of the universe, in a range of masses where they are relativistic well after the CMB decoupling.

Bounds on neutrino lifetimes are usually quoted in terms of the rest-frame lifetime to mass ratio τ/m. Given a measurement in the time interval t using neutrinos with Lab energy E, the naive bound which one can put is $\tau/m \gtrsim t/E$. Using then the longest timescale available, the universe lifetime $t_0 \simeq H_0^{-1}$ (where H_0 is the Hubble constant), and the lowest energy neutrinos, the ones of the cosmic background which are at least partially non-relativistic, a bound of the order of ($m_{50} \equiv m/50\,\mathrm{meV}$)

$$\frac{\tau}{m} \gtrsim \frac{1}{m\,H_0} \simeq 10^{19}\,m_{50}^{-1}\,\mathrm{s/eV}\,, \tag{8}$$

is the strongest constraint attainable in principle. This is to be compared with the strongest direct bound available at present given by the observation of solar MeV neutrinos, of the order of $\sim 10^{-4}\,\mathrm{s/eV}$ [24]. A bound of the order of $\tau/m \gtrsim 4 \times 10^{11} m_{50}^2\,\mathrm{s/eV}$, has been claimed to follow already from the requirement that the neutrinos are free-streaming at the time of the photon decoupling, as deduced by precise measurements of the CMB acoustic peaks [25]. Yet, the robustness of this conclusion has been questioned in [26]. We shall see that the proposed bound based on cosmological neutrino mass detection would be much closer to the maximal theoretical bound of Eq. (8), thus superseding by several orders of magnitude the previous ones. More importantly, it is not based on a model for secret neutrino interactions, but on the "observation" of neutrino survival, and it applies whatever the final state light particles are.

Let us define the fraction of matter energy density in neutrinos as $f_\nu \equiv \rho_\nu/(\rho_{\rm m} + \rho_\nu)$, where $\rho_{\rm m}$ is the average cold dark matter (CDM) plus baryon density, and ρ_ν is the neutrino energy density. Keeping e.g. $\rho_{\rm m} + \rho_\nu$ constant, a non-vanishing f_ν today would change the redshift of matter radiation equality $z_{\rm eq}$ with respect to the massless neutrino case, as well as attenuate the growth of structures. The combined effect of the shift in the time of equality and of the reduced CDM fluctuation growth during matter domination produces an attenuation of perturbations for modes $k > k_{\rm nr}$, where $k_{\rm nr}$ is the minimum of the comoving free-streaming wavenumber attained when neutrinos turn non-relativistic, and given by [13]

$$k_{\rm nr} \simeq 1.5 \times 10^{-3} m_{50}^{1/2}\, {\rm Mpc}^{-1}. \tag{9}$$

An instantaneous decay of the massive neutrinos at a redshift z_d in the matter era can be thought as replacing the neutrino fluid with one having the same energy content at z_d, but whose energy density scales from that moment on as $(1+z)^4$, since the daughter particles are relativistic. Let us estimate how large a value of z_d, or equivalently of the proper time $t_d(=\tau$ if the neutrino is non-relativistic), can be probed cosmologically. Quickly after the neutrino decay one has formally $f_\nu \to 0$, provided that $t_d \ll t_0 \simeq H_0^{-1}$; from that moment on, the cosmological effects of the decaying neutrino scenario are analogous to the ones of a massless neutrino universe. The condition $t_d \ll t_0$ is required by the fact that when $t_d \to t_0$, the radiation content of the relativistic daughters of the massive neutrino has no time to decline to zero with respect to the matter density. This condition is necessary to change appreciably the energy budget of the universe, thus affecting the predicted growth of the structures and the time of equality with respect to a massive neutrino scenario. Clearly, for a given sensitivity to the effect of neutrino masses there is a maximum value $t_d^{\rm max}$ which would result in a detectable change of cosmological observables. A precise estimate of this parameter would imply a detailed forecast analysis, which goes beyond the purpose of this paper. Yet, a simple argument shows that, relying on the existing forecasts, a conservative lower limit is $t_d^{\rm max} \gtrsim t_{\rm nr}$, where $t_{\rm nr}$ is the epoch at which the heavier neutrinos become non-relativistic, whose redshift is defined by $m = 3\,T_{\nu,0}(1+z_{\rm nr})$, $T_{\nu,0}$ being the present temperature of the neutrino gas. Indeed, when the decay epoch satisfies $t_{\rm d} \lesssim t_{\rm nr}$, the energy content of the products is the same of a relativistic neutrino fluid, and it redshifts the same way. So, all physical effects of this scenario are basically the same of the case where neutrinos are massless. In Fig. 3, from top to bottom as seen from the left side of the plot, we show $f_\nu(z)$ for the following cases: (i) a massive neutrino cosmology, where we assume an IH neutrino mass pattern and the lightest neutrino is massless; (ii) as in (i), but for NH; (iii) a decaying neutrino cosmology, where massive neutrinos have IH; (iv) as in (iii), but for NH; (v) a massless neutrino cosmology. For the decaying cases, we assume that all massive neutrinos decay at $t_d = t_{\rm nr}$, where $t_{\rm nr}$ is the time of non-relativistic transition of the heaviest neutrino state ($m \simeq 0.05\,{\rm eV}$) . For simplicity we have assumed a matter-dominated cosmology with the matter density parameter $\Omega_{\rm m} = 0.24$ and the reduced Hubble constant $h = 0.73$ [27].

Clearly, the cases (iii), (iv), and (v) are very similar (exactly degenerate if $t_d \ll t_{\rm nr}$) and, as long as $t_{\rm d} \lesssim t_{\rm nr}$, if the massless neutrino case can be disproved, the decaying neutrino bound immediately follows. The improvement in the bound on the neutrino lifetime is tremendous. In particular, neutrinos turn non-relativistic at $z_{\rm nr} \simeq m/3\,T_{\nu,0} \simeq 100\,m_{50}$, i.e. when the universe has about $(100\,m_{50})^{-3/2} \sim 10^{-3} m_{50}^{-3/2}$ of its present age, and the bound is about $10^{-3} m_{50}^{-3/2}$ of the maximum attainable limit reported in Eq. (8),

$$\frac{\tau}{m} \gtrsim 10^{16}\, m_{50}^{-5/2}\, {\rm s/eV}\,. \tag{10}$$

Obviously, the previous argument does not exclude that an accurate forecast analysis may reveal a sensitivity to a somewhat larger $t_d^{\max}$. Note also that we do not require that cosmological data need to distinguish between NH and IH: if future observations will suggest e.g. $\Sigma = 0.08\,\mathrm{eV}$ with a 1σ error of 0.02 eV, the two neutrino mass patterns would be both consistent within 1 σ with the best fit, yet a complete decay of neutrinos into relativistic particles with lifetime lower than the value reported in Eq. (10) could be excluded at $4\,\sigma$. Of course, for a given cosmological sensitivity, the significance of the above bound increases if the inverted hierarchy is realized in nature: in that case $\Sigma \simeq 0.1\,\mathrm{eV}$ holds, and the cosmological effects of neutrino masses are larger. Note that accelerator neutrino experiments, magnetized detectors of atmospheric neutrinos, direct mass searches, and the serendipitous observation of neutrinos from a galactic supernova may all be used to determine the mass hierarchy. It is thus possible that by the time cosmology will be sensitive to $\Sigma \lesssim 0.1\,\mathrm{eV}$, the hierarchy information may be available independently.

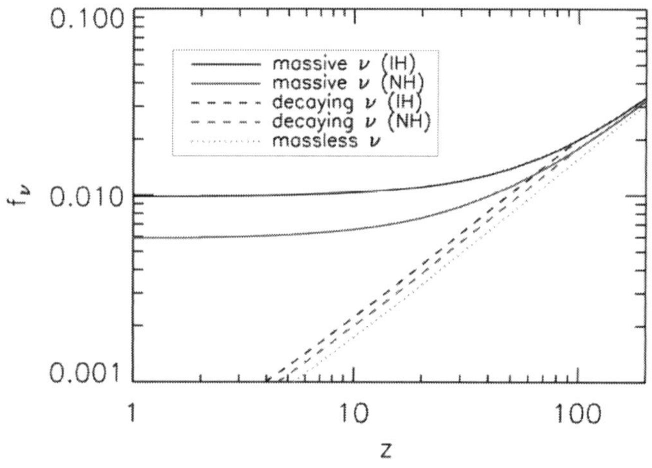

Figure 3. The function $f_\nu(z)$ for the relevant cosmological cases considered in the text.

To appreciate how strong the bound of Eq. (10) would be, let us consider a model of a "secret" neutrino interaction with a (quasi-)massless majoron field ϕ of the kind $\mathcal{L} = g\bar{\nu}_i\nu_j\phi +$ h.c, i, j labeling different mass eigenstates. The total decay rate for a hierarchical neutrino mass pattern and summing over neutrino and antineutrino final state channels is [24, 25]

$$\Gamma_{\rm d} = t_d^{-1} = \frac{g^2}{16\pi} m\,. \tag{11}$$

This holds in the neutrino rest frame, but in our case this is also the Lab decay width, give or take a factor $\mathcal{O}(1)$, since the neutrino is just turning non-relativistic. The constraint of Eq. (10) leads to the stringent bound

$$g \lesssim 4 \times 10^{-14}\, m_{50}^{1/4}\,. \tag{12}$$

This has to be compared with traditional bounds found in the literature in the range $g \lesssim 10^{-4} \div 10^{-5}$ (see e.g. [28]). Even the extremely stringent bound reported in [25] is more than two orders of magnitude weaker. Note that the tiny couplings which may induce the decay are not sufficient to thermalize extra degrees of freedom in the early universe. So, this model does not predict departure from the standard expectation for the effective number of neutrinos $N_{\rm eff}$ which can be consistently fixed in deriving the bound.

5. Discussion and Conclusions

In this paper, we have revisited the bound on the neutrino radiative lifetime coming from cosmology, deriving updated constraints from the high precision CMB spectrum data collected by the FIRAS instrument on board of COBE, within the presently allowed range of mass parameters suggested by neutrino oscillation physics and tritium endpoint experiments. The updated bounds are reported in Fig. 2. The improvement over pre-existing bounds depends both on narrowing the uncertainty on the neutrino mass spectrum and on the observational test of the blackbody nature of the spectrum at better than 1 part in 10^4 [5, 6]. This provides and explicit example of the continuing interplay of Lab neutrino experiments and cosmology.

Looking at the future, it is very interesting that current forecast analyses suggest that future cosmological surveys may attain the sensitivity to detect the effects of a sum of neutrino masses as small as $\sim$0.06 eV, the lower limit predicted by oscillation data. Provided that the systematics can be controlled to that level, we have discussed in this paper how such a detection would have profound consequences for the particle physics of the neutrino sector, besides providing a way to measure the absolute neutrino mass scale. In particular, when taking into account the expectations from the Lab, excluding the $\Sigma = 0$ case would improve by many orders of magnitude the existing limits on neutrino lifetime, and as a consequence on neutrino secret interactions with (quasi-)massless particles as in majoron models. Strictly speaking these bounds apply to the heaviest (or the two heaviest, in IH) mass eigenstate, but naturaleness and phase-space considerations suggest that the lifetime of the lightest state(s) is longer, and its coupling with a majoron field weaker, than for the heavier one(s). It also applies to any possible invisible decay channel, provided that the total mass of the final state particles is much smaller than Σ. In particular, this bound applies to 3-ν final state decays $\nu_i \to \bar{\nu}_j\nu_j\nu_k$, as well as to decays $\nu_i \to \nu_j + \phi$ in majoron-like models. We think that this idea summarized here provides another beautiful example of interplay between particle physics and cosmological arguments and motivates further the efforts to fully exploit the potential of future cosmological surveys.

Acknowledgments

I would like to thank the organizers for the kind invitation to celebrate the 75th birthdays of such two prominent members in the field of neutrino and low background physics, Frank Avignone and Ettore Fiorini. I would also like to dedicate this paper to the memory of S. Peter Rosen, an inspirational figure for all neutrino physicists. It is a pleasure to acknowledge A. Mirizzi and D. Montanino for collaborating on part of the topics summarized here. Support by the US Department of Energy and by NASA grant NAG5-10842 is acknowledged. Fermilab is operated by Fermi Research Alliance, LLC under Contract No. DE-AC02-07CH11359 with the United States Department of Energy.

References

[1] A. Mirizzi, D. Montanino and P. D. Serpico, Phys. Rev. D **76**, 053007 (2007).
[2] P. D. Serpico, Phys. Rev. Lett. **98**, 171301 (2007).

[3] M.T. Ressell and M.S. Turner, Comm. Astrophys. **14**, 323 (1990).
[4] S. D. Biller *et al.*, Phys. Rev. Lett. **80**, 2992 (1998).
[5] D. J. Fixsen *et al.*, Astrophys. J. **473**, 576 (1996).
[6] J. C. Mather *et al.* Astrophys. J. **512**, 511 (1999).
[7] G. Mangano *et al.*, Nucl. Phys. B **729**, 221 (2005).
[8] P. D. Serpico and G. G. Raffelt, Phys. Rev. D **71**, 127301 (2005).
[9] G. L. Fogli, E. Lisi, A. Marrone and A. Palazzo, Prog. Part. Nucl. Phys. **57**, 742 (2006).
[10] C. Kraus *et al.*, Eur. Phys. J. C **40**, 447 (2005).
[11] E. Massò and R. Toldrà, Phys. Rev. D **60**, 083503 (1999).
[12] H. Dole *et al.*, astro-ph/0603208.
[13] J. Lesgourgues and S. Pastor, Phys. Rept. **429**, 307 (2006).
[14] S. Hannestad, Ann. Rev. Nucl. Part. Sci. **56**, 137 (2006).
[15] J. Dunkley *et al.* [WMAP Collaboration], arXiv:0803.0586 [astro-ph].
[16] U. Seljak, A. Slosar and P. McDonald, JCAP **0610**, 014 (2006).
[17] G. L. Fogli *et al.*, Phys. Rev. D **75**, 053001 (2007).
[18] J. Lesgourgues, L. Perotto, S. Pastor and M. Piat, Phys. Rev. D **73**, 045021 (2006).
[19] Y. S. Song and L. Knox, Phys. Rev. D **70**, 063510 (2004).
[20] S. Hannestad, H. Tu and Y. Y. Y. Wong, JCAP **0606**, 025 (2006).
[21] S. Wang, Z. Haiman, W. Hu, J. Khoury and M. May, Phys. Rev. Lett. **95**, 011302 (2005).
[22] F. B. Abdalla and S. Rawlings, astro-ph/0702314.
[23] R. E. Lopez, S. Dodelson, R. J. Scherrer and M. S. Turner, Phys. Rev. Lett. **81**, 3075 (1998); S. Hannestad, Phys. Rev. D **59**, 125020 (1999); M. Kaplinghat, R. E. Lopez, S. Dodelson and R. J. Scherrer, Phys. Rev. D **60**, 123508 (1999); M. J. White, G. Gelmini and J. Silk, Phys. Rev. D **51**, 2669 (1995).
[24] J. F. Beacom and N. F. Bell, Phys. Rev. D **65**, 113009 (2002).
[25] S. Hannestad and G. Raffelt, Phys. Rev. D **72**, 103514 (2005); See also A. Basboll, O. E. Bjaelde, S. Hannestad and G. G. Raffelt, arXiv:0806.1735 [astro-ph].
[26] N. F. Bell, E. Pierpaoli and K. Sigurdson, Phys. Rev. D **73**, 063523 (2006).
[27] W. M. Yao *et al.* [Particle Data Group], J. Phys. G **33**, 1 (2006).
[28] Y. Farzan, Phys. Rev. D **67**, 073015 (2003).

The Mysterious Disappearance of Ettore Majorana

Barry R. Holstein
Department of Physics-LGRT, University of Massachusetts, Amherst, MA 01003
E-mail: holstein@physics.umass.edu

Abstract. We review the short but interesting life of Ettore Majorana.

1. Introduction

I would like to add my voice today to those honoring Frank Avignone and Ettore Fiorini. I am only ten years away from *my* seventy-fifth birthday and it is an inspiration to me to see their level of activity in the field has scarcely diminished. I also wish to eulogize Peter Rosen. Not only did he and I share an interest in much of the same physics, but we also shared a job! Thirty years ago I was a program officer in theoretical physics at the National Science Foundation, and Peter took that position after I left.

Deciding what to speak about this afternoon was difficult, since I have not done any recent neutrino work. Instead I thought it would be interesting to review the life of one whose name has been oft-mentioned at this meeting — Ettore Majorana — for his is an interesting story. I have extracted much of the material below from two books — one by the Sicilian writer Leonardo Sciascia[1] and the second a volume produced by the Italian Physical Society to honor the centennial of his birth, which includes a retrospective by Edoardo Amaldi[2]. I have also used material given by Amand Faessler in interesting talks about Majorana's life.

2. Majorana as a Youth

Ettore Majorana was born 5 August 1906 in Catania, which is a province on the eastern side of the island of Sicily. His family was highly respected and quite well to do, and owned homes in Catania, in Palermo, in Rome, etc. His father Fabio Majorana (b. Catania 1875-d. Rome 1934) was an engineer and for many years served as the Director of the Catania telephone company. The family "moved" to Rome in 1921 and in 1928 Fabio was appointed to the position of General Inspector of the Italian Ministry of Communication in Rome. (Interestingly his father had an older brother Quirino Majorana (1871-1957) who was an experimental physicist at the University of Bologna). His mother Dorina Corso Majorana (b. Catania 1876-d. Rome 1965), was said to be somewhat overbearing. The union produced five children — Rosina, who later married Werner Schultze, Salvatore, who became a Doctor of Law, Luciano, who became a civil engineer, Ettore, and Maria, who became a musician and piano instructor[2].

Ettore was very shy as a child and stories are told by his family of him being able to multiply three-digit numbers or to extract square or cube roots in his head. However, an indication of this shyness is that when presented with such a problem, he would disappear underneath a table in order to do the calculation and later would call out the answer. (I wonder whether today we might classify him as autistic.) He was also cited in the local newspaper for his chess prowess.

Ettore was schooled in Catania for his first few years, but then was sent to Rome first as a boarding student in classical studies at the Jesuit Instituto Parificato Massimiliano Massimo. Above I mentioned that his family "moved" to Rome in 1921, when Ettore was fifteen, so that he finished his studies as a day student. I put the word "moved" in quotes because this is not how you or I would move our family — putting our house for sale, hiring some movers, buying a house in a new city. Rather for the Majoranas, it was a matter of telling the servants in Catania and Palermo to close up the house for a while and notifying the servants in Rome to open up the apartment[3]. In any case, Majorana earned his final diploma "licenz Liciale" at the Liceo Statale Torquato Tasso in 1923.

Apparently Ettore intended to follow in his father's engineering footsteps, for on November 3, 1923 he entered the two year preparatory course for engineers, held at the Faculty of Sciences at the University of Rome, and he earned his final diploma in the autumn of 1925. He then took the next step, entering the three year School of Applications for Engineers. His work was fine, except for his study of hydraulics — he passed the hydraulics exam 7 December 1927 with a grade of only 75/100 — but he did not graduate. That is because he had befriended a young student Emilio Segre, who convinced him that his future lay in the field of physics. Segre himself had begun his studies as an engineer, but in the summer of 1927 followed the advice of Franco Rasetti, who was a lecturer in Physics at Florence, and transferred to the Institute of Physics where he studied under the direction of the young (27 year old) Enrico Fermi, who had just been appointed as extraordinary Professor of Theoretical Physics at the University of Rome in in a 1926 national competition. The board of examiners had met in November of that year and chose Fermi for this position in Rome, Enrico Persico for a professorship in Florence and Aldo Potremoli for a position in Milan.

3. Majorana at the University of Rome

Majorana was admitted to the physics program at the beginning of 1928. How he joined the program involves an interesting story by his friend Edoardo Amaldi. Amaldi, Segre, and others had urged him to set up a meeting with Fermi about transferring to physics and when this came about and he discussed this possibility with Fermi, Ettore was given presentations concerning the various activities in the institute[12]. One such activity had to do with what we now call the Thomas-Fermi statistical model of the atom and Fermi showed him a numerical table which he had developed in this regard using a simple hand calculator. Majorana listened without comment and left, but returned the next day with his own sheet of paper on which he had generated his own figures. He asked to look at Fermi's table and, after comparing them, announced Fermi's work to be correct. He then left the Institute for the day and within a week had begun his study thesis work with Fermi. He submitted his thesis "On the Quantum Mechanics of Radioactive Nuclei" on July 7, 1929 and it was accepted with a grade of 110/110 with distinction. The subject of the work was a generalization of the Gamow and Houterman theory of alpha decay to the case of particles in excited states and with higher angular momentum.

After obtaining his degree, Majorana continued to attend the Institute and on 12 November 1932 he obtained his university teaching diploma in theoretical physics. The examination board, consisting of Fermi, Antonino Lo Surdo and Enrico Persico, was unanimous in their recommendation and reported that the candidate possessed a "complete mastery of theoretical physics." During this period he generated a modest output of publications involving quantum mechanics. His papers included work

i,ii) on the Thomas-Fermi model of atomic structure[4],[5]:

G. Gentile and E. Majorana, "On the splitting of Röntgen and optical terms due to the rotating electron, and on the intensities of Cesium lines," Rend. dell'Acc. dei Lincei **8**, 229-33 (1928).
E. Majorana, "Search for a general expression of Rydberg corrections, valid for neutral atoms or positive ions," Nuovo. Cim **6**, 16-16 (1929).

iii,iv) on chemical bonding:
E. Majorana, "On the formation of the molecular helium ion," Nuovo Cim. **8**, 22-28 (1931).
E. Majorana, "Pseudopolar reaction between hydrogen atoms," Rend. dell'Acc. dei Lincei **13**, 58-61 (1931).

v,vi) on spectroscopic lines:
E. Majorana, "The possible anomalous terms of helium", Nuovo Cim. **8**, 78-83 (1931).
E. Majorana, "Theory of incomplete P'triplets," Nuovo Cim. **8**, 107-113 (1931).

vii) on polarized atoms in the presence of changing magnetic fields (a phenomenon which is now called "Majorana flip")
E. Majorana, "Oriented atoms in a variable magnetic field," Nuovo Cim. **9**, 43-50 (1932).

viii) on the form of the relativistic wave equation for arbitrary spin[6] (infinite dimensional unitary representations for the relativistic Lorentz group):
E. Majorana, "Relativistic theory of particles with arbitrary intrinsic angular momentum," Nuovo Cim. **9**, 335-344 (1932).

ix) and on nuclear theory:
E. Majorana, "On the theory of nuclei," Zeit. für Physik **82**, 137-45 (1933).

(As an aside, I should mention that after his death his brother found an unpublished manuscript "The value of statistical laws in physics and social sciences" which was subsequently published in Scientia **36**, 58 (1942) through the influence of Majorana's friend Giovanni Gentile, Jr. This paper was recently republished in Quantitative Finance **5**, 133 (2005).)

Then in January of 1933 he received a grant from the Italian National Research Council in the amount of 12,000 Lire, which allowed him to travel to Leipzig to work with Heisenberg. His application was supported by Fermi. Leipzig at this time was a center for theoretical physics and besides Heisenberg the theory group included Hund, Peierls, and Bloch. He arrived in Germany during January of 1933 and stayed until the beginning of the summer, when he went to Copenhagen for a three month visit with Bohr at the Institute. After a brief time back in Leipzig, he returned to Rome in the autumn of 1933.

After coming back from this trip, however, his publications ceased. It is not clear why this was the case, but it is known that he developed a serious case of gastritis while in Germany, which his family doctors attributed to nervious exhaustion. Whatever the cause, this illness stayed with him and Majorana became a very reluctant author. Once he had discovered something, he considered his own work to be commonplace, so that he was very disinclined to write anything about it. An example is the work by Joliot and Curie in France, which had discovered a neutral particle that could enter matter and expel a proton. The experimenters assumed that the particle must be a photon, since that was the only neutral particle known at the time. However, Majorana immediately saw that it must be a particle with a mass near that of the proton, in order to move something as heavy as the proton. When he heard this Fermi urged him to write it up immediately, but nothing happened and soon thereafter Chadwick was (properly) given the credit for the discovery of the neutron for his work using Beryllium[9].

4. Majorana at the University of Naples

Because of this writer's block, very little is known about Majorana's career during this time except that he ceased his regular presence at the physics institute and took up an interest

Journal of Physics: Conference Series **173** (2009) 012019 doi:10.1088/1742-6596/173/1/012019

in philosophy, especially the work of Schophauser. During this period he also developed an interest in economics and politics. In 1937, however, a new competition for physics chairs was announced — the first which had been held since 1926. Fermi was appointed chair of the examination committee, which consisted on himself, O. Lazzarino, E. Persico, G. Polvani, and A. Carelli. The committee met on October 25, 1937 and they selected

a) Gian Carlo Wick, who went to Palermo, then to Padova in 1938 and then to Rome in 1940
b) G. Racah, who went to Pisa but lost the chair in 1938 due to the fact that he was Jewish. He moved from there to Jerusalem.
c) G. Gentile, Jr., who was the son of the politician G. Gentile, Sr. obtained a position in Milan, but died there in 1942 from what is described as a "minor" infection.

Fermi had urged Majorana to apply for this position, and in addition to the list of publications given above the candidate was asked to supply a new research paper. The paper which is the genesis of the two-component neutrino (the "Majorana" neutrino)

E. Majorana, "Symmetrical theory of the electron and the positron," Nuovo Cim. **5**, 171-184 (1937).

was written for this competition. It is said that when Fermi approached him, Majorana pulled some research notes from a drawer in his desk but that possibly Fermi helped to write the paper. In any case, his application for a position was considered by the committee, but they did not offer him any of the three positions. Instead, at the recommendation of Senator Giovanni Gentile, Sr. (who perhaps saw that Ettore's presence on the list would push his son to fourth position and out of the running) Majorana was considered above comparison with any in this group and was recommended for a position at Naples on the grounds of "alta fama di singolare perizia cui e pervenuto nel campo degli studi de Fisica teorica" (high fame of singular expertise reached in the field of theoretical physics). He thus began his career in January 1938 teaching quantum mechanics at an annual salary of 26,000 Lire. Some notes from these lectures are available[7],[8],[13] and there are some like N. Cabibbo who say that Majorana was on track to discover the Feynman path integral. However, having read these notes, I think that this is a bit of a stretch.

Obviously, Fermi had a very high opinion of his former thesis student in order to justify the award of this special position. In this regard, speaking to Guiseppi Cocconi, Fermi has been quoted as stating that[11] "Because you see, in the world there are various categories of scientists: people of a secondary or tertiary standing, who do their best but never go very far. There are also those of high standing, who come to discoveries of great importance, fundamental for the development of science [Fermi considered himself to be in this category.] But then there are the geniuses, like Galileo and Newton. Well, Ettore was one of these. Majorana had what no one else in the world had ..."

Also during the Manhattan project, Oppenheimer recalled that at one time a critical issue arose, whereupon Fermi turned to Wigner and said "If only Ettore were here." Another time he said "This calls for Ettore." After hearing this General Groves asked Fermi who this "Ettore" was and was told that it was Majorana. Groves asked where he was so that they could bring him to work on the project. Wigner replied "Unfortunately he disappeared many years ago."

For several months things went well with his teaching career. In Naples, Majorana moved first into the Hotel Terminus and then in February to the fancier Hotel Bologna. He began his lectures on quantum theory 13 January 1938 to a group of students.[1] However, the level of these lectures was very high, and there were only a few who showed much understanding.

[1] Note that ten of these lectures have been already published[13].

During this period he did not touch any of his salary money, which was deposited into an account in a bank in Naples. In late March, however, he took all of this money out of the bank and on March 23, 1938 took the night boat to Palermo. While in Palermo he sent a note to the Director of the Naples Physics Institute, Antonio Carrelli, hinting at dramatic plans. However, as soon as he had posted it he apparently changed his mind and so telegraphed a message to Carrelli telling him to disregard his previous note. The original note that had been dispatched to Carrelli states

March 25 1938
Dear Carrelli, I made a decision that has become unavoidable. There isn't a bit of selfishness in it, but I realize what trouble my sudden disappearance will cause you and the students. For this as well, I beg your forgiveness, but especially for betraying the trust, the sincere friendship and the sympathy you gave me over the past months. I ask you to remind me to all those I learned to know and appreciate in your Institute, especially Sciuti: I will keep a fond memory of them all at least until 11 pm tonight, possibly later too. E. Majorana

and Carrelli was naturally concerned when he received it. Later he received a second letter from Palermo

March 26, 1938
Dear Carrelli, I hope you got my telegram and my letter at the same time. The sea rejected me and I'll be back tomorrow at the Hotel Bologna traveling perhaps with this letter. However, I have the intention of giving up teaching. Don't think I'm like an Ibsen heroine, because the case if different. I'm at your disposal for further details. E. Majorana

In any case, Carrelli was concerned enough by these events to contact the Majorana family in Rome, and Ettore's brother Luciano was dispatched to Naples to scope out what had happened. It was determined that Majorana had traveled back to Naples on the steamer on the night of March 25, 1938, and he had shared a compartment with Professor Michele Strazzeri of the University of Palermo. However, after this no trace was found, even after a reward of 30,000 Lire was offered by the family.

On December 6, 1938 the minister issued a decree where Majorana was considered as having resigned from his public office, starting on March 25, 1938 because he "abandoned the duty of his office for more than ten days, without any justified reason."

Fermi wrote a letter to Mussolini to undertake a search, saying "I have no hesitation to state to you, and I am not saying this as an hyperbolic statement, that of all Italian and foreign scholars that I have had the opportunity to meet, Majorana is among all of them the one that has most struck with for his deep brilliance." The family also entreated Mussolini to find Ettore but despite a police search, nothing was turned up.

5. Majorana's Fate

So what happened? There have been many speculations.

i) The most common theory is suicide, jumping into the Mediterranean on the boat trip from Palermo. His note to Carrelli can certainly be read in this way. Also, Sciascia reports that Majorana left a note at the hotel in Palermo for his family stating[1]:
I have only one wish: do not wear black. If you must conform to custom just wear some emblem of mourning, but not for more than three days. After that remember me in your hearts, if you can, and forgive me.

On the other hand, why would he take all his money out of the bank? Also, his family insisted that he had a very strong Catholic faith and would certainly *not* have taken his own life.

ii) After examining all the evidence, Leonardo Sciascia became convinced that Majorana decided to disappear because he foresaw that nuclear forces would lead to nuclear weapons[1]. In this regard, Laura Fermi said that her husband had said that "Ettore was too intelligent. If he has decided to disappear, no one will be able to find him."

iii) Some (including his family) say he had a crisis of spiritual faith and retreated into a monastery. Monsignor Riccieri, who was Bishop of Trapani after serving in Catania, said that Majorana had experienced "mystical crises," and agreed that suicide was to be discounted. He cited the sanctity of the confessional to say no more[9].
The monastery theory is given some additional traction by reports that a few days before Majorana sailed for Palermo a young man who looked like him and appeared upset had shown up at the Chiesa del Gesu Nuovo near the Hotel Bologna in Naples and asked to go into retreat. Upon being informed that a long term accomodation would be required before he could join the novitiate, he thanked the brothers and left.

iv) Others say that he became a beggar in Naples and report a disheveled person helping local students with their mathematics.

v) There was a report that he went to South America and that a "Majorana" had checked into a hotel in Buenos Aires. Detectives were dispatched but found the registration page missing from hotel register.

vi) Finally, the Sicilian writer Salerno says "A theory of suicide has always been advanced, but the politics of the Fascist and Nazi years leading up to the Second World War suggest homicide. As some of Majorana's work might have facilitated development of an atomic bomb, a project with which Heisenberg was eventually involved for Nazi Germany, there exists a strong possibility that he was murderded, probably by Nazi agents[10]."

So what really happened? Since I am a theorist, I will present my theory this afternoon as to Majorana's fate, which is that we shall never know.

Acknowledgements

This work was supported in part by the National Science Foundation under award PHY05-53304. It is a pleasure to thank Amand Faessler for interesting me in this subject and for providing information about the life of Majorana. I also wish to thank Laura Catanati for translating the note sent to Carrelli.

Bibliography

[1] L. Sciascia 1987 *The Moro Affair and the Mystery of Majorana* (New York: Carcanet Press).
[2] E. Amaldi 2006 *Ettore Majorana: Scientific Papers*, ed. G.F. Bassani (New York: Springer).
[3] A. Faessler, talk given at the 2007 Erice school of nuclear physics.
[4] S. Esposito 2006 *Elec. J. Theor. Phys.* **3** 265.
[5] F. Guerra and N. Robotti 2008 *Phys. in Pers.* **10** 1422.
[6] D.M. Fradkin 1966 *Am. J. Phys.* **34** 314.
[7] S. Esposito 2007 *Am. J. Phys.* **75** 781.
[8] S. Esposito 2006 *Eur. J. Phys.* **27** 1147.
[9] A. Zichichi, 2006 *Cern Courier* Jul. 25.
[10] V. Salerno 2006 *Best of Sicily Magazine*.
[11] E. Recami 2006 *Elec. J. Theor. Phys.* **10** 1.
[12] E. Segre 1971 *Enrico Fermi, Fisico* (Bologna: Zanichelli); 1965 *Autobigrafia di un Fisico* (Rome: Il Mulino).
[13] E. Majorana 1987 *Ettore Majorana — Lezioni all'Universita di Napoli*, Ed. B. Preziosi (Napoli: Bibliopolis).

Some aspects of classical and quantum phases

Yakir Aharonov[1], **Tirza Kaufherr**, **Shmuel Nussinov**[1]

Tel Aviv University, Sackler School Faculty of Sciences, Ramat Aviv, Tel Aviv 69978, Israel

[1] Chapman University, Schmid College of Science, Dept. of Physics, Computational Sciences and Engineering, One University Drive, Orange CA 92266, USA

E-mail: nussinov@post.tau.ac.il

Abstract.
We study classical and quantum phases in the adiabatic Born-Oppenheimer context. These include a classical astronomical case, the general dual description of the phases, a new "Paradox" connected to scattering Berry phase and its resolution and various elaboration of topological/geometrical/non-abelian phases.

1. Introduction

Adiabatic Born-Oppenheimer settings are common in classical and quantum physics. Thus envision a point-like massive object (or strongly charged particle)—the "Strong Probe"—passing by a binary star (or an atom)—the "System".

If the probe's motion is slow compared with characteristic periods of the system (or $h/(E_i - E_0)$ for the atom in a non-degenerate ground state), we can use an adiabatic approximation and solve the dynamics of the system—or the Schrödinger's Equation—for each "instantaneous" location $R(t)$ of our probe. The resulting energy shift $E_0(R) - E_0(R = \infty)$ defines the effective "B.-O. potential" that our external probe sees. Interfering parts of a quantum mechanical wave function following different paths allows sensing potential differences even in cases where the test particle is always in force-free regions.

In the following we focus on the general principle that phases be understood from both the point of view of the external probe particle and that of the dynamical system with which it interacts. This principle affords a unified point of view of Q.M. phases yielding, in particular, a new "Paradox" for scattering type adiabatic set-ups and a very amusing resolution thereof.

2. A classical (astronomical) Berry phase

The adiabatic approximation was used by Gauss for celestial motions. Also here we have on top of the dynamical effect of changed period an extra Berry phase (or time lapse). The latter resembles a spin-half dynamical system which attempts tracking the location of the external probe, and the accumulated effect of it failing to do so instantaneously is the analog of the geometric phase.

Consider a test mass (Earth) moving around a very heavy fixed center (Sun) with another heavy planet (Jupiter) present. We will assume that Jupiter's orbit is coplanar with Earth's orbit. The question of interest is how does the presence of Jupiter effect the period (year) of the earth? Since $M_{\rm Jupiter} = 10^{-}3\ M_{\rm Sun}$, $R_{\rm Jupiter} = 5.2\ R_{\rm Earth}$ and $T_{\rm Jupiter} = 11.9\ T_{\rm Earth} = 11.9$

years, both a perturbative (in force ratio $F_{\text{Jupiter}}/F_{\text{Sun}} \sim 5.10^{-5}$) *and* adiabatic (in the sense that $T_{\text{Jupiter}} > T_{\text{Earth}}$) approaches are feasible. The latter suggests solving the problem for an instantaneous frozen position of Jupiter. In particular, we need the new period T'_{Earth} in the presence of the Sun and Jupiter. Remarkably, that special planar two-center problem is exactly solvable.[1] (One can also use perturbative approximation in this case.)

There is, however, a finite extra correction; namely, the Berry-like delay—the very analog of the geometric Berry phase arising from the fact that the dynamical system, the orbiting earth, fails to instantaneously adjust itself (the orbit's period in the case of interest here) to the new position of the probe (Jupiter)—the sum of these over one Jupiter period, 'the astronomical Berry phase' can be computed.[2] Unlike the huge ordinary Gauss/Born-Oppenheimer effect, it is barely observable.

3. The two complementary ways of viewing quantum phases

When the external probe describes a closed path and the dynamical system returns to its initial position, its wave function (and that of the system as a whole) can only pick up a phase. It is very instructive to understand how this phase arises from two complementary points of view: that of the motion and interactions of the external probe alone or, alternatively, by considering the "Internal" dynamics of the system.

This applies to all cases. For the electric (or magnetic) potential effects the A.B. phase emerges via relative linear momentum (or angular momentum) imparted by the electron to the plates (solenoid). This is particularly transparent in the electric case.

An electron, playing here the role of an external *weak* probe, stays for time t in the force-free region outside two parallel condenser plates. During this time the plates are at a fixed relative distance x with a uniform field E existing between them. The standard (electric) potential modifies the wave function of an electron passing to the left-L of the plates by the phase $eV_L \cdot t$ = eExt (or $eV_R \cdot t$ = -eExt for electrons passing on the right. The relative A.B. phase 2eExt then manifests in the interference of these two paths.

The forces $F_{L,R} = +/-$ eE between the electron and the L/R plate, cancel out for the electrons and for the motion of the center of mass of the whole capacitor—i.e., the two plates jointly. However, insofar as the *relative* motion of the two plates is concerned (i.e., the internal dynamics of the "system" in the present case), the effects add up: F_L imparts a momentum $P_x(L) = Ft$ =eEt to the left plate and F_R imparts the opposite momentum $P_x(R)$ = -eEt to the right plate at a relative distance x away. The phase picked up by the internal (relative x) wave function of the system is thus 2eEt$\cdot x$—the precise A.B. phase.

Localizing the electron to the left *or* to the right of the capacitor—namely, within $\Delta(X) < x$, induces by the uncertainty principle an uncertainty $\Delta(P_x) > 1/x$ in its momentum. The A.B. phase $\sim$ eExt can be taken as $< \pi$. The uncertainty in the imparted momentum does then exceed the momentum itself. Consequently the A.B. phase becomes uncertain and, as expected, knowing which path (L or R) the electron took destroys the interference.

For complex "Systems" the task of identifying the internal dynamical change accounting for the phase—otherwise simply explained by an appropriate (albeit somewhat mysterious at times) external potential, becomes increasingly difficult. Already for the magnetic A.B. effect, one needs to use a simplified "solenoid"—say, counter-rotating almost overlapping concentric cylinders with opposite charges—in order to have the analog of the above simple reasoning as in the electric A.B. case. Still, the general principle that such alternative, complementary, interpretations of the phase are possible is quite powerful.

In Sects. 4 and 5 we briefly discuss two examples illustrating this. Implementing the above general principle appear at times to lead to a "Paradox". Such a case is presented along with its amusing resolution. The second example involves a discussion of well-known Berry phases.

4. A paradox related to scattering Berry phase and its resolution

The "system" with which our external weak and or strong probe interacts need not be static or bound as often implicitly assumed, but can involve a scattering state.

Thus let our system consist of a quantum particle (electron, photon, neutron, etc.) incident from the right ($x = +\infty$) along the x axis, moving in a semi-infinite channel and reflecting from a mirror at the origin ($x = 0$). The small channel width w, prevents excitation of transverse (y) modes rendering the system one dimensional.

The particle's wave function is then

$$\psi(x) \sim \exp(-ipx) - \exp(ipx) \tag{1}$$

so that it vanishes at $x = 0$, the location of our mirror. In reality we can have an incident wave packet, which encounters the mirror at time $t \sim 0$ and becomes the reflected wave packet. Our external, strong probe is located outside the channel at (X, Y), $X > 0$ and $Y > w$. Its short-range repulsive interaction with the internal particle $U(X, Y; x) = U(|r - R|)$ is strong at short distances $|r - R| = [(x - X)^2 + Y^2]^{1/2} <$ few w, and decreases to almost zero for $|r - R| > Y_0$. Hence when at small Y the probe reflects the internal particle at $x = X$ before it reaches the origin introducing an extra phase $\sim 2pX$. On the other hand, when the probe is far from the channel at $Y \sim Y_0$ it will not effect the incident internal particle which will reflect from the mirror with no extra phase. A heavy probe which hardly moves during the collision will pick (after a time $T \sim Y_0/v$ with $v = p/m$ the velocity of the internal particle required for the latter to reflect and stop interacting) a phase $\Phi(X, Y)$.

The overall change of the phase as Y changes between w and Y_0 $\Delta\Phi \sim 2pX$ implies a phase gradient or momentum in the y direction of $2pX/Y_0$ and the average force obtained by dividing by the effective duration of the collision, T, is $F \sim 2pXv/Y_0^2 \sim 2p^2X/mY_0^2$. However, it is clearly impossible for the force to indefinitely increase with the distance X from the distant mirror at the origin.

The paradox arises due to a serious omission. Even when the external probe is very near to the x axis the internal particle does not see an infinite barrier. Consequently there is some small probability, $\sim \epsilon$, that the particle will not reflect from the probe at $x = X$, imparting to it a momentum $2p$, but rather tunnel through the barrier. It will then be reflected from the mirror at $x = 0$ and subsequently by symmetry, will reflect back with a high probability, $1 - \epsilon$ from the left end of the barrier imparting to it a momentum $-2p$ (with probability $\epsilon \cdot (1 - \epsilon)$. This backward and forward reflections keep going on. The nth reflection with probability $f_n = \epsilon \cdot (1 - \epsilon)^{(n+1)}$ and will then impart a momentum $-2pf_n$. The geometric series sum of all the many small negative momentum kicks exactly cancels the original positive $2p(1 - \epsilon)$. In physical terms the particle will be trapped, albeit with a small probability ϵ on the left-hand side of the barrier. The trapping, however, lasts for a long time, $\sim 1/\epsilon$, thereby equalizing the pressures exerted on the right- and left-hand side of the barrier (namely, the external probe) at $x = X$. Thus our probe suffers no net momentum kick and the internal particle reflects just from the end mirror as if nothing much happened in between!

The latter suffer no net momentum kick and the internal particle reflect just from the end mirror as if nothing much happened in between!

Is it possible to have an actual realization of this using laser beam of light and a tiny mirror? In such a case one could envision reflection by the probe of one transverse polarization and ensuing extreme sensitivity to tiny rotations of the plane of polarization of the trapped photons during their many traversals. We will not speculate any further on this here. Experiments of this type are, in fact, being done at MIT by Nervis Paluvalaha and her group in connection with the ultimate quantum noise limitations on the measurements of gravitational waves utilizing multiple reflections of laser light from mirrors.

5. Berry-A.B. geometric/topological phases

Returning to the initial external slow probe polarizing an atom, consider the case of just a two-level system-"Atom" arising when we have not just one isolated ground state but two, nearby states well separated from all other levels.

The atom can then be replaced by a spin-half system at the origin and its interaction with the probe is $H = \vec{A} \cdot \hat{\sigma}$ with the vector consisting of the three Pauli matrices, appropriate for spin-1/2. For time reversal invariant interaction and no external B fields, only the real $\sigma(1)$ and $\sigma(3)$ appear.

The two-level spin system becomes degenerate when $|A| = 0$. For the time reversal invariant case this reduces to

$$A_1(\vec{R}) = A_3(\vec{R}) = 0. \tag{2}$$

Two equations define a one-dimensional closed (or infinite) curve C*. A basic result—reminiscent of the magnetic A.B. effect with a singular B field vortex—is the following: When the probe adiabatically describes a closed circuit C and the system returns to its initial non-degenerate ground state it picks a topological Berry phase, $\Phi_{Berry} = \pi$, if and only if the curves C and C* interlock.

Generally an $A_2\ \sigma(2)$ term is present. Requiring that A_2 vanish as well reduces the degeneracy manifold in R space to a single point R^*. In the famous example described next, this point $R^*{=}0$ is at the location of the system-"Atom" itself. A "Geometric" Berry phase manifests for probe system interaction: $A\ \hat{\sigma} \cdot \hat{n}(t)$ with A a constant and $\hat{n}(t)$ the unit vector pointing towards $\vec{R}(t)$—the instantaneous probe's location. At the origin $\vec{R} = 0$ and hence the direction $\hat{n}$ and the Hamiltonian $H = A\hat{s}\cdot\hat{n}$ are ill defined. We define H—via a procedure similar to Schwinger's point splitting—by averaging the interaction over a spherical neighborhood around each $\vec{R}$. This does not change H for any $|R| > 0$ but makes $H = 0$ at $R = 0$. Hence $\vec{R}^*{=}0$ as claimed above.

As $\vec{R}(t)$ moves around the energy of the system, its interaction energy, $\sim |A|$, does *not* change and there are no B.O. forces. Yet the probe picks up after describing closed curve C a "Geometric phase" equal to half the solid angle substended by C at the origin. This and the independence of the phase on $|A|$, the interaction strength seems surprising!

We next consider this phase from both points of view: That of the probe and of the two-level system.

The first is well known and elaborated in many reviews: There *is* a "Lorentz"-like force acting on the probe which, however, is at all times perpendicular to its velocity so that no work is done and no energy change incurred. This force requires introducing a fictitious monopole of strength g at the origin, endowing our probe with a fictitious charge e and imposing Dirac quantization, $e \cdot g \sim h/2$. The standard A.B. phase, $\Phi_{A.B.} = \int_C e\vec{A} \cdot d\vec{\ell}$ with $\vec{A}$ the vector potential, is then the flux of the monopole's $\vec{B} = g\,\hat{r}/\vec{r}^2$ through surfaces bounded by C, namely, the solid angle above.

Note that this geometric Berry phase due to the fictitious monopole nicely ties in with the previous topological "A.B.-like" Berry phase. Thus consider imposing on the Hamiltonian $A \cdot \hat{n}$ the condition $\vec{A} \cdot \hat{n}_0 = 0$ for a specified direction $\hat{n}_0$. In this case H vanishes only on the infinite line directed along $\pm\hat{n}_0$. This line then becomes the degeneracy manifold—the curve C* mentioned above. A Probe encircling this line collects a topological Berry phase—which in terms of the formal e.m. analogy becomes an "A.B." phase due to a "fluxon" aligned along $\pm\hat{n}_0$. It is as if some symmetry breaking phase transition, like in superconductivity, forced the total unit flux of the monopole to equally divide to half fluxons going along $\pm\hat{n}_0$. In this case (and ONLY for such a flux!) is the A.B. phase equal to π, independent of the $\pm\hat{n}_0$ orientation.

A simple topological argument for such a phase emerges when we ask what the A.B. phase is accumulated in a 2-d setup when a point vortex encircles a portion of the bound state of a quantum charge (electron). For the case of half fluxon, time reversal invariance allows only for a vanishing phase (e.g., for a path which does not encircle any charge) or a phase of π, e.g., for paths encircling all charges. Therefore there is some special point (and in general an odd number of such points) inside the wave function support, the encircling of which entails a jump of the phase by π. This is the point where a half fluxon stationed there make the ground state level in question[3] degenerate.

This indeed is the Berry phase for the case of two fold degeneracy. The case of just one degeneracy point at the origin of the 3-d space is special. The geometric phase is here the average phases of π arising if the flux was directed along any one of the rays included inside the solid angle encircled, with total weight Ω, and the 0 phase due to the rest, $4\pi - \Omega$ of the sphere.

Remarkably the E.M. analog tends to reconstruct the original spin-1/2 system. The magnetic and electric fields of a monopole at the origin and a charge at $\vec{R}$ generate angular momentum $\sim \int d^3\vec{r}\vec{r} \times \vec{E} \times \vec{B} \sim 1/2\, eg\hat{r}1/2$, namely, spin-1/2, along the unit vector $\hat{n}$ between them, precisely the $\sigma \cdot \hat{n}$ we started from!

SU(2) $\rightarrow$ SU(N) extensions appear straightforward but are rather nontrivial. In the above SU(2) we had U(1) $\sim$ O(2) subgroup corresponding to rotations around the preferred $\hat{n}_0$ "Degeneracy" axis above (or around half fluxons along the $\pm\hat{n}_0$ direction in the E.M./A.B analog). Formally this subset of SU(2) was selected by time reversal symmetry and the need to avoid imaginary pieces in the Hamiltonian. The SU(3) analog of the Pauli matrices are the eight Gell-Mann l_k in the lie algebra: $[l_k, l_m] = i\, f\,(kmn) l_n$; the $f(kmn)$ are real. Hence the three imaginary l_k span a SU(2)= O(3) subgroup. The latter can be embedded in many ways in the SU(3)/O(3) manifold like in the choice of $\hat{n}_0$ above on the sphere.

Note that for imaginary l_k the finite group elements exp $i(\alpha_k \cdot l_k)$ are real. In general, SU(N) has upon restriction to real transformations O(N) subgroups. The analog of the heavy probe at $R(t)$ in three-dimensional R space is now a point in the $n^2 - 1$ dimensional parameter space $A_k(t) = 1...n^2 - 1$, with the n level hamiltonian $H = A_k\, G_k$, with G_k the $n^2 - 1$ generators in the fundamental representations the SU(n). The infinite or close line degeneracy domain C* in the SU(2) case becomes here a closed (or infinite surface) S* of dimensionality $n(n+1)/2 - 1$ in the SU(n)/O(n) manifold. The topological Berry phases are now picked up when the system describes a closed path in $n^2 - 1$ dimensional parameter space which threads the degeneracy surface S*. These are "Genuine" n-fold degeneracies of *all* n levels corresponding to the complete vanishing of the Hamiltonian and cannot be reduced to lower degeneracies of, say, $n - 1$ out of the n levels and trivial SU(n) $\rightarrow$ SU(n-1) reductions.

Next we introduce all A_k components—in which case we can write just like before: $i\,\Sigma A_k/|\vec{A}| \cdot G_k$, with the sum extending over all the group generators. Interesting questions—not addressed here—relate to the analog of the geometric Berry phase and higher dimensional analog of the monopole's charges and angular momentum—now constructed presumably as exterior products of higher-order forms with many different coordinates:

How does the (ordinary 3-d) geometric Berry phase arise from the point of view of the atom/spin-1/2 system? We answer this for a particular probe trajectory, namely, an equatorial circle of radius R traversed at constant (angular) velocity (w) and speed $v = Rw$. This provides a simple illustration of the complementary point of view where small deviations from true adiabaticity integrate to the final, finite, geometric-like phases.

The Hamiltonian for this case is $H(t) = A[cos(wt)\sigma(1) + sin(wt)\sigma(2)]$. Transforming via $U = \exp(iwt\ \sigma(3)/2)$ to the "rotating frame", the Hamiltonian becomes $H' = U^{+}HU + iU^{+}\frac{\partial}{\partial t}U = A\,\sigma(1) + w\,\sigma(3)/2$. The new H' implies a non-vanishing expectation value of $\sigma(3) : <\sigma(3)> = w/(2A)$. This E.V. measures the (small) deviation from adiabaticity: $|A|$ is the level spacing and w the rate of the probe's periodic motion. With $w/A = \epsilon << 1$ the system stays in its lowest state anti-aligned with the instantaneous $\hat{n}$ direction with a *crucial* tiny correction: It tracks it not in the equatorial plane (as it should in the perfectly adiabatic case) but is slightly shifted up to an almost flat cone of opening angle $\pi - \epsilon$! Since $\sigma(3)$ is invariant under the $U = \exp(i\,\sigma(3)wt)$ rotations the phase acquired due to it keeps accumulating. The (long!) time required for the probe to complete the cycle is $T = 2\pi/w$. The corresponding accumulated phase is then: $A\epsilon < \sigma > \cdot T = A(w/A)(2\pi/w) \cdot 1/2 = \pi$. Thus both the long period time and strong coupling A cancel out and we are left with just the geometric phase corresponding to the solid angle of half a sphere! It is relatively simple to generalize this to smaller "latitude" circles at polar angle θ of radius $R \sin\theta$ enclosing a solid angle $\Omega = \pi\cdot(1-\cos\theta) = 2\pi \sin^2\theta/2$. Finally, any part of the sphere can be exhausted by infinitely many such tangent circles generalizing the proof to any closed path.

6. Adiabatic complete MSW switching of solar neutrinos: a non-abelian Berry phase

To tie in SCINSP neutrinos consider adiabatic $\nu_e \to \nu_\mu$ MSW conversion in the sun.

As a ν_e produced at the solar core moves outwards it encounters eventually a layer where matter effects make it degenerate with ν_μ. For an extended crossover region even tiny μ, e mixings cause a complete ν_e to ν_μ conversion as in the famous figure in J. Bahcall's book.[4] An entertaining classical analog I showed in a colloquium using a demo that Robert Sproul and George King arranged is the following: We start with two long weakly coupled "e" and "μ" pendulums. Initially only the "e" pendulum is excited. It is a bit longer with lower natural frequency: $\Delta = (g/l_\mu)^{1/2} - (g/l_e)^{1/2} < 0$. A slow motor—mimicking the changing matter effects—adiabatically shortens it so that in the end the μ pendulum is longer and Δ flips sign. Since the (small!) coupling ϵ of the two pendulums stays constant, we have resonant beating with many back and forth energy transfers between the two pendulums.

Remarkably at the end when $|\Delta| >> \epsilon$ *all* the energy is transferred from one pendulum to the other, just as all the initial ν_e wave becomes a pure ν_μ in the sun.

To explain this using the Berry phase we use the Hamiltonian: $H = \epsilon\sigma(1) + \Delta\sigma(3)$. The above experiments performed half a circuit in the $\Delta - \epsilon$ plane. Had we completed the circuit with Δ changing sign again regaining the initial configuration, then the system would return to the initial pure "e" (i.e., only the "e" pendulum excited) state possibly with a 1/-1 overall topological Berry phase.[5] The second lengthening stage is—by symmetry—the same as the first. Hence the transformation T in the $e - \mu$ Hilbert space corresponding to half a circuit satisfies $T^2 = -1$ implying that T effects a "μ" $\leftrightarrow$ "e" transposition.

Acknowledgments

The subject of my presentation differs from most others (Freedman's excepted!). Yet it closely ties with another, less familiar, aspect of Frank: his passion for and contributions to theoretical physics. This manifested not only in the theoretical papers that he has coauthored. Rather, Frank got Yakir Aharonov, the late Jeeva Anandan (whose true passion were these phases), Pawel Mazur, and the "Quantum Summer Institute". Thanks to him I have been visiting the University of South Carolina for the last 22 years—a joyful experience for which I am truly grateful.

References

[1] See Withakker *Dynamics of Point Particles and Rigid Bodies* p 76
[2] Berry M, private communication
[3] Aharonov Y, Coleman S, Goldhaber A S, Nussinov S, Popescu S, Reznik B, Rohrlich D, Vaidman L 1994 "Aharonov-Bohm and Berry Phases for a Quantum Could of Charge" *Phys. Rev. Lett.* **73** 918
[4] J N Bahcall 1989 *Neutrino Astrophysics* (Cambridge, MA: Cambridge University Press) figure 9.4 p 266
[5] To close the path of an elongated rectangular shape in the $\Delta - \epsilon$ plane, we need to change $\epsilon \to -\epsilon$ for, say, $\Delta > 0$ and $\Delta >> \epsilon$ and again -$\epsilon \to \epsilon$ at the opposite point $\Delta < 0$ but $|\Delta| >> |\epsilon|$. Since this is done far away from resonance (namely, $\Delta = 0$), no further change in the pendulum's state ensues.

Journal of Physics: Conference Series **173** (2009) 012021 doi:10.1088/1742-6596/173/1/012021

IceCube Science

Francis Halzen

Department of Physics, University of Wisconsin, Madison, WI 53706, USA

E-mail: halzen@icecube.wisc.edu

Abstract. We discuss the status of the kilometer-scale neutrino detector IceCube and its low energy upgrade Deep Core and review its scientific potential for particle physics. We subsequently appraise IceCube's potential for revealing the enigmatic sources of cosmic rays. After all, this aspiration set the scale of the instrument. While only a smoking gun is missing for the case that the Galactic component of the cosmic ray spectrum originates in supernova remnants, the origin of the extragalactic component remains as inscrutable as ever. We speculate on the role of the nearby active galaxies Centaurus A and M87.

1. The First Kilometer-Scale High Energy Neutrino Detector: IceCube

A series of first-generation experiments[1] have demonstrated that high energy neutrinos with $\sim$ 10 GeV energy and above can be detected by observing the Cherenkov radiation from secondary particles produced in neutrino interactions inside large volumes of highly transparent ice or water instrumented with a lattice of photomultiplier tubes. The first second-generation detector, IceCube, is under construction at the geographic South Pole[2]. IceCube will consist of 80 kilometer-length strings, each instrumented with 60 10-inch photomultipliers spaced by 17 m. The deepest module is located at a depth of 2.450 km so that the instrument is shielded from the large background of cosmic rays at the surface by approximately 1.5 km of ice. The strings are arranged at the apexes of equilateral triangles 125m on a side. The instrumented detector volume is a cubic kilometer of dark, highly transparent and totally sterile Antarctic ice. The radioactive background is dominated by the instrumentation deployed into the natural ice. A surface air shower detector, IceTop, consisting of 160 Auger-style 2.7m diameter ice-filled Cherenkov detectors deployed pairwise at the top of each in-ice string, augments the deep-ice component by providing a tool for calibration, background rejection and cosmic ray studies.

Each optical sensor consists of a glass sphere containing the photomultiplier and the electronics board that digitizes the signals locally using an on-board computer. The digitized signals are given a global time stamp with residuals accurate to less than 3 ns and are subsequently transmitted to the surface. Processors at the surface continuously collect the time-stamped signals from the optical modules; each functions independently. The digital messages are sent to a string processor and a global event trigger. They are subsequently sorted into the Cherenkov patterns emitted by secondary muon tracks that reveal the direction of the parent neutrino[3].

IceCube detects neutrinos with energies in excess of 0.1 TeV. An upgrade of the detector, dubbed Deep Core, consists of an infill of 6 strings with 60 DOMs with high quantum efficiency. They are mostly deployed in the highly transparent ice making up the bottom half of the

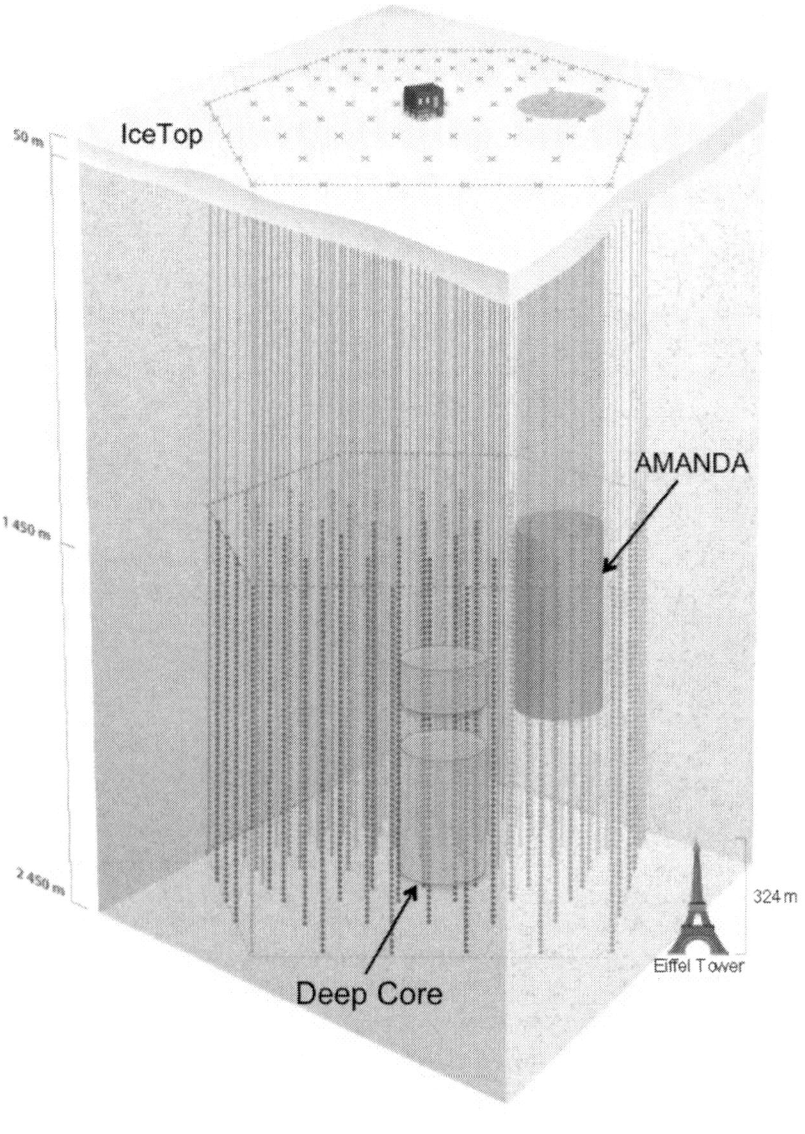

Figure 1. The IceCube detector, consisting of IceCube and IceTop and the low-energy sub-detector DeepCore. Also shown is the first-generation AMANDA detector.

IceCube detector. Deep Core will decrease the threshold to $\sim 10\,\text{GeV}$ over a significant fraction of IceCube's fiducial volume and will be complete by February 2010; see Fig. 1.

The main scientific goals of IceCube fall into broad categories:

(i) Detect astrophysical neutrinos produced in cosmic sources with an energy density comparable to the energy density in cosmic rays[4]. Supernova remnants satisfy this requirement if they are indeed the sources of the galactic cosmic rays as first proposed by Zwicky; his proposal is a matter of debate after more than seventy years. The sources of the extragalactic cosmic rays naturally satisfy the prerequisite when particles accelerated near black holes, possibly the central engines of active galaxies or gamma ray bursts, collide with photons in the associated radiation fields. While the secondary protons may remain trapped in the acceleration region, approximately equal numbers of neutrons, neutral and charged pions escape. The energy escaping the source is therefore distributed between cosmic rays, gamma rays and neutrinos produced by the decay of neutrons and neutral and charged pions, respectively. We will elaborate on the potential of IceCube to reveal the sources of the cosmic rays; this goal is of primary importance as it sets the scale of the detector.

(ii) As for conventional astronomy, neutrino astronomers observe the neutrino sky through the atmosphere. This is a curse and a blessing; the background of neutrinos produced by cosmic rays in interactions with atmospheric nuclei provides a beam essential for calibrating the instrument. It also presents us with an opportunity to do particle physics[5]. Especially unique is the energy range of the background atmospheric neutrino beam covering the interval $1-10^5$ TeV, energies not within reach of accelerators. Cosmic beams of even higher energy may exist, but the atmospheric beam is guaranteed. IceCube is expected to collect a data set of order one million neutrinos over ten years with a scientific potential that is only limited by our imagination.

(iii) The passage of a large flux of MeV-energy neutrinos produced by a galactic supernova over a period of seconds will be detected as an excess of the background counting rate in all individual optical modules[6]. Although only a counting experiment, IceCube will measure the time profile of a neutrino burst near the center of the Galaxy with a statistics of about one million events, equivalent to the sensitivity of a 2 megaton detector.

(iv) IceCube will search for neutrinos from the annihilation of dark matter particles gravitationally trapped at the center of the Sun and the Earth[7]. In searching for generic weakly interacting massive dark matter particles (WIMPs) with spin-independent interactions with ordinary matter, IceCube is only competitive with direct detection experiments[8] if the WIMP mass is sufficiently large. For spin-dependent interactions IceCube already has improved the best limits on spin-dependent WIMP cross sections by two orders of magnitude[9].

Construction of IceCube and other high-energy neutrino telescopes is mostly motivated by their potential to open a new window on the Universe using neutrinos as cosmic messengers; more about this in the rest of the talk. The IceCube experiment nevertheless appeared on the U.S. Roadmap to Particle Physics[10]. As the lightest of fermions and the most weakly interacting of particles, neutrinos occupy a fragile corner of the Standard Model and one can realistically hope that they will reveal the first and most dramatic signatures of new physics.

Besides its potential to detect dark matter, IceCube's opportunities for particle physics include[11]

(i) The search for signatures of the unification of particle interactions, possibly including gravity at the TeV scale. In this case neutrinos approaching TeV energies would interact gravitationally with large cross sections, similar to those of quarks and leptons; this increase yields dramatic signatures in a neutrino telescope including, possibly, the production of black holes[12].

(ii) The search for modifications of neutrino oscillations that result from non-standard neutrino interactions[13].

(iii) Searching for flavor changes or energy-dependent delays of neutrinos detected from cosmic distances as a signature for quantum decoherence.

(iv) The search for a breakdown of the equivalence principle as a result of non-universal interactions with the gravitational field of neutrinos with different flavor.

(v) Similarly, the search for a breakdown of Lorentz invariance resulting from different limiting velocities of neutrinos of different flavors. With energies of 10^3 TeV and masses of order 10^{-2} eV or less, even the atmospheric neutrinos observed by IceCube reach Lorentz factors of 10^{17} or larger.

(vi) The search for particle emission from cosmic strings or any other topological defects or heavy cosmological remnants created in the early Universe. It has been suggested that they may be the sources of the highest energy cosmic rays.

(vii) The search for magnetic monopoles, Q-balls and the like.

The Deep Core upgrade of IceCube will significantly extend IceCube's scientific potential as an atmospheric neutrino detector. It will accumulate atmospheric neutrino data covering the first oscillation dip near 20 GeV with unprecedented statistics. Its instrumented volume is of order 10 Mton. Buried deep inside IceCube, Deep Core will use the surrounding strings as a veto in order to observe the tracks of contained events; see Fig. 1. It has been shown that the event statistics is sufficient to determine the mass hierarchy to at least 90% confidence level assuming the current best-fit values of the oscillation parameters, and for values of θ_{13} close to the present bound[14]. A positive result will also require a sufficient understanding of the systematics of the measurement. This is under investigation and although a result is at this point not guaranteed, the good news is that the relevant data are forthcoming in the next few years.

The physics behind the measurement is the same as for long baseline experiments[15]; the key is to measure the Earth matter effects associated with the angle θ_{13} which governs the transitions between ν_e and $\nu_{\mu,\tau}$. The effective θ_{13} mixing angle in matter in a two-flavor framework is given by:

$$sin^2 2\theta_{13}^m = \frac{sin^2 2\theta_{13}}{sin^2 2\theta_{13} + (cos 2\theta_{13} \pm \frac{\sqrt{2}G_F N_e}{\Delta_{13}})^2}, \quad (1)$$

where the plus (minus) sign refers to (anti) neutrinos. N_e is the electron number density of the Earth, $\sqrt{2}G_F N_e\,(eV) = 7.6 \times 10^{-14} Y_e \rho\,(g/cm3)$ and Y_e, ρ the electron fraction and the density of the Earth's interior. The critical quantity is $\Delta_{13} = \Delta_{13}^2/2E$; its sign determines the mass hierarchy. The resonance condition is satisfied for neutrino energies of order 15 GeV for the baselines of thousands of kilometers studied in atmospheric neutrino experiments. Deep Core extends the threshold of IceCube to this energy. Both the disappearance of muon neutrinos and the appearance of tau and electron neutrinos can be observed.

In the presence of Earth matter effects the neutrino (antineutrino) oscillation probability is enhanced if the hierarchy is normal (inverted). Long baseline detectors, unlike IceCube, measure the charge of the secondary muon thus selecting the sign associated with each event in above equation. The hierarchy is determined by simply looking in which channel, neutrino or antineutrino, the signal is enhanced by matter effects. Provided that the mixing parameter $sin^2 2\theta_{13}$ is not too small and the statistics sufficient, the magnitude of the Δ_{13} term can be measured even without charge discrimination. This is in principle possible with Deep Core[14] but cannot be guaranteed until the systematics of the detector has been studied in the unexplored low energy range.

We next return to IceCube's prospects to detect cosmic neutrinos and, possibly, reveal the sources of the cosmic rays prior to the one hundredth anniversary of their discovery by Victor Hess in 1912. Although impressive progress has been made, driven by the commissioning of a new generation of air Cherenkov telescopes and the Auger air shower array, the origin of comic rays is as enigmatic as ever. Observations show that candidate cosmic accelerators such as supernova remnants, active galaxies and gamma ray bursts emit roughly equal energies in cosmic rays and gamma rays. We will argue that a similar amount of energy should be radiated in neutrinos and that, in this case, IceCube will reveal the sources.

2. The energetics of cosmic ray sources

Cosmic accelerators produce particles with energies in excess of 10^8 TeV; we still do not know where or how [16]. The flux of cosmic rays observed at Earth is shown in Fig. 2. The energy spectrum follows a broken power law. The two power laws are separated by a feature dubbed the "knee" at an energy[1] of approximately 3 PeV. There is evidence that cosmic rays up to this energy are Galactic in origin. Any association with our galaxy disappears in the vicinity of a second feature in the spectrum referred to as the "ankle"; see Fig. 2. Above the ankle, the gyroradius of a proton in the Galactic magnetic field exceeds the size of the Galaxy and it is routinely assumed that we are witnessing the onset of an extragalactic component in the spectrum that extends to energies beyond 100 EeV. Direct support for this assumption now comes from two experiments [17] that have observed the telltale structure in the cosmic ray spectrum resulting from the absorption of the particle flux by the microwave background, the so-called Greissen-Zatsepin-Kuzmin cutoff[2]. The origin of the flux in the intermediate region covering PeV to EeV energies remains a complete mystery. Although the routine assumption is that it results from some high energy extension of the reach of the Galactic accelerators, no convincing mechanism for this has been identified.

It is assumed that cosmic rays originate in cosmic accelerators[3]. Acceleration to TeV energy and above requires massive bulk flows of relativistic charged particles. These are likely to originate from the exceptional gravitational forces in the vicinity of black holes or neutron stars. Gravity powers large currents of charged particles that are the origin of high magnetic fields. These create the opportunity for particle acceleration by shocks. It is a fact that black holes accelerate electrons to high energy; astronomers observe them indirectly by their synchrotron radiation. Some must accelerate protons because we detect them as cosmic rays.

The Swiss astronomer Fritz Zwicky suggested as early as 1933 that supernova remnants could be sources of the Galactic cosmic rays. It is assumed that the accelerators are powered by the conversion of 10^{50} erg of energy into particle acceleration by diffusive shocks associated with young ($\sim$ 1000 year old) supernova remnants expanding into the interstellar medium [18]. Like a snowplough, the shock sweeps up the $\sim$ 1 proton/cm^3 density of hydrogen in the Galactic plane. The accumulation of dense filaments of particles in the outer reaches of the shock, clearly visible as sources of intense X-ray emission, are the sites of high magnetic fields. It is theorized that particles crossing these structures multiple times can be accelerated to high energies following an approximate power-law spectrum $dN/dE \sim E^{-2}$. The mechanism is familiar from solar flares where filaments of high magnetic fields accelerate nuclear particles to tens of GeV. The higher energies reached in cosmic ray accelerators are the consequence of particle flows of much larger intensity powered by the gravitational energy of collapsed objects such as neutron stars and black holes.

From a myriad of ideas, speculations on the sites for the acceleration of extragalactic cosmic rays have converged on the supermassive black holes at the centers of active galactic nuclei (AGN) or the primary engines of gamma ray bursts (GRB). It should surprise nobody, however, if the final answer turned out to be neither of these.

AGN are of special interest because some emit most of their luminosity at TeV energy and above. Their inferred isotropic luminosities can be as high as $10^{45} - 10^{49}$erg s^{-1}. They produce a typical spectrum $dN/dE_\gamma \propto E^{-2}$ in the MeV-GeV range. The energetics requires mass accretion

[1] We will use energy units TeV, PeV and EeV, increasing by factors of one thousand from GeV-energy.

[2] The possibility has not been eliminated that the "cutoff" is simply the maximum energy reached by the accelerator(s).

[3] Speculations on the origin of the highest energy cosmic rays fall into two categories, top-down and bottom-up. In top-down models it is assumed that the cosmic rays are the decay products of cosmological remnants or topological defects associated with, for instance, Grand Unified theories with unification energy $M_{\rm GUT} \sim 10^{24}$ eV. These models predict, besides cosmic rays, large fluxes of gamma rays and neutrinos that have not been observed.

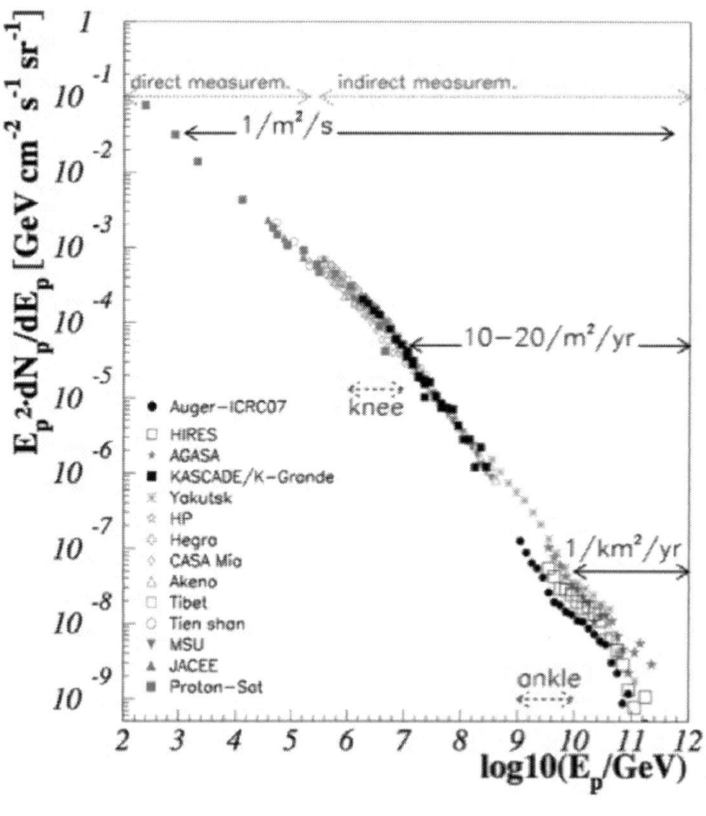

Figure 2. At the energies of interest here, the cosmic ray spectrum follows a sequence of 3 power laws. The first two are separated by the knee, the second and third by the ankle. The cosmic rays beyond the ankle are a new population of particles produced in extragalactic sources.

on a black hole that is up to one billion times more massive than our Sun. Some of the matter falling onto the black hole is deflected and accelerated in highly beamed jets aligned along its rotation axis. Both the inflow onto the central engine and the jet provide opportunities for the buildup of large magnetic fields and shock acceleration. For all these reasons AGN were pinpointed by Ginzberg and Syrovatskii [19] as far back as 1964 as candidate cosmic ray accelerators. A subset, called blazars, emit high-energy radiation in collimated jets pointing at the Earth and are the sources of photons with energies of tens of TeV. Their emission is highly variable over several time scales. TeV-energy bursts as short as minutes have been observed. Exceptionally, the nearest active galaxies, Centaurus A (Cen A) and M87 have been detected in TeV gamma rays even though their jets are not pointing at us.

Some argue instead that gamma ray bursts (GRB), outshining the entire Universe for the duration of the burst, are the best motivated sources of high energy cosmic rays [20]. The collapse of a massive star to a black hole has emerged as the origin of "long" GRB with durations of tens of seconds. In the collapse a fireball is produced of electromagnetic plasma that expands with a highly relativistic velocity powered by radiation pressure. The fireball eventually runs into the stellar material that is still accreting onto the black hole. If it successfully punctures through this stellar envelope the fireball emerges to produce a GRB display. While the energy transferred to highly relativistic electrons is observed in the form of synchrotron radiation, it is a matter of speculation how much energy is transferred to protons. The assumption that GRB are the sources of the highest energy cosmic rays determines the energy of the baryons in the fireball. Accommodating the observed energy spectrum of extragalactic cosmic rays leads to the requirement of roughly equal efficiency for conversion of fireball energy into the kinetic energy of protons and electrons.

It is routinely emphasized that the flux of cosmic rays, especially at the highest energies, is very low. For example, at the onset of the extragalactic component near 10 EeV the flux is only at the level of one particle per kilometer squared per year for a typical array with a steradian acceptance in angle. This can be translated into an energy flux

$$E\left\{E\frac{dN}{dE}\right\} = \frac{10^{19}\,\mathrm{eV}}{(10^{10}\,\mathrm{cm}^2)(3\times 10^7\,\mathrm{s})\,\mathrm{sr}} = 3\times 10^{-8}\,\mathrm{GeV\,cm^{-2}\,s^{-1}\,sr^{-1}}\,. \tag{2}$$

The particles may be few, but each carries an enormous energy that can be expressed in macroscopic units, tens of Joules for the highest energies. We can derive the average energy density ρ_E of cosmic rays in the Universe using the relation that the total flux = velocity × density, or

$$4\pi \int dE \left\{E\frac{dN}{dE}\right\} = c\rho_E\,. \tag{3}$$

We obtain

$$\rho_E = \frac{4\pi}{c}\int_{E_{\min}}^{E_{\max}} \frac{3\times 10^{-8}}{E} dE\,\frac{\mathrm{GeV}}{\mathrm{cm}^3} \simeq 10^{-19}\,\frac{\mathrm{TeV}}{\mathrm{cm}^3}\,, \tag{4}$$

taking the extreme energies of the accelerator(s) to be $E_{\max}/E_{\min} \simeq 10^3$. This is the value for cosmic rays corresponding to the energy density of, for instance, microwave photons of 410 photons of 2.7 K per cubic centimeter.[4]

The energy content derived "professionally" by integrating the observed extragalactic spectrum in Fig. 2, including the GZK feature, is $\sim 3\times 10^{-19}\,\mathrm{erg\,cm^{-3}}$ [21]. This is within a factor of our back-of-the-envelope estimate (1 TeV = 1.6 erg).

It has been realized for a long time that the corresponding quantity for the Galactic component of the spectrum may be a revealing number. If one repeats the integration that we just introduced for the Galactic flux in Fig. 2, from TeV-energy up to the knee near 3 PeV, one obtains an energy density of the cosmic rays in our Galaxy of

$$\rho_E \sim 10^{-12}\,\mathrm{erg\,cm^{-3}}.$$

This happens to be very close to the energy density of light in our Galaxy and to the energy density $B^2/8\pi$ in its microgauss magnetic field. This density, as well as the one for extragalactic cosmic rays in the Universe, represent informative benchmarks for speculating on the sources.

3. Multi-wavelength Astronomy: Cosmic rays, gamma rays and neutrinos

Although the origin of cosmic rays remains a matter of speculation, particle astrophysicists have developed ambitious instrumentation that probes their sources with increased sensitivity detecting cosmic rays, gamma rays and neutrinos. In the 3-prong attack on the cosmic ray problem, TeV-astronomy is by far the most mature [22]. A new generation of ground based air Cherenkov detectors has revealed plausible candidate cosmic ray sources, some, surprisingly, had not been previously identified in other wavelengths. However, the basic hurdle to conclusively identify the observed TeV gamma rays as the decay products of pions produced by a cosmic ray accelerator has not been overcome. Synchrotron radiation by energetic electrons, routinely

[4] We note that the energy density of photons changes by less than 5 orders of magnitude over the electromagnetic spectrum, from radio waves to GeV-photons, while the flux drops by almost 20 orders of magnitude.

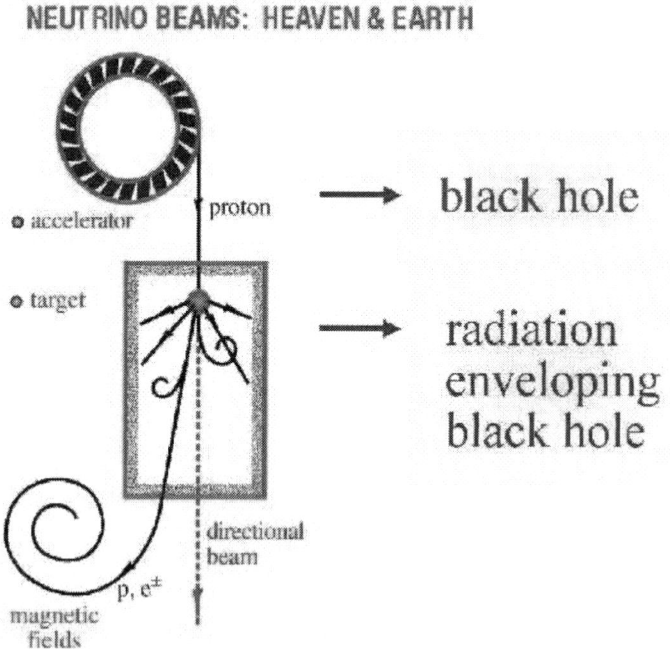

Figure 3. Sketch of a cosmic ray accelerator producing photons and neutrinos. Cosmic rays accelerated in the vicinity of a black hole produce pions in interactions with the radiation and gases surrounding it. Neutral and charged pions decay into photons and neutrinos, respectively.

observed in non-thermal sources such as supernova remnants, AGN and GRB, cannot be excluded as their origin. Neutrinos from the decay of charged pions accompanying pionic gamma rays can provide incontrovertible evidence for cosmic ray acceleration. The predicted fluxes are small and difficult to detect [23]. Let's quantify the problem.

How many gamma rays and neutrinos are produced in association with the cosmic ray beam? For orientation, consider the neutrino beam produced at an accelerator laboratory in Fig. 3. The accelerated protons interact with a target, referred to as a beam dump, producing pions that decay into gamma rays and neutrinos.[5] Generically, a cosmic ray source should also be a beam dump. Cosmic rays accelerated in regions of high magnetic fields near black holes inevitably interact with the radiation surrounding it, for instance, UV photons in active galaxies or the MeV-photons in GRB fireballs. In these interactions they generate neutral and charged pions by the processes

$$p + \gamma \rightarrow \Delta^+ \rightarrow \pi^0 + p \text{ and } p + \gamma \rightarrow \Delta^+ \rightarrow \pi^+ + n.$$

While the secondary protons may remain trapped in the high magnetic fields, neutrons and the decay products of neutral and charged pions escape. The energy escaping the source is therefore distributed among cosmic rays, gamma rays and neutrinos produced by the decay of neutrons, neutral pions and charged pions, respectively.

In reality the relation between the secondary fluxes is more complex. The cosmic rays may interact with gas besides radiation producing equal numbers of pions of all three charges in

[5] For earthbound accelerators the dump is designed to reabsorb all secondary electromagnetic and hadronic showers. Only neutrinos exit the dump. If Nature constructed such a "hidden source" in the heavens, conventional astronomy would not reveal it.

hadronic collisions $p+p \rightarrow n\,[\pi^0+\pi^++\pi^-]+X$. Pionic gamma rays may cascade in the source steepening their spectrum and, as already pointed out, there may be an additional contribution to the gamma ray flux originating from purely electromagnetic processes.

The secondary flux of neutrinos and gamma rays in astrophysical sources can be related to the initial accelerated proton spectrum by specifying the final-state multiplicities of the pions and their average fractional energies x_i relative to the proton. For instance, in hadronic collisions of cosmic rays interacting with gas in or near the source, equal numbers of each pion charge $\pi^0+\pi^++\pi^-$ are produced. The neutral pions decay into two gamma-rays $\pi^0 \rightarrow \gamma+\gamma$ and the charged pions into leptons and neutrinos $\pi^\pm \rightarrow e^\pm+\nu_\mu+\nu_\mu+\nu_e$. The effect of neutrino oscillations is to equalize the number of neutrinos of each flavor arriving at Earth, resulting into one neutrino of each flavor per charged pion. In summary, we obtain two-thirds of a charged pion per interacting proton and one neutrino (of each flavor) per charged pion; and one-third of a neutral pion per interacting proton and two photons per neutral pion. Therefore,

$$\frac{dN_\nu}{dE}(E) = 1 \times \frac{2}{3} \times \frac{1}{x_\nu} \times \frac{dN_p}{dE}\left(\frac{E}{x_\nu}\right) \tag{5}$$

$$\frac{dN_\gamma}{dE}(E) = 2 \times \frac{1}{3} \times \frac{1}{x_\gamma} \times \frac{dN_p}{dE}\left(\frac{E}{x_\gamma}\right) \tag{6}$$

The fractional energies x_i are determined by the average pion inelasticity measured to be approximately 0.2 at accelerators. In the approximation that the pion decay products carry equal energies, we obtain that $x_\nu \simeq 0.25 \times x_\pi \simeq 0.05$ and $x_\gamma \simeq 0.5 \times x_\pi \simeq 0.1$.

Notice that one can eliminate the initial cosmic ray flux to obtain a relation between secondary gammas and neutrinos. It just follows from the fact that there are 2(1) charged pions for every neutral pion in pp(pγ) interactions and, after taking into account oscillations, 2(8) photons for every ν_μ. The flux referred to sums over neutrinos and anti-neutrinos which the experiments cannot separate. For instance, for the case of a E^{-2} spectrum elimination of the cosmic ray flux from above equations yields:

$$\frac{dN_\nu}{dE}(E) = \frac{1}{4} \times \frac{x_\gamma}{x_\nu}\frac{dN_\gamma}{dE}(E) \simeq \frac{1}{2}\frac{dN_\gamma}{dE}(E)\,. \tag{7}$$

We expect one muon neutrino (and antineutrino) for every two TeV gamma ray of pion origin. In the photoproduction case the neutrino flux is reduced by a factor 4.

4. The Sources of Galactic Cosmic Rays

In 1934, Baade and Zwicky[24] pointed out that supernovae could be the sources of the Galactic cosmic rays provided that a substantial fraction of the energy released in the explosion is converted into the acceleration of relativistic particles. Their proposal has been commonly accepted despite the fact that to date no source has been conclusively identified, neither of cosmic rays, nor of accompanying gamma rays and neutrinos produced when they interact with Galactic hydrogen. Galactic cosmic rays reach energies of at least several PeV, the "knee" in the spectrum; therefore their interactions should generate gamma rays and neutrinos from the decay of secondary pions reaching hundreds of TeV. Such sources are referred to as PeVatrons. Straightforward energetics arguments are sufficient to conclude that present air Cherenkov telescopes have the sensitivity to detect TeV photons from PeVatrons.

A first step into pinpointing the sources of the Galactic cosmic rays may have been taken when an all-sky survey in the 10 TeV energy region revealed a subset of sources not readily associated with known supernova remnants or with non-thermal sources observed at other wavelengths [25]. They are associated with nearby star forming regions in Cygnus and in the vicinity of galactic latitude $l = 40$ degrees. Subsequently air Cherenkov telescopes were pointed at 3 of the sources revealing them as potential PeVatrons with a very hard gamma ray energy spectrum that extends to tens of TeV without evidence for a cutoff [26, 27]. This is in sharp contrast with the best studied supernova remnants RX J1713-3946 and Vela Junior.

Pions are produced when the cosmic rays in the expanding remnant interact with the hydrogen in the Galactic plane. In the star forming regions where supernova are more likely to occur, the cosmic rays can interact with dense molecular clouds. Expecting an association of supernova and molecular clouds, the Milagro sources are suspected to be molecular clouds illuminated by the comic ray beam from remnants within $\sim 100\,\mathrm{pc}$. One expects that multi-PeV cosmic rays are accelerated only during a short period when the remnant transitions from the free-expansion to the beginning of the Sedov phase and the shock velocity is high. The high energy particles can produce photons and neutrinos over much longer times as they diffuse through the interstellar medium to interact with nearby molecular clouds; for a detailed discussion see reference [28].

Despite directed observations to reveal plausible candidate TeV sources, the basic hurdle to conclusively associate the observed TeV gamma rays with the decay of pions produced by a cosmic accelerator has not been overcome. Synchrotron radiation by energetic electrons followed by inverse Compton scattering, routinely observed in non-thermal sources, cannot be excluded as their origin. Neutrinos from the decay of charged pions accompanying pionic gamma rays can provide incontrovertible evidence for cosmic ray acceleration in the source. It has been argued for some time that the brightest sources produce TeV neutrino rates comparable to the background of atmosheric neutrinos[31]. It is likely that a neutrino signal should emerge after several years from the data of a kilometer-scale detector as the predicted fluxes are at the level of the background from atmospheric neutrinos. Especially for the case of molecular clouds, the neutrino flux should be predictable at a quantitative level as any confusion of pionic gamma rays with gamma rays of electromagnetic origin should be minimal[28]. Let's try to quantify these claims.

The energy density of the cosmic rays in our Galaxy is $\rho_E \sim 10^{-12}\,\mathrm{erg\,cm^{-3}}$. Galactic cosmic rays are not forever; they diffuse within the microgauss fields and remain trapped for an average containment time of 3×10^6 years. The power needed to maintain a steady energy density requires accelerators delivering 10^{41} erg/s. This happens to be 10% of the power produced by supernovae releasing 10^{51} erg every 30 years (10^{51} erg correspond to 1% of the binding energy of a neutron star after 99% is initially lost to neutrinos.) This coincidence is the basis for the idea that shocks produced by supernovae exploding into the interstellar medium are the accelerators of the Galactic cosmic rays.

A generic supernova remnant releasing an energy of $W \sim 10^{50}$ erg into the acceleration of cosmic rays will inevitably generate TeV gamma rays by interacting with the hydrogen in the Galactic disk. The emissivity in pionic gamma rays Q_γ is simply proportional to the density of cosmic rays n_{cr} and to the target density n of hydrogen atoms. Here $n_{cr} \simeq 4 \times 10^{-14}\,\mathrm{cm^{-3}}$ by integrating the spectrum for energies in excess of 1 TeV. For a E^{-2} spectrum,

$$Q_\gamma \simeq c\left\langle \frac{E_\pi}{E_p} \right\rangle \lambda_{pp}{}^{-1}\, n_{cr}\ (>1\,\mathrm{TeV}), \tag{8}$$

$$\simeq 2cx_\gamma \sigma_{pp}\,[n\, n_{cr}]. \tag{9}$$

or

$$Q_\gamma(>1\,\mathrm{TeV}) \simeq 10^{-29}\,\frac{\mathrm{photons}}{\mathrm{cm^3\,s}}\left(\frac{n}{1\,\mathrm{cm^{-3}}}\right). \tag{10}$$

The proportionality factor is determined by particle physics; x_γ is the average energy of secondary photons relative to the cosmic ray protons and $\lambda_{pp} = (n\sigma_{pp})^{-1}$ is the proton interaction length ($\sigma_{pp} \simeq 40\,\mathrm{mb}$) in a density n of hydrogen atoms. The corresponding luminosity is

$$L_\gamma(>1\,\mathrm{TeV}) \simeq Q_\gamma \frac{W}{\rho_E} \simeq 10^{33}\,\mathrm{photons\,s^{-1}}, \tag{11}$$

where W/ρ_E is the volume occupied by the supernova remnant. We here made the approximation that the volume of the young remnant is approximately given by W/ρ_E or, that the density of particles in the remnant is not very different from the ambient energy density $\rho_E \sim 10^{-12}\,\mathrm{erg\,cm^{-3}}$ of Galactic cosmic rays.

We thus predict a rate of TeV photons from a supernova at a nominal distance d of order 1 kpc of

$$E\frac{dN_{\mathrm{events}}}{dE}(>1TeV) = \frac{L_\gamma}{4\pi d^2} \simeq 10^{-12} - 10^{-11}\left(\frac{\mathrm{photons}}{\mathrm{cm^2\,s}}\right)\left(\frac{W}{10^{50}\,\mathrm{erg}}\right)\left(\frac{n}{1\,\mathrm{cm^{-3}}}\right)\left(\frac{d}{1\,\mathrm{kpc}}\right)^{-2}. \tag{12}$$

Such sources must emerge in an all-sky TeV gamma ray survey performed with an instrument with the sensitivity of the Milagro experiment [25].

The estimate above ignores the fact that supernovae are not uniformly distributed but are associated with regions of star formation such as the Cygnus region and the center of the Galaxy. Dense molecular clouds, common in star forming regions, should be efficient at converting cosmic rays into pions that decay into gamma rays and neutrinos. To date, the Milagro collaboration has identified 6 such PeVatron candidates. Not surprisingly, they cluster in star forming regions in the nearby spiral arms, four in the Cygnus region (MGRO J2019+37, MGRO J2031+41, MGRO J2043+36 and MGRO J2032+37) and two more (MGRO J1908+06 and MGRO J1852+01) near galactic longitude of $l = 40$ degrees. One has to realize that, after subtracting the sources considered here, an excess of TeV gamma rays persists in the Milagro's skymap from the general direction of the Cygnus region [29]. This "diffuse" flux almost certainly originates in unresolved sources that contribute additional neutrinos.

The spectrum of 3 of the sources supports their identification as a PeVatron. H.E.S.S. observations of MGRO J1908+06 reveal a spectrum consistent with a E^{-2} dependence from 500 GeV to 40 TeV without evidence for a cutoff [26]. In a follow-up analysis the MILAGRO collaboration [30] showed that its own data are consistent with an extension of the H.E.S.S. spectrum to at least 90 TeV. This is suggestive of pionic gamma rays from a PeVatron whose cosmic ray beam extends to the knee in the cosmic ray spectrum at PeV energies. MGRO J2031+41, has been observed [27] by the MAGIC telescope with a spectrum that is also consistent with E^{-2}. The lower flux measured by MAGIC, which we will conservatively adopt in our calculations, is likely attributed to the problem of differentiating the source from the background in a high density environment like the Cygnus region. Finally, the failure of Veritas to observe MGRO J2019+37 at lower energies implies that the slope of the spectum must be larger than -2.2.

In the end, despite the suggestive evidence, conclusively tracing the observed gamma rays to pions produced by cosmic-ray accelerators has so far been elusive. It is one of the main missions of neutrino telescopes to produce the smoking gun for cosmic-ray production by detecting neutrinos associated with the charged pions.

As previously emphasized, particle physics is sufficient to compute the neutrino fluxes associated with pionic gamma rays. The neutrino flux associated with MGRO J1908+06 is shown in Fig. 4 along with the TeV gamma ray observations from which it is derived. Note the approximate factor of 2 between the gamma ray and neutrino flux previously argued for. The sensitivity of IceCube to the Milagro sources has been evaluated assuming that the 6 sources represent the imprint of the Galactic cosmic-ray accelerators on the TeV sky [31]. While the number of events with energies of tens of TeV is relatively low, it is clear that this is the energy

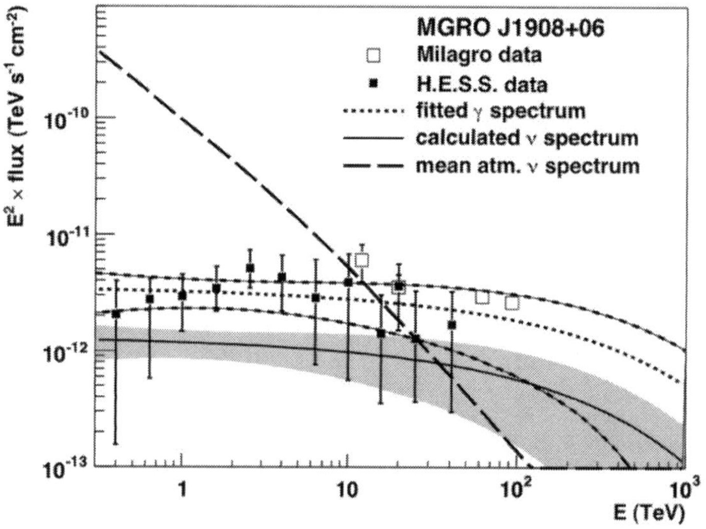

Figure 4. The gamma ray and neutrino fluxes from MGRO J1908+06. The shading surrounding the neutrino flux represents the range in spectra consistent with the statistical and systematic uncertainties on the input gamma ray flux. Also shown is the flux of atmospheric neutrinos at the same zenith angle as the source (dashed line), taking into account the source size and angular resolution.

region where the atmospheric neutrino background, also shown in Fig. 4, is suppressed and an excess from these sources can be statistically established. While observing individual sources may in some cases be challenging, establishing a correlation between the Milagro and IceCube sky maps should be conclusive after several years; see Fig. 5. A "stacked" source search that will look for correlations between all six Milagro sources and the IceCube sky map shown in Fig. 5 has a Poisson probability of 3–5 σ after 5 years; for a sample calculation see Fig. 6. The range reflects the imprecise knowledge of the gamma ray fluxes. The use of optimised methods using unbinned searches beyond the simple binned method considered here will further increase IceCube's sensitivity.

In conclusion, it is important to emphasize again that the photon flux from the Milagro sources is consistent with the flux expected from a typical cosmic-ray-generating supernova remnant interacting with the interstellar medium. In other words, the TeV flux is consistent with the energetics that is required to power the cosmic ray flux in the Galaxy. Alternative sources such as microquasars have been theorized to contribute to the Galactic cosmic rays. If this were indeed the case, cosmic ray energetics would require that they leave their imprint on the Milagro skymap, but none have been observed so far.

5. The Sources of the Extragalactic Cosmic Rays

We first return to the observable that seemed so revealing in the case of the Galactic component, the energy density which is $\sim 3 \times 10^{-19}$ erg cm^{-3} for extragalactic cosmic rays. The power required for a population of sources to generate this energy density over the Hubble time of 10^{10} years is $\sim 3 \times 10^{37}$ erg s^{-1} per (Mpc)^3 or, as often quoted in the literature, $\sim 5 \times 10^{44}$ TeV per (Mpc)^3 per year.

As previously discussed, a GRB fireball converts a fraction of a solar mass into the acceleration

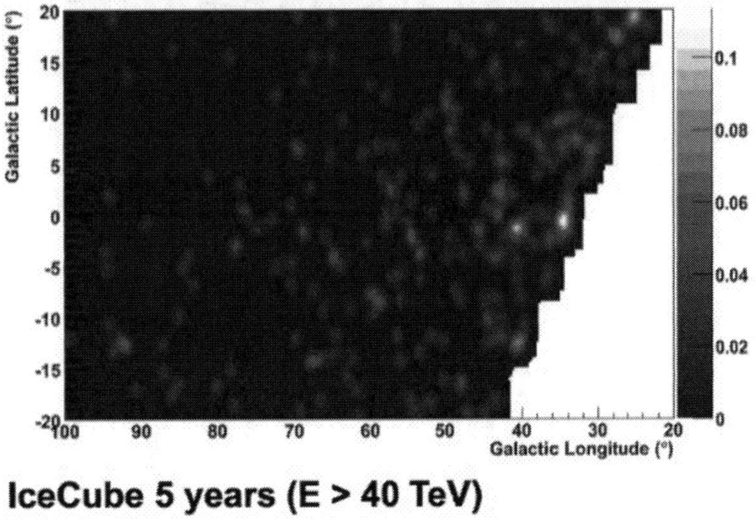

Figure 5. Simulated sky map of IceCube in Galactic coordinates after 5 years of operation of the completed detector. Two of the Milagro sources are visible "by eye" with 4 events for MGRO J1852+01 and 3 events for MGRO J1908+06 with energy in excess of 40 TeV. These as well as the background events have been randomly distributed according to the resolution of the detector and the size of the sources.

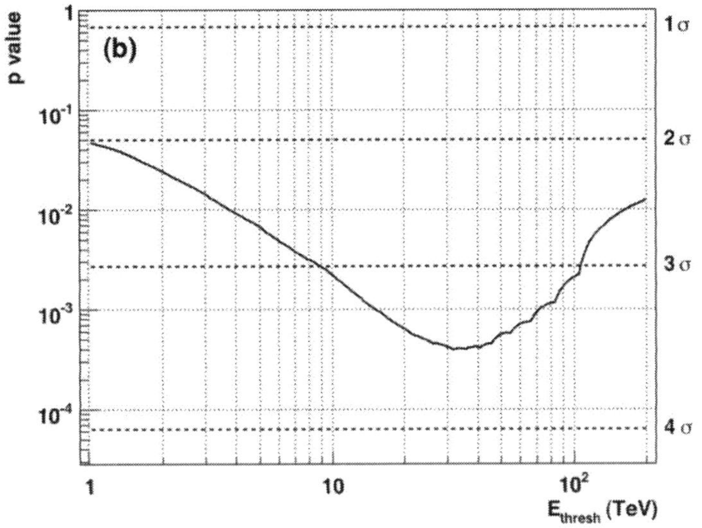

Figure 6. Poisson probability of the excess from the 6 Milagro hotspots after 5 years as a function of the energy of the neutrino-induced muons and a gamma-ray cut-off of the source at 300TeV. Clearly most sensitivity is associated with muons with energies in excess of 40 TeV.

of electrons, seen as synchrotron photons. The energy in extragalactic cosmic rays can be accommodated with the reasonable assumption that shocks in the expanding GRB fireball convert roughly equal energies into the acceleration of electrons and cosmic rays. It so happens that $\sim 2 \times 10^{52}$ erg per cosmological gamma ray burst will yield the observed energy density in cosmic rays after 10^{10} years given that their rate is of order 300 per Gpc^3 per year. Hundreds of bursts per year over Hubble time produce the observed cosmic ray density, just like 3 supernova per century accommodate the steady flux in the Galaxy. Problem solved? Not really, it turns out that the same result can be achieved with active galaxies.

The energy density of $3 \times 10^{-19}\,\text{erg cm}^{-3}$ works out to not only [21]

- $\sim 2 \times 10^{52}$ erg per cosmological gamma ray burst, but also to
- $\sim 3 \times 10^{42}\,\text{erg s}^{-1}$ per cluster of galaxies,
- $\sim 2 \times 10^{44}\,\text{erg s}^{-1}$ per active galaxy.

The coincidence between above numbers and the observed output in electromagnetic energy of these sources explains why they have emerged as the leading candidates for the cosmic ray accelerators. Whether GRB or AGN, the observational evidence that the sources radiate similar energies in photons and cosmic rays is consistent with the beam dump scenario previously discussed. In the interaction of cosmic rays with radiation and gases near the black hole, roughly equal energy goes into the secondary neutrons, neutral and charged pions whose energy ends up in cosmic rays, gamma rays and neutrinos, respectively.

Can IceCube reveal the extragalactic cosmic ray sources? Naively, the neutrino flux should be the same as the observed flux of cosmic rays of Eq. 1. Not so naively, it is about 5 times smaller[6]; we will refer to the energetics estimate for the neutrino flux accompanying the extragalactic cosmic ray accelerators as the band shown in Fig. 7, or

$$E_\nu^2 dN/dE_\nu = 1 - 5 \times 10^{-8}\,\text{GeV}\,\text{cm}^{-2}\,\text{s}^{-1}\,\text{sr}^{-1} \tag{13}$$

As illustrated in the figure, after 7 years of operation AMANDA's sensitivity is approaching the interesting range, but it takes IceCube to explore it.

If AGN are indeed the sources, the proximity of Cen A and M87 single them out as potential accelerators. We will review the information on their spectral energy distibution and argue that the observations are consistent, within large ambiguities, with the neutrino flux estimated above. In fact, Cen A and M87, the next nearest FRI whose jet is not aligned to our line of sight, have previously been singled out as potential cosmic ray accelerators on the basis of gamma-ray data [34].

Interpreting TeV gamma ray observations is challenging because the high energy emission of AGN is highly variable and it is difficult to compare multi-wavelength data taken at different times. Extragalactic TeV sources observed so far are all BLLacs, a subclass of Fanaroff-Riley I (FRI) active galaxies whose jets are oriented along our line of sight. The data have yielded no hints of cosmic ray acceleration so far; the observed spectra can be readily accommodated by synchrotron radiation by electrons, followed by inverse Compton scattering of the photons to TeV energy. Exceptionally, the nearby FRI M87 was observed in the late 1990's by the HEGRA stereoscopic system of five imaging atmospheric Cherenkov telescopes despite the fact that its jet is angled to more than 30° from our line of sight [35].

Where Cen A is concerned, there exists archival data of TeV emission collected in the early 1970's with the Narrabri optical intensity interferometer of the University of Sydney [36]. At the time the Sydney interferometer unsuccessfully searched the sky for gamma ray sources detecting the highest fluctuation in the direction of Cen A. As a followup they exclusively observed the

[6] For experts, it is the Waxman-Bahcall "bound" adjusted downward because only 20% of the proton energy is transferred to pions [32].

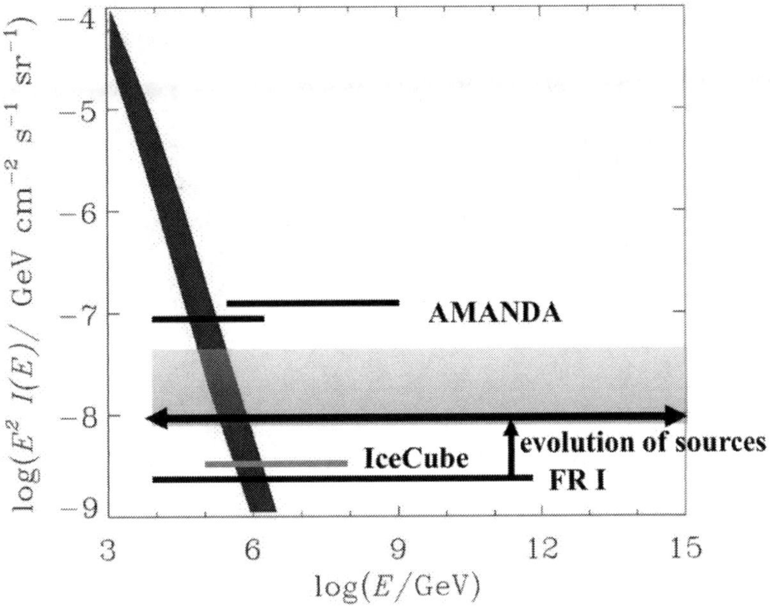

Figure 7. Our energetics estimate of the flux of neutrinos associated with the sources of the highest energy cosmic rays (the shaded range) is compared to the limits established by the AMANDA experiment and the sensitivity of IceCube [33]. Also shown is the flux derived from the assumption that AGN are the sources which we model using spectral energy information on the nearby active galacties Cen A and M87; see Fig. 8. Integration of AGN to larger redshifts can reconcile the two estimates which differ by a factor 3. Also shown is the background flux of atmospheric neutrinos.

source for a 3 year period. A variable flux was observed in two periods of heightened activity lasting roughly one year, pointing at a region of coherent emission of size of order 0.3 pc. This is consistent with the idea that the high energy emission is from an isotropic region near the base of the jet or the central black hole of mass 2×10^8 solar masses, about two orders of magnitude more massive than the one at the center of our Galaxy [34]. The fact that the interferometer beam did not include the radio lobes at the end of the jets further supports the idea of a central engine at the base of the jets. An average flux of

$$dN_\gamma/d(\ln E)(E_\gamma > 0.3\,\mathrm{TeV}) = 4.4 \pm 1.0 \times 10^{-11}\,\mathrm{cm}^{-2}\,\mathrm{s}^{-1} \tag{14}$$

was reported for the period of 3 years over which the source was variable.

The integrated flux of Cen A is close to the one observed by HEGRA [37] from M87

$$dN_\gamma(E_\gamma)/d(\ln E)(E_\gamma > 0.73\,\mathrm{TeV}) = 0.96 \pm 0.23 \times 10^{-12}\,\mathrm{cm}^{-2}\,\mathrm{s}^{-1}, \tag{15}$$

after scaling the flux of M87 at 16 Mpc to the distance to Cen A and adjusting for the different thresholds of the experiments. The observations are therefore consistent with identical source luminosities of roughly $7 \times 10^{40}\,\mathrm{erg\,s^{-1}}$, assuming an E^{-2} gamma-ray spectrum. This suggests that they may be generic FRI, a fact we will exploit to construct the diffuse neutrino flux from all FRI. Recently, MAGIC reported short day-long bursts of M87 [38] pointing at acceleration of particles even closer to the black hole. This, as well as the absence of associated X-ray activity expected in the case of electromagnetic processes, further supports the possibility of cosmic ray acceleration.

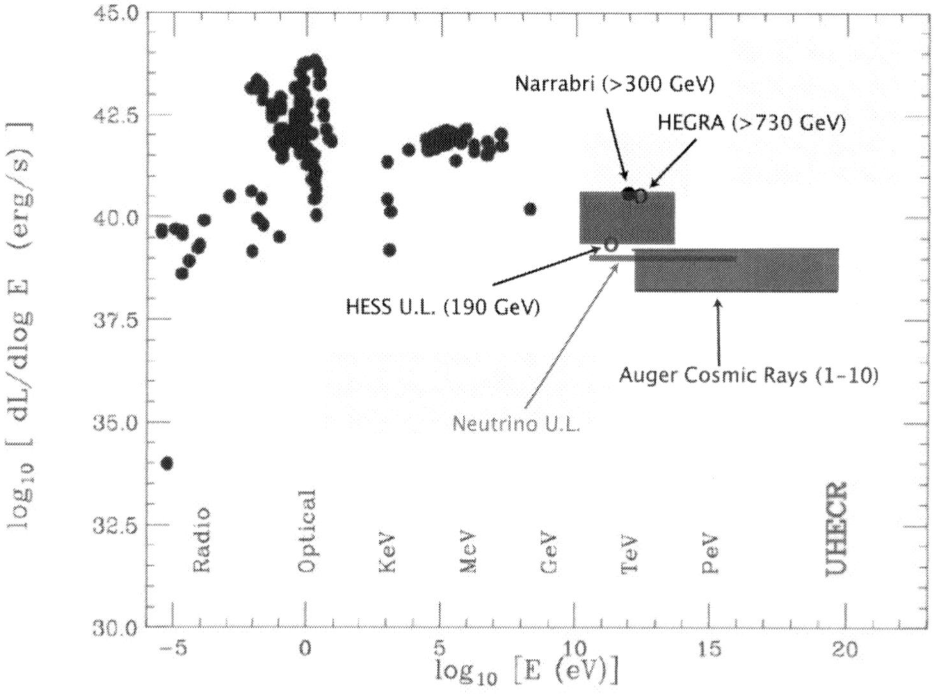

Figure 8. Spectral energy distribution of Cen A (black dots). Having in mind that the source is variable, we show our estimates for the flux of TeV gamma rays (gray shading) and cosmic rays assuming that between 1 and 10 events observed by Auger originated at Cen A (blue shading). We note that the cosmic ray and TeV gamma ray fluxes estimated in this paper are at the level of the electromagnetic component shown from radio waves to GeV photons. Our estimate for the neutrino flux (labeled upper limit) is shown as the red line.

For completeness, the H.E.S.S. experiment obtained a limit on Cen A [39] of 10^{-12} in the same units, a flux smaller than the one just argued for. Given the burst nature of the data the disagreement may be acceptable. We have compiled the TeV gamma ray data in Fig. 8 along with other observations on the multi-wavelength emission spectrum of Cen A recently compiled by Lipari [40].

The same conversion of TeV gamma rays to neutrinos, exploited for Galactic sources, yields a neutrino flux of

$$\frac{dN_\nu}{dE} \leq 5 \times 10^{-13} \left(\frac{E}{\mathrm{TeV}}\right)^{-2} \mathrm{TeV}^{-1}\,\mathrm{cm}^{-2}\,\mathrm{s}^{-1}, \tag{16}$$

conservatively normalized not to exceed the contemporary limit of the H.E.S.S. experiment.

Finally, having previously argued for a relation between the energetics of TeV gamma rays, cosmic rays and neutrinos, we ask the question what Auger data may reveal about the flux of the source

. The answer depends on the number of events in their sky map, N_{events}, that actually correlate with Cen A, a number that depends on the angular broadening of the source by the deflection of the particles by magnetic fields [41]. N_{events} is therefore a matter of speculation. We estimate the flux assuming a power-law spectrum of the form $dN/dE = N_0(E/E_0)^\alpha$, where the normalization N_0 is fixed by the number of events observed N_{events}. For $\alpha = -2.0$, we have N_{events} = field of view × time × efficiency × N_0/E_{thresh}. This gives us $dN_{\mathrm{cr}}/dE =$

$N_{\text{events}} \times 10^{-13}(E/\text{TeV})^{-2}\,\text{TeV}^{-1}\,\text{cm}^{-2}\,\text{s}^{-1}$. The number of events with energy in excess of 6×10^7 TeV may be as many as 10, thus obtaining the range of cosmic ray fluxes shown in Fig. 8. It is close to, but below the flux of TeV gamma rays and suggests an even smaller neutrino flux than the one derived from TeV gamma ray information. The variability of the source and the possibility of a more complicated shape of the spectrum clearly prevent us from reaching quantitative results on the basis of the present data.

The neutrino flux from a single source such as Cen A is clearly small: repeating the calculation for power-law spectra between 2.0 and 3.0, we obtain, in a generic neutrino detector of effective muon area $1\,\text{km}^2$, between 0.8 and 0.02 events/year only. Having estimated the neutrino flux from a point cosmic ray source, we calculate next the total diffuse flux from all such sources within our horizon. Given an FRI density of $n \simeq 8 \times 10^4\,\text{Gpc}^{-3}$ within a horizon of $R \sim 3\,\text{Gpc}$ [42], the total diffuse flux from all $4\pi\,\text{sr}$ of the sky is simply the sum of the luminosities of the sources weighted by their distance:

$$\frac{dN_\nu}{dE_{\text{diff}}} = \sum \frac{L_\nu}{4\pi d^2} = L_\nu\, n\, R = 4\pi d^2 n R \frac{dN_\nu}{dE}, \tag{17}$$

where dN_ν/dE is given by Eq. 1. We performed the sum by assuming that the galaxies are uniformly distributed. This evaluates to:

$$\frac{dN_\nu}{dE_{\text{diff}}} = 2 \times 10^{-9} \left(\frac{E}{\text{GeV}}\right)^{-2} \text{GeV}^{-1}\,\text{cm}^{-2}\,\text{s}^{-1}\,\text{sr}^{-1}, \tag{18}$$

approximately a factor 3 below the flux estimated on the basis of source energetics. Varying the spectral indices as before, we obtain an event rate per km^2 year from the northern sky of between 19 and 0.5 neutrinos per year. Considering sources out to 3 Gpc, or a redshift of order 0.5 only, is probably conservative. Extending the sources beyond $z \sim 1$ increases the flux by a factor 3 or so as was discussed in connection with the results in Fig. 7. We can thus bridge the gap between this and the previous estimate of the cosmic neutrino flux based solely on the energetics of the cosmic rays.

The predicted flux of Eq. 16 should be within reach of IceCube and a future Mediterranean kilometer-scale neutrino telescope. The flux is close to IceCube's 90% confidence level sensitivity in a single year. It follows that detection at the 5σ level should be possible within ~ 5 years.[7]

It is not challenging to produce models that generate a neutrino flux at the level proposed here. For instance [34], the central engine in FRI such as Cen A and M87 may feed a beam dump consisting of the gas surrounding the supermassive black hole[8].

In summary, while the road to the identification of the sources of the Galactic cosmic ray has been mapped, the origin of the extragalactic component remains as mysterious as ever.

Acknowledgments

I would like to thank my collaborators C. Gonzalez-Garcia, D. Grant, Alexander Kappes and Aongus Ó Murchadha as well as John Beacom, Julia Becker, Peter Biermann, Steen Hannestad and Stefan Westerhoff for valuable discussions. This research was supported in part by the National Science Foundation under Grant No. OPP-0236449, in part by the U.S. Department of Energy under Grant No. DE-FG02-95ER40896, and in part by the University of Wisconsin Research Committee with funds granted by the Wisconsin Alumni Research Foundation.

[7] We here used the atmospheric background flux from charm particles predicted by [43] rather than the larger flux conservatively implemented in the sensitivity calculation presented in the IceCube publication. The enhancement in the charm flux relative to the one used here is a direct consequence of the assumption that charm fragmentation functions scale and therefore clearly unphysical.

[8] Given the paucity of data, models can be invoked where, alternatively, the gamma rays are produced in photohadronic ($p\gamma$) rather than hadronic (pp) interactions [44] and in the jet rather than the black hole.

References

[1] C. Spiering, arXiv:0811.4747 [astro-ph].

[2] A. Achterberg *et al.* [IceCube Collaboration], Astropart. Phys. **26**, 155 (2006), arXiv:astro-ph/0604450; T. Montaruli *et al.* [IceCube Collaboration], in Proc. of Topics in Astroparticle and Underground Physics (TAUP07), Sendai, Japan, 2007; S. R. Klein [IceCube Collaboration], arXiv:0807.0034 [physics.ins-det].

[3] F. Halzen, Eur. Phys. J. C **46**, 669 (2006) [arXiv:astro-ph/0602132].

[4] T. K. Gaisser, F. Halzen and T. Stanev, Phys. Rept. **258**, 173 (1995) [Erratum-ibid. **271**, 355 (1996)] [arXiv:hep-ph/9410384].

[5] M. C. Gonzalez-Garcia, F. Halzen and M. Maltoni, Phys. Rev. D **71**, 093010 (2005) [arXiv:hep-ph/0502223].

[6] F. Halzen, J. E. Jacobsen and E. Zas, Phys. Rev. D **53**, 7359 (1996) [arXiv:astro-ph/9512080].

[7] G. Bertone, D. Hooper and J. Silk, Phys. Rept. **405**, 279 (2005), arXiv:hep-ph/0404175.
[8]

[8] B. Sadoulet, Science **315**, 61 (2007).

[9] A. Rizzo, Seventh International Heidelberg Conference on Dark Matter in Astro- and Particle Physics, Christchurch, New Zealand (2009); C. DeClercq, PANIC 2008, Eilat, Israel (2008).

[10] www.pha.jhu.edu/ bagger/talks/HEPAP.pdf.

[11] L. Anchordoqui and F. Halzen, arXiv:hep-ph/0510389.

[12] J. L. Feng, A. Rajaraman and F. Takayama, Int. J. Mod. Phys. D **13**, 2355 (2004), arXiv:hep-th/0405248.

[13] For a recent review see, S. Pakvasa and J. W. F. Valle, Proc. Indian Natl. Sci. Acad. **70A**, 189 (2003).

[14] O. Mena, I. Mocioiu and S. Razzaque, Phys. Rev. D **78**, 093003 (2008) [arXiv:0803.3044 [hep-ph]].

[15] V. Barger *et al.*, arXiv:0705.4396 [hep-ph].

[16] For a recent review of the cosmic ray problem, see P. Sommers and S. Westerhoff, arXiv:0802.1267 [astro-ph]; A. M. Hillas, arXiv:astro-ph/0607109; V. Berezinsky, arXiv:0801.3028 [astro-ph].

[17] R. Abbasi *et al.* [HiRes Collaboration], Phys. Rev. Lett. **100**, 101101 (2008) [arXiv:astro-ph/0703099]; J. Abraham *et al.* [Pierre Auger Collaboration], Phys. Rev. Lett. **101**, 061101 (2008) [arXiv:0806.4302 [astro-ph]].

[18] The idea has been revisited over time by many aythors; see, for instance, F. Aharonian, L. O'C. Drury, & H.J. and Völk, Astron. Astrophys. **285**, 645 (1994).

[19] V. L. Ginzberg and S.I. Syrovatskii, Origin of Cosmic Rays, Pergamon Press (1964).

[20] E. Waxman, Phys. Rev. Lett. **75**, 386 (1995), arXiv:astro-ph/9701231; M. Vietri, Phys. Rev. Lett. **80**, 3690 (1998), arXiv:astro-ph/9802241; M. Bottcher and C. D. Dermer, arXiv:astro-ph/9801027.

[21] T. K. Gaisser, OECD Megascience Forum, Taormina, Italy, 1997, arXiv:astro-ph/9707283; the discussion was recently revisited in Markus Ahlers, Luis A. Anchordoqui, Haim Goldberg, Francis Halzen, Andreas Ringwald, Thomas J. Weiler, Phys.Rev. D72, 023001 (2005) [arXiv:astro-ph/0503229].

[22] H. Krawczynski *et al.*, arXiv:0709.0704 [astro-ph].

[23] T. K. Gaisser, F. Halzen and T. Stanev, Phys. Rept. 258 (1995) 173 [Erratum-ibid. 271 (1996) 355]. F. Halzen and D. Hooper, Rept. Prog. Phys. 65 (2002) 1025. J. G. Learned and K. Mannheim, Ann. Rev. Nucl. Part. Sci. 50 (2000) 679; J. Becker, Phys. Rept. 458, 173 (2008).

[24] W. Baade and F. Zwicky, Phys. Rev. D **46**, 76 (1934).

[25] A. A. Abdo *et al.*, Astrophys. J. **658** L33-L36 (2007) arXiv:astro-ph/0611691.

[26] A. Djannati-Atai *et al.* [H.E.S.S. Collaboration], 30th ICRC, Merida, Mexico, 2007, arXiv:0710.2418.

[27] J. Albert *et al.*, arXiv/astro-ph 0801.2391.

[28] For a recent discussion; see S. Gabici, F. .A. Aharonian, arXiv:0705.3011.

[29] A. A. Abdo *et al.*, arXiv:0805.0417 [astro-ph].

[30] S. Valenti [Milagro Collaboration], Non-thermal Hadronic Processes in Galactic Sources Workshop, Heidelberg, Germany, 2008.

[31] For details of the calculation; see F. Halzen, A. Kappes and A. O'Murchadha, Phys. Rev. D 78, 063004 (2008), arXiv:astro-ph/0803.0314; see also L. Anchordoqui, F. Halzen, T. Montaruli and A. O'Murchadha, Phys. Rev. D **76**, 067301 (2007), arXiv:astro-ph/0612699; J. F. Beacom and M. D. Kistler, Phys. Rev. D **75** 083001 (2007), arXiv:astro-ph/0701751; F. Halzen and A. O'Murchadha, Phys. Rev. D **76**, 123003 (2007), arXiv:astro-ph/0705.1723.

[32] E. Waxman and J. N. Bahcall, Phys. Rev. D 59, 023002 (1999) [arXiv:hep-ph/9807282].

[33] G. C. Hill [IceCube Collaboration], *Prepared for 30th International Cosmic Ray Conference* (ICRC 2007), Merida, Yucatan, Mexico, 3-11 Jul 2007; L. Gerhardt [IceCube Collaboration], *Prepared for 30th International Cosmic Ray Conference* (ICRC 2007), Merida, Yucatan, Mexico, 3-11 Jul 2007.

[34] L. A. Anchordoqui *et al.*, Phys. Lett. B 600 (2004) 202; for an earlier suggestion using beamed emission see R. J. Protheroe, A. C. Donea and A. Reimer, Astropart. Phys. **19**, 559 (2003) [arXiv:astro-ph/0210249].

[35] S. Horiuchi *et al.*, Publ. Astron. Soc. Jap. 58 (2006) 211.

[36] J. E. Grindlay, H. F. Helmken, R. Hanbury Brown, J. Davis, and L. R. Allen, Astrophys. J. 197, L9 (1975);

see also R.W. Clay, B. R. Dawson and R. Meyhandan, Astropart. Phys. 2, 347 (1994); W. H. Allen *et al.* [JANZOS Collaboration], Astropart. Phys. 1, 269 (1993).

[37] See e.g., M. Punch *et al.*, Nature 358, 477 (1992); S. M. Bradbury *et al.* [HEGRA Collaboration], Astron. Astrophys. 320, L5 (1997) [arXiv:astro-ph/9612058]; F. Aharonian *et al.* [HEGRA Collaboration], Astron. Astrophys. 393, 89 (2002) [arXiv:astro-ph/0205499]; D. Petry *et al.* [VERITAS collaboration], Astrophys. J. 580, 104 (2002) [arXiv:astro-ph/0207506]; F. Aharonian, A. Akhperjanian and M. Beilicke [HEGRA Collaboration], Astron. Astrophys. 406, L9 (2003) [arXiv:astro-ph/0305275]; J. Holder *et al.*, Astrophys. J. 583, L9 (2003).

[38] J. Albert *et al.*, arXiv/astro-ph 0806.0988.

[39] F. Aharonian *et al.* [H.E.S.S. Collaboration], Astron. Astrophys. 441 (2005) 465.

[40] P. Lipari, Proceedings of NO-VE, IV International Workshop on "Neutrino Oscillations in Venice", M. Baldo-Ceolin Editor.

[41] J. Abraham *et al.*, Science 318 (2007) 939.

[42] P. Padovani and C. M. Urry, Astrophys. J. 356, 75 (1990).

[43] P. Gondolo, G. Ingelman and M. Thunman, Nucl. Phys. Proc. Suppl. 48, 472 (1996) [arXiv:hep-ph/9602402].

[44] A. Cuoco and S. Hannestad [arXiv:0712.1830]; F. Halzen and A. O'Murchadha, Proceedings of NO-VE, IV International Workshop on "Neutrino Oscillations in Venice", M. Baldo-Ceolin Editor, arXiv:0802.0887 [astro-ph]; J. K. Becker and P. L. Biermann, arXiv:0805.1498 [astro-ph]; M. Kachelriess, S. Ostapchenko and R. Tomas, arXiv:0805.2608 [astro-ph].

Journal of Physics: Conference Series **173** (2009) 012022 doi:10.1088/1742-6596/173/1/012022

New Approaches to Dark Matter and Neutrino Detection

J.I. Collar

Enrico Fermi Institute and Kavli Institute for Cosmological Physics, University of Chicago, Chicago, IL 60637, USA

E-mail: `collar@uchicago.edu`

Abstract. Two new initiatives in astroparticle and neutrino physics are briefly reviewed. COUPP employs ultra-stable heavy liquid bubble chambers to search for WIMP dark matter. First results leading to improved limits on spin-dependent WIMP couplings are presented, together with the most recent progress and prospects. CoGeNT aims at the detection of very faint (~100 eV) signals in detectors massive enough (~1 kg) to allow searches for rare processes, using recently developed p-type point contact (PPC) germanium detectors. The broad range of applications (coherent neutrino scattering, light WIMP searches, double-beta decay) for these semiconductor devices is succinctly described. The implications from their first underground operation are discussed, as well as the status of an ongoing deployment at the San Onofre nuclear generation plant.

1. Preface

There is nothing new under the sky. The old adage fits quite nicely when applied to the "new" methods for particle detection briefly reviewed below. In the first case, the Chicagoland Observatory for Underground Particle Physics (COUPP), we have revisited that old shoe, the bubble chamber, to find new applications for it at the forefront of modern astroparticle physics. In CoGeNT (Coherent Germanium Neutrino Technology) we have taken devices tested and shelved by germanium detector manufacturers a quarter of a century ago (we find out only now!), realizing a plethora of exciting applications for them. Pausing to acknowledge that these projects could have been undertaken decades ago is a humbling exercise. But not nearly as much as being in Carolina celebrating the pioneering efforts of Frank Avignone and Ettore Fiorini. Their accomplishments, when paraded together as during the week of the CISNP, pretty much define the early days of what we would now call the intersection of neutrino, astroparticle and underground physics (Fig. 1).

2. COUPP: not your daddy's bubble chamber

COUPP uses CF_3I bubble chambers to look for particle dark matter, in particular Weakly Interacting Massive Particles (WIMPs) (Fig. 2). The operation of COUPP chambers at a modest superheat permits such searches for exotica in a detector otherwise known for its instability. The advantages of this approach are many, including an optimal target maximally sensitive to both spin-dependent and -independent WIMP couplings, an extreme intrinsic insensitivity to minimum ionizing backgrounds (Fig. 2), low cost and room-temperature operation.

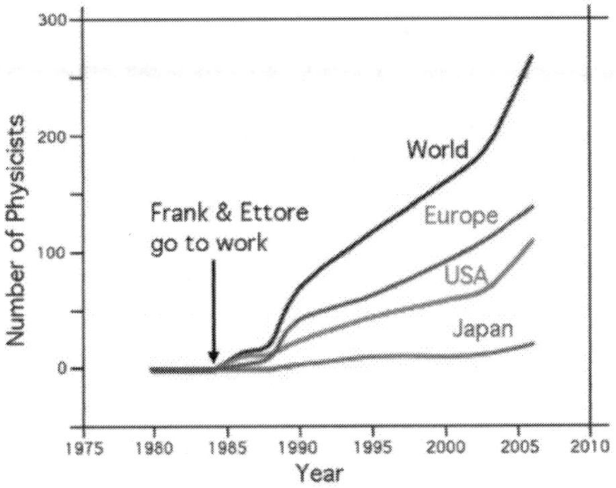

Figure 1. And this is just for the field of Dark Matter detection. Figure shamelessly stolen and adapted from an original by B. Sadoulet.

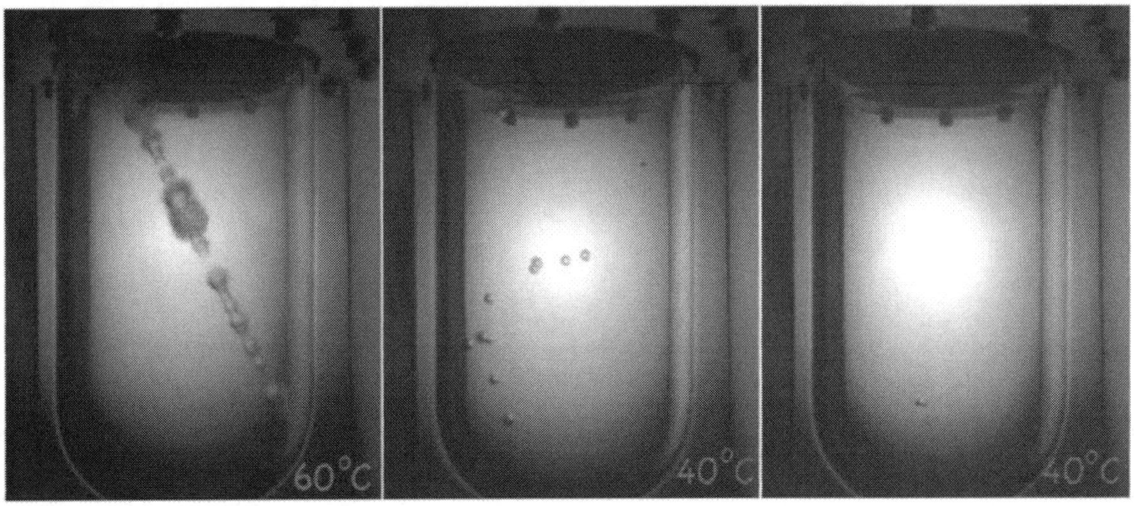

Figure 2. Left: at a high degree of superheat this bubble chamber containing 2 kg of CF_3I displays the typical sensitivity to minimum ionizing particles. The track from a cosmic muon traversing it looks rather robust, due to the relatively slow triggering of recompression in this device. Center: closer to equilibrium, only particles with high stopping powers such as nuclear recoils from WIMPs or neutrons can produce nucleations. The intrinsic background rejection against gamma-induced events has been shown to be better than 10^{-10} (the best in the field) in conditions that nevertheless lead to a full response to low energy recoils [1]. Simultaneous bubbles mark the sites of multiple neutron scattering. A WIMP can, on the other hand, produce single bubbles only (right panel).

COUPP released its first results during 2008 [1]. This analysis used engineering runs, where a majority of events was caused by alpha-recoils from radon penetration into the chamber. No effort was made at the time to control this source of background. The spin-dependent limits obtained are nevertheless some of the best anywhere (Fig. 3). These bounds exclude the possibility that the DAMA annual modulation effect [2] might originate in spin-dependent WIMP couplings. Radon control has since been implemented, resulting in a considerable reduction in event rate. The residual signal agrees with the expectations from muon-induced neutrons at the shallow depth of the NUMI gallery at FNAL, where COUPP currently operates (Fig. 4). A muon veto has been commissioned to tag these events, further reducing the background.

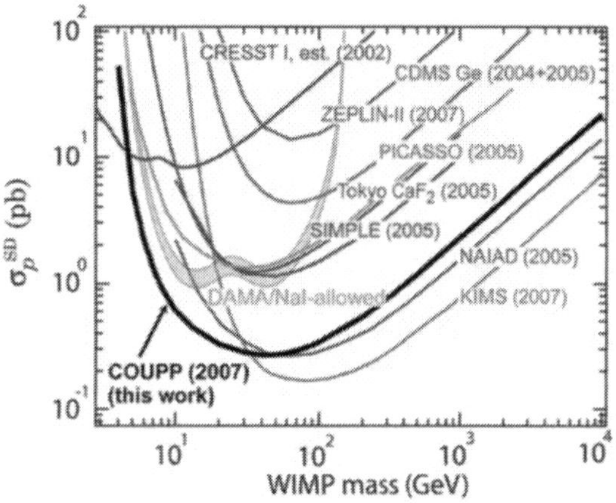

Figure 3. Improved spin-dependent limits from COUPP [1]. The last region of phase space allowing an interpretation of the DAMA annual modulation effect on the basis of spin-dependent WIMP interactions is eliminated by these bounds.

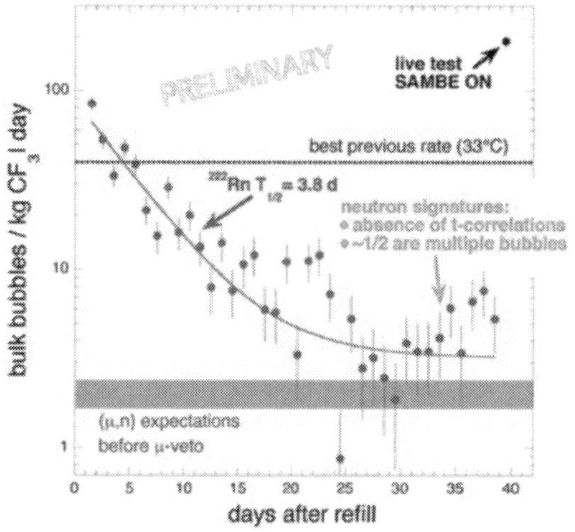

Figure 4. Observed reduction in event rate from radon control. The residual neutron-like rate is now tagged with a 4π muon veto, recently commissioned. The increase in rate resulting from a switchable Am/Be calibration source (SAMBE) is also visible in the figure.

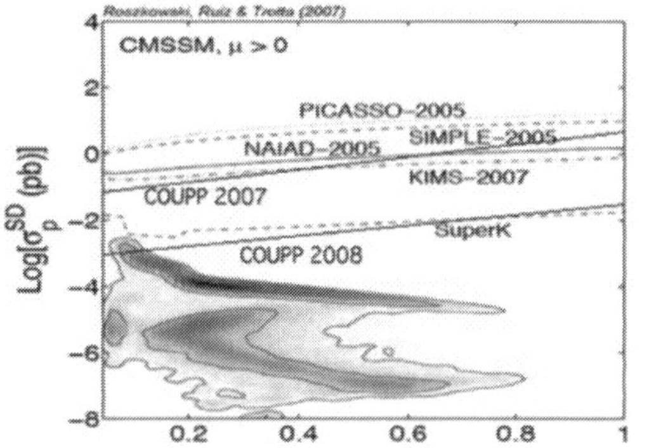

Figure 5. Sensitivity expected from COUPP during 2008, based on a preliminary inspection of recent data. The colored region shows theoretically favored spin-dependent couplings [3].

Figure 6. Images from a new 60 kg COUPP bubble chamber under construction at Fermilab.

Based on a preliminary analysis of recent data, we expect to reach the region of spin-dependent WIMP couplings favored by supersymmetric models very soon (Fig. 5). This would be a first, and particularly important since the supersymmetric phase space probed is complementary and different from that explored by CDMS and XENON [4]. The collaboration plans to commission two new chambers during the present year, one containing 20 kg of CF_3I in a new (windowless) simplified design, and a second conventional model containing 60 kg of active mass (Fig. 6). The first chamber is already operational and undergoing tests above ground. Footage of the "first light" (first bubble) from this large device can be found at [5]. After a short period of tests at 300 m.w.e., the chambers will be installed deep underground, Soudan and SNOlab being presently under consideration. A collaborative effort with Canadian members of SNO and PICASSO, specialists in fluid purification, has started. The collaboration is presently considering a preliminary design for a 500 kg chamber, a direct technical extrapolation from the devices already built and under test.

3. CoGeNT: the "neutrino radio" at last?

We recently reported on a new type of germanium detector [6] (Fig. 7). PPCs (p-type point contact), as we refer to them, feature an unprecedented marriage between large mass and low electronic noise, opening up more than a decade of low-energy (sub-keV) range to exploration with ultra-low background devices (Fig. 8). We have recently learned that Canberra Industries, our partners in crime in this (ad)venture, had actually experimented with such devices (minus the emphasis on ultra-low electronic noise) as early as in 1983, finding no immediate use for them. To make the situation even more strange, a quasi-planar version of this design, the so-called BEGe detector, has been commercially available from Canberra for about 15 years, finding a niche in very specific markets only. Good things can happen when industry and academia finally sit down to listen to each other. While our initial interest in the design was motivated by low-energy applications, we soon realized their strong potential for high-energy gamma background discrimination (Fig. 9) in next-generation double-beta decay projects such as MAJORANA.

The first PPC prototype [6] was operated early in 2008 in a shallow underground site (300 m.w.e.). This new laboratory is part of Chicago's Tunnel And Reservoir Project (TARP), sited 20 minutes from the UC campus (yes folks, we are talking them sewers). A full shield featuring a triple active veto was built around the detector. These runs revealed a low-energy background internal to the cryostat, now in the process of being addressed. The collected data have nonetheless allowed us to impose new limits on light WIMPs (Fig. 10), ruling out the very last possibility remaining for the DAMA modulation [2] to be caused by a standard WIMP halo. These results can be found in [7]. Besides their unmatched sensitivity to light WIMPs, PPCs are presently the only existing detector capable of testing the alternative hypothesis that light axion-like particles might be the cause for the DAMA effect. While TARP runs already constrain this hypothesis, the ongoing cryostat upgrade should allow us to confirm or deny this exciting possibility [7]. The detector is now taking data 25 m away from one of the reactors at the San Onofre Nuclear Generation Station (SONGS, Fig. 11). We expect to perform in this site a first observation of coherent neutrino-nucleus scattering, a process of great interest in fundamental and applied neutrino physics [6]. The large cross-section expected from this process can lead to a miniaturization in the mass of low-energy neutrino detectors by up to three orders of magnitude, enabling technological applications such as neutrino reactor safeguards.

PPC detectors are becoming increasingly popular within MAJORANA. The planned MAJORANA 60 kg demonstrator will feature at least two-thirds of its target mass as PPCs. Several PPCs have been successfully developed within MAJORANA during 2007, demonstrating reproducibility in their manufacture. One more, featuring upgraded electronics and based on a BEGe crystal, is under construction for the University of Chicago group. A vigorous research program on improved front-end electronics is in progress: we envision realistic future energy thresholds lower than 100 eV. In other words, the energy resolution typical of a CCD, embodied in a large ($\sim$1 kg) germanium gamma spectrometer.

Journal of Physics: Conference Series **173** (2009) 012022 doi:10.1088/1742-6596/173/1/012022

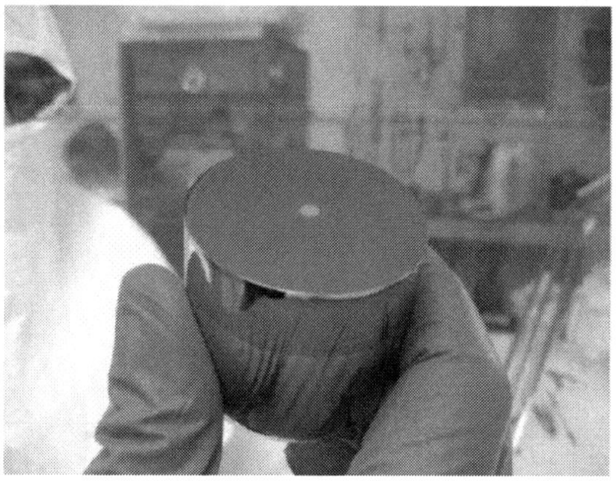

Figure 7. A view of the point contact region in a PPC germanium detector. Minutes after this photograph was taken, the author and his friends proceeded to rather nonchalantly maim this contact, "bricking" the detector.

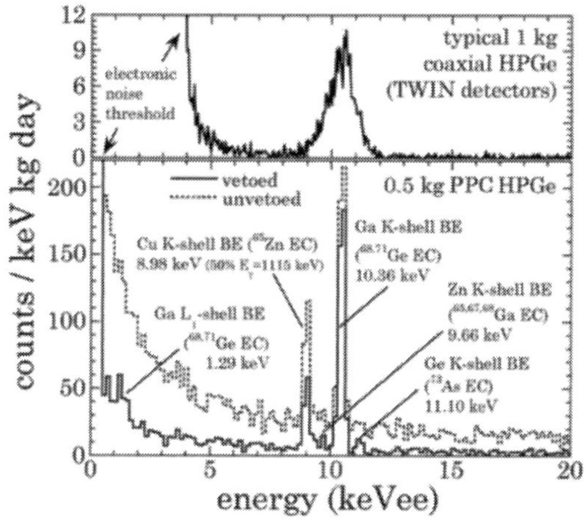

Figure 8. (from [7]) Improvements in threshold and resolution in a PPC design (bottom), compared to a typical coaxial HPGe (top). Cosmogenic peaks are clearly resolved in the PPC spectrum. BE stands for binding energy, EC for electron capture.

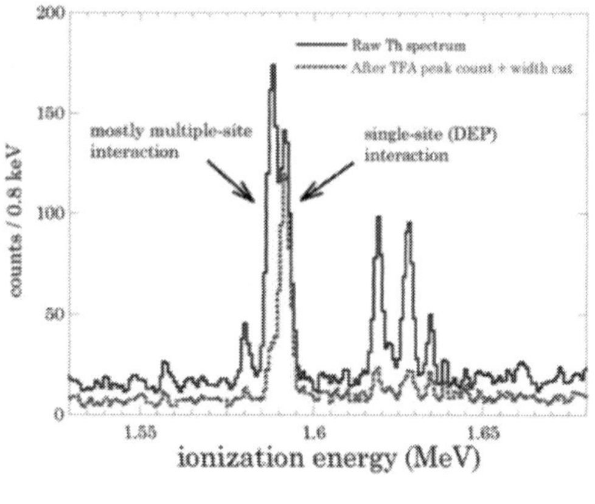

Figure 9. (from [6]) Natural Thorium irradiation of a PPC detector and effect of pulse-shape cuts on rejection of multiple-site events (see text). The energy resolution appears slightly degraded due to the limited resolution of the digitizer employed (8 bit).

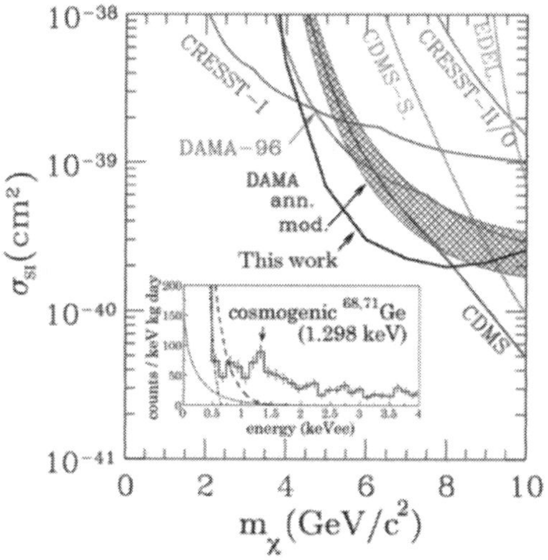

Figure 10. (from [7]) New limits on spin-independent couplings from light WIMPs, from a shallow underground run of the first PPC prototype (see text).

4. Afterword

Here's to Frank and Ettore, with the desire, perhaps certainty, that the many exciting fronts of research they opened up for the rest of us should bear fruit very soon. May we all get to meet during the next CISNP to discuss some discoveries. Exclusion plots and improved limits are nice, but how about a little signal now and then? Patience comes to those who wait...

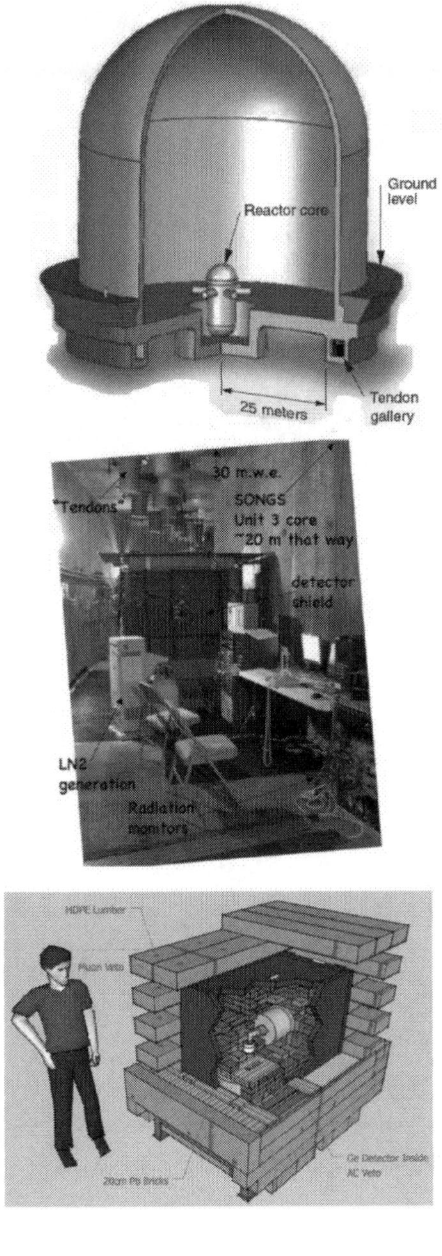

Figure 11. Top: "Tendon" galleries are commonly found along the periphery of power reactor containment domes. The term makes reference to the endings of the steel reinforcement cables that criss-cross the structure. Middle: the experimental setup for our ongoing attempt at coherent neutrino detection. Galleries like these combine a high (anti)neutrino flux ($\sim 10^{13}$ ν/cm^2 s) with the background reduction benefits from a ~30 m.w.e. overburden. Liquid nitrogen is generated from air in situ. Bottom: shield design. A triple active veto is combined with several layers of passive shielding. A low-efficiency external muon veto is not shown in the diagram, but visible in the photograph.

References

[1] Behnke E *et al.* 2008 *Science* **319** 933.
[2] Bernabei R *et al.* 2008 First results from DAMA/LIBRA and the combined results with DAMA/NaI *Preprint* arXiv:0804.2741.
[3] Roszkowski L, Ruiz de Austri R and Trotta R 2007 Implications for the Constrained MSSM from a new prediction for b to s gamma *Preprint* arXiv:0705.2012.
[4] Bertone G, Cerdeno D G, Collar J I and Odom B 2007 *Phys. Rev. Lett.* **99** 151301.
[5] `http://kicp.uchicago.edu/~collar/f1.gif`
[6] Barbeau P S, Collar J I and Tench O 2007 Large-Mass Ultra-Low Noise Germanium Detectors: Performance and Applications in Neutrino and Astroparticle Physics *J. Cosmol. Astropart. Phys.* JCAP09(2007)009.
[7] Aalseth C E *et al.* 2008 Experimental constraints on a dark matter origin for the DAMA annual modulation effect *Preprint* arXiv:0807.0879.

Double Chooz and the search for short range anti-neutrino oscillations

Francis X Hartmann

Division of Particle and Astroparticle Physics, Max-Planck-Institut fuer Kernphysik, Postfach 10 39 80, D-69029 Heidelberg, Germany

f.hartmann@mpi-hd.mpg.de

Abstract. The Double Chooz Experiment seeks to search for short range antineutrino oscillations from the nuclear reactors at the Chooz Nuclear Power Station operated by Electricite de France in Northeastern France. The measurements are of interest to constraining the value for θ_{13} in current neutrino oscillation models. New scintillator types based on beta-diketone and pH stabilized carboxylic acid chemistry are described. New results from the study of these scintillators in the context of the detector design are reported.

1. Physics and Concept of the Experiment

1.1. The physics

The nature of the neutrino is not well understood. The neutrino state arising from electron capture or nuclear beta decay is currently viewed [1] as a superposition of three neutrino mass eigenstates (i=1,2,3). The "electron neutrino" is the neutrino state of the three mass eigenstate particles produced in association with the charged lepton of the electron "flavor" (α). As such, the states of the neutrino types are defined in relationship to the charged current. The electron's loss of electric charge through the weak interactions leads not to an elementary particle, in a sense to which we are accustomed to defining such particles, but a mixture of three entities.

In the current model of three neutrino mass eigenstates, the state of the neutrino associated with the electron can be found from the matrix equation $\langle \nu_e | j \rangle$ whose three factors are:

$$\begin{pmatrix} \nu_e \\ \nu_\mu \\ \nu_\tau \end{pmatrix} = \begin{pmatrix} 1 & 0 & 0 \\ 0 & \cos\theta_{23} & \sin\theta_{23} \\ 0 & -\sin\theta_{23} & \cos\theta_{23} \end{pmatrix} \begin{pmatrix} \cos\theta_{13} & 0 & \sin\theta_{13} e^{i\delta} \\ 0 & 1 & 0 \\ -\sin\theta_{13} e^{-i\delta} & 0 & \cos\theta_{13} \end{pmatrix} \begin{pmatrix} \cos\theta_{12} & \sin\theta_{12} & 0 \\ -\sin\theta_{12} & \cos\theta_{12} & 0 \\ 0 & 0 & 1 \end{pmatrix} \begin{pmatrix} \nu_1 \\ \nu_2 \\ \nu_3 \end{pmatrix}$$

(1)

The angle θ_{23} is known from atmospheric neutrino studies [1] to be greater than 37 degrees (to a maximum of 45 degrees) and the angle θ_{12} is known from solar neutrino and reactor experiments [1] to be 30 to 35 degrees. The neutrino state is given by:

$$\nu_e = \left(\tfrac{\sqrt{3}}{2}\cos\theta_{13},\ \tfrac{1}{2}\cos\theta_{13},\ \sin\theta_{13} \right) \bullet \mathbf{B} \quad (2)$$

Journal of Physics: Conference Series **173** (2009) 012023 doi:10.1088/1742-6596/173/1/012023

where **B** is the basis vector of the three massive neutrinos ν_j (one mass can be zero). The numbers are simplified by using $\theta_{23} = 45$ degree, $\theta_{12} = 30$ degree and setting $\delta = 0$ (when CPT holds, a non-zero value indicates CP violation). The propagation of the neutrino state through space and time involves many interesting physics issues which have some complicating aspects [2]. In the simple situation here, we can summarize that the electron neutrino state will propagate in distance L, such that the probability of detecting an electron neutrino state is given by:

$$\begin{aligned} P_{\nu_e \to \nu_e} &= \left| \sum_{j=1}^{3} \langle \nu_e | \langle j | \nu_e \rangle \langle j | \exp[-i(E_j t - p_j L)] | j \rangle \langle \nu_e | j \rangle | \nu_e \rangle \right|^2 \\ &= \tfrac{5}{8} + \tfrac{3}{8}[\sin^4\theta_{13} + (1-\sin^4\theta_{13})\cos\Phi_{12}] - \tfrac{1}{2}\sin^2 2\theta_{13}[(\tfrac{5}{8} + \tfrac{3}{8}\cos\Phi_{12}) - \tfrac{3}{4}\cos\Phi_{13} - \tfrac{1}{4}\cos\Phi_{23}] \end{aligned} \tag{3}$$

where the phase factor differences $\Phi_{ij} = (E_i - E_j)t - (p_i - p_j)L \to (\Delta m_{ij}^2 L/2E)$ per [1,2].

Consider the case of a 2 MeV neutrino. Close to the source, the $\cos\Phi_{12}$ terms approach unity since the experimental value [1] $\Delta m_{12}^2 = 0.00008$ eV implies an oscillation length of 62 km. Also $\Phi_{13} \approx \Phi_{23}$ since $\Delta m_{23}^2 = 0.00250$ and $\Delta m_{13}^2 = 0.00258$ eV [1] imply nearly identical oscillation lengths of 1.98 km and 1.92 km respectively. In equation 3 the first two terms sum to *ca.* 1.0; the second term oscillates such that at the minimum (where the $\cos\Phi_{13}, \cos\Phi_{23} \cong -1$) the neutrino deficit is directly equal to $\sin^2 2\theta_{13}$. The minimum occurs close to 1.0 km.

1.2. Concept of the experiment

Motivated by the foregoing discussion, a suitable experiment [3] would be where one can search for anti-neutrinos from a nuclear reactor at a distance of about 1.0 km from the core. A suitable situation exists in the town of Chooz in the northeastern Departement of Ardennes in France. Two cores, for a total thermal power P of 8.4 GW are operated by Electricite de France. The number of neutrinos per second is $N_\nu \approx 2 \times 10^{11} P(watts)$. The site of the first Chooz experiment [4] lies 1.05 km from the midpoint of the two reactor cores. A first "Far Detector" is sited here. A second "Near Detector" is planned 400 meters from the same point where the oscillation probability is much less (about 15 % of the maximum) so to minimize uncertainties having to do with the reactors themselves. The observable deficits are summarized in figure 1.

For the specific case of $\theta_{13} = 5$ degrees, figure 1, there is a 3% reduction of the neutrino flux at 1.0 km. The second detector permits lowering the systematic uncertainties by monitoring the actual neutrino spectrum in the near absence of oscillations. To see such a signal is still challenging. The first Chooz experiment achieved a limit of $\theta_{13} \leq 11.4$ deg for $\Delta m_{23}^2 = 0.0025$. To improve upon the situation the detector will incorporate a number of new features as described next.

2. Description of the detector

The Far Detector scintillators are described here. The Near Detector is planned to be identical, except it will have different background considerations. The Far Detector is located in a tunnel with an overburden of 300 meters water equivalent. The central region consists of a blue fluorescent scintillator containing Gd.

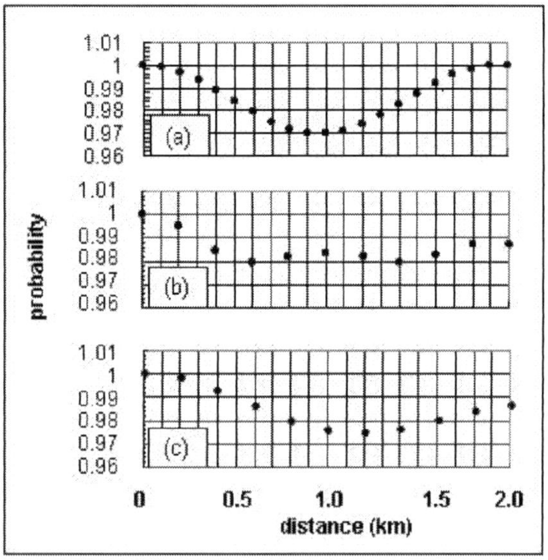

Figure 1. Neutrino Oscillations. The figure at the left (a) shows the oscillation minimum at about 1 km for a monoenergetic 2 MeV neutrino. The deficit is about 3%. A continuum of neutrinos is emitted by the U-235 fuel [5] which, in a simple histogram calculation (b), shows the reappearance of the lower energy neutrinos below 2 MeV leading to some reduction in the magnitude of the deficit. The neutrino detection method imposes a threshold (modeled as 2 MeV), so in (c), the deficit grows while the minimum broadens due to the presense of higher energy neutrinos.

The target scintillator, figure 2, gives a prompt signal following deposition of the positron energy and its annihilation; microseconds later, a delayed signal due to neutron capture on Gd follows. The follow-on gammas emitted from the Gd are detected in the concentric gamma catcher region. The target and gamma catcher regions involve new scintillator formulations. A non-scintillating buffer region shields the central targets from background gammas. The phototubes are mounted on the wall of the stainless steel buffer tank. These regions are contained in a larger tank which serves as an inner veto shield. Above these tanks is an outer plastic veto scintillator to "tag" cosmic ray events.

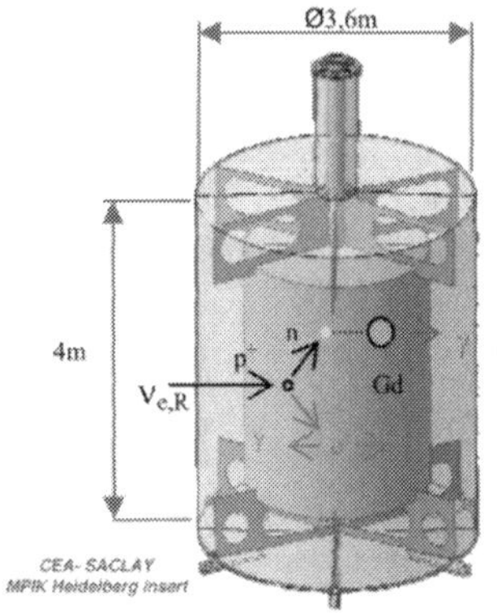

Figure 2. Depiction of a Neutrino Intereraction in the Gamma Catcher and the Neutrino Target. The anti-neutrino interaction with the proton of the Neutrino Target (10 m^3, 20% ortho-phenylxylylethane, o-PXE, and 80% normal-dodecane, n-C12) yields a prompt signal due to a positron summed with its own annihilation energy, thus recording the reactor neutrino energy. The microsecond scale delayed capture of the neutron on Gd leads to higher energy multi-MeV gammas that are captured in the Gamma Catcher (22 m^3, o-PXE, n-C12 and a C12-C14 mineral oil, MO). The Gamma Catcher is surrounded by an optically transparant liquid buffer region (102 m^3, MO and tetra-methyldecane, TMD). An inner veto (muon) detector (87 m^3, TMD and a mixture of linear alkyl benzenes, LAB) encases the main detector.

The neutrino rate is about 70 nu/day. Correlated background signals due to stopping and reacting muons or fast neutrons are expected to be about 2 /day. Accidental backgrounds originating from gammas from the rock or a fast neutron are expected to be < 5 /sec in the energy window of interest (E > 0.7 MeV). In addition errors are reduced not only due to background reduction but improved target weighing (no. of protons) and the use of two reactors to factor out reactor specific quantities. Reductions of the systematic uncertainty from 2.7% (Chooz) to 0.6% (Double Chooz) and better statistics is expected from a number of improvements: larger detector volume, longer exposure, two detectors, lower e+ threshold, improved e+ and n detection efficiency, background shielding, radiopurity and overburden. One key improvement is in the scintillator which we describe next.

3. New Gd scintillators

3.1 Neutrino target chemistry

A carboxylic acid (CBX) is shown in figure 3 where R represents an arbitrary hydrocarbon of choice. A base will remove the proton of the –OH leaving a negative anion of "triangular" shape. The inclusion of an additional two carbons allows one to achieve a "hexagonal" shape seen in the center of figure 3 which is better configured to grab the Gd^{3+} ion. The choice of a double bond in the extension leads to a sp^2 planar hybridization of the complex such that all atoms of the hexagon lie in the same plane. The resulting enol is a tautomer and rapidly rearranges by movement of a proton to the third structure of figure 3. Here, the C=O is a ketone, the carbon alpha to it holds the moveable proton and the carbon beta to it is a second ketone – thus a "beta-diketone" (BDK). Removal of the proton leads to a negative anion of hexagonal shape that bonds to the Gd with the added feature of resonance stabilization via the out-of-plane p orbitals. Such complexes are much more stable than carboxylic acids, in fact they are stable to temperatures just above 200 deg C. They have considerable vapour pressures that can be used in their purification.

The original scintillator of Reines [6] was based on the use of propionic acid (R having three carbons). The cation was Cd++. The first complexes of Gd utilized ethylhexanoic acid [7]; Palo Verde made significant work with ethyl-2-hexanoic acid [8], and describe well their observed problems. The first Chooz experiment used gadolinium nitrate in hexanol dissolved in the organic scintillator. In the Double Chooz project, the desire is to have a very stable scintillator that lasts out to at least five years. As a consequence we developed, at MPIK, metal based scintillator systems based on the specific acid TMHA. These are produced in a pH stabilized manner [9]. The approach keys on providing a defense against hydrolysis reactions.

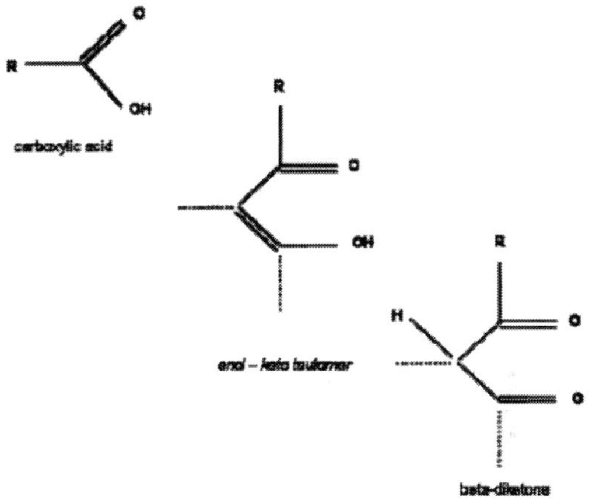

Figure 3. Carboxylic acid (CBX *e.g.* TMHA, section 3.3) and beta-diketone (BDK *e.g.* THD, section 3.2) ligands used in the Gd molecular investigations.

The first Gd-TMHA scintillators were made at MPIK in 2003 and first presented in detail in [10]. Investigations over the course of four years are reported in Section 3.3. The BDK scintillators [11] are described next in 3.2.

3.2. BDK scintillators

Beta-diketone systems are interesting from the viewpoint of "neutrino physics chemistry", that is "putting the right nucleus in the right place at the right time". It is useful to summarize some background information. The first use of beta-diketones in this regard was to produce gaseous forms of Ho-163 in measurements [12] that currently set the best limit on the neutrino mass (electron based) [1]. These experiments followed some ideas communicated to us by Davis who had tried to put Be in the gas phase as part of a Li solar neutrino experiment [13] using pentane-2,4-dione (acetylacetonate). In the LENS project [14] two scintillator systems were tested in prototype cells for an Indium solar neutrino experiment at Gran Sasso, one of which was the BDK form of Indium and the other a form of indium hydroxide suspended with the help of added carboxylic acid. From such experience, we investigated and developed the beta-diketone versions of Gd for a number of ligand species. The final system is Gd(III)-tris-(2,2,6,6-tetramethyl-hepane-3,5-dionate) or $Gd(THD)_3$. The structure takes the basic backbone of the acetylacetonate and adds to it two tert-butyl groups whose steric effects accomplish improved properties in outer and inner sphere coordination chemistry and in vapor pressure. Stability of such Gd complexes was observed at Princeton when radioactively labelled complexes of $Gd(THD)_3$ were first studied by us [15] at elevated temperatures during testing of the Ho-163 neutrino mass experiment.

The six main areas that concern the development of a scintillator of this type are: molecular stability, light yield, optical absorbance, solubility, overall system stability and radopurity.

3.2.1. Molecular stability and light yield. The structure of the Gd complex is shown in the inset to figure 4 [16]. The central region is strongly bonded via resonance stabilized coordination. The resonance structures have some absorbance in the ultra-violet. We develop the scintillator so as to construct energy transfer pathways. The pathways are studied using a model based on [17] that has now been extended to this more complicated situation [18]. As a consequence we attain a light yield for the project of about 45 % relative to anthracene (recorded versus BC505).

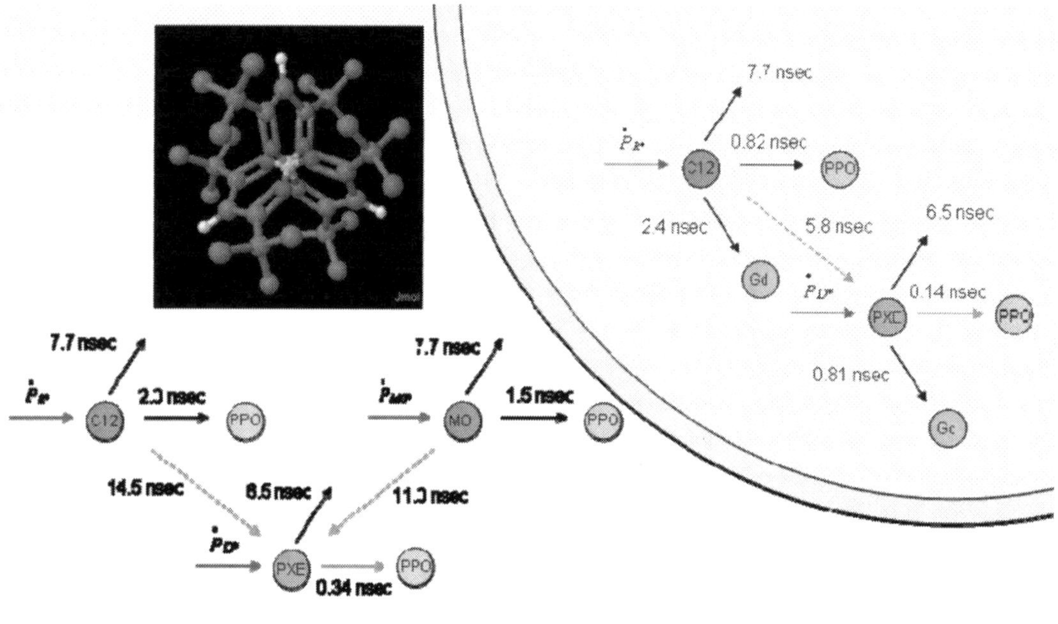

Figure 4. Molecular structure and light yield. The target in the upper right creates light via dodecane C12 and PXE channels, with one cross-transfer from C12 to PXE. The light yield in the gamma catcher is designed to match the target light yield and match as well in density. The degrees of freedom allow for one free parameter, the ratio of PPO to PXE in the gamma catcher. This freedom allows us to tune the pulse shapes so as to achieve a neutrino target – gamma catcher difference. At issue is adjusting the energy transfer pathway of the C12→PPO versus C12→PXE by PPO adjustment above and below a Forster-Dexter effective critical concentration as found for C12. Time constants shown are determined from the model after experimental determination of pairwise transfer coefficients. Pulse shape measurements are also made.

3.2.2. Absorbance, solubility and overall scintillator stability. Figure 5 summarizes the absorbance characteristics of the neutrino target. The issue of solubility over five years is a complicated subject that requires an understanding of the fundamental aspects of the inner and outer coordination sphere chemistry of Gd which is itself influenced by the nature of the coordinated ligands and their bonding.

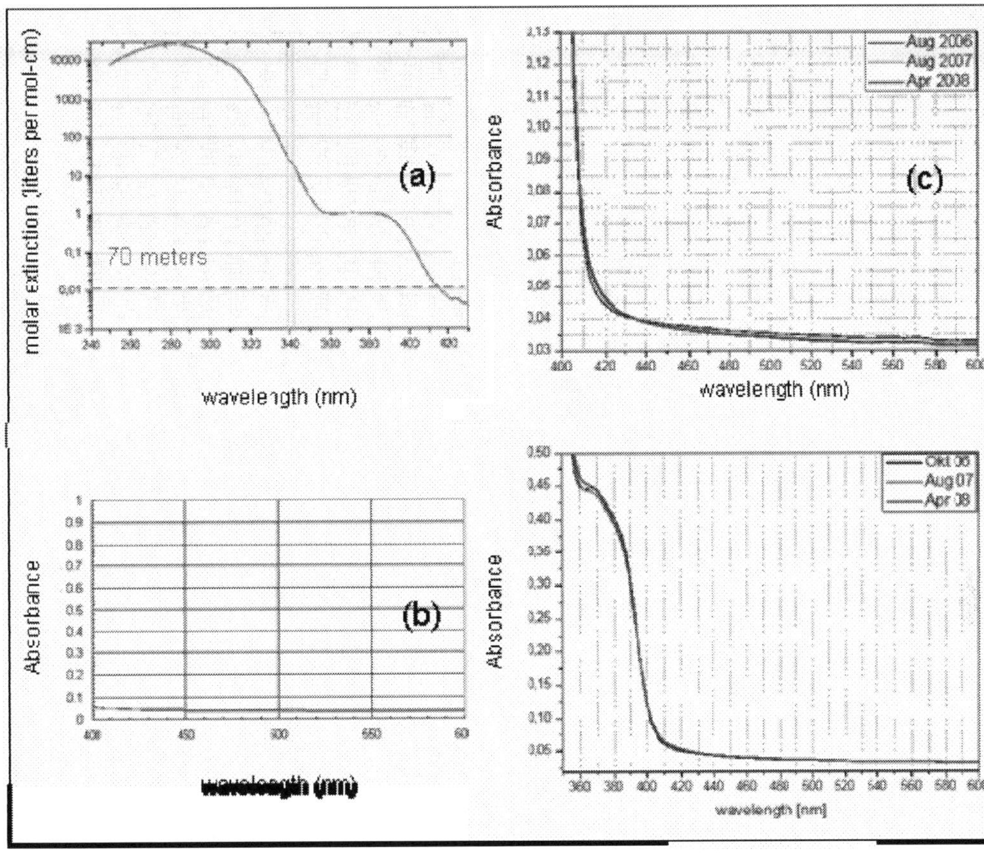

Figure 5. Absorbance, Solubility and Overall Stability. In the upper left (a), we report a precise measurement of the molar extinction coefficient of the Gd material [18]. In the visible region >410 nm, the Gd component, at 1 g/l, has an absorbance length well in excess of 70 meters. In the lower left (b) the scintillator minus fluors is highly transparent; at 440 nm it changes 10.4 meters to 8.0 meters (99.0 %T to 98.8 %T) in three years. In the upper right (c), a solution containing 5 g/l Gd is stable. After almost two years the absorbance change at 430 nm is about 0.0015 a.u. (30 m component). At 400 nm the change is only about 0.0043 (10 m component); there is no solid formation. The full scintillator mixture prepared by Buck [19], in the lower right (d), has an absorbance change at 430nm of 0.0022 (20 meter component) in just less than two years.

3.2.3 Radiopurity. One hundred kilograms of Gd-BDK were vaporized and re-crystalized into molecular crystals in order to achieve a high radiopurity. Table 1 shows the radiopurity results following these sublimations as measured in the MPIK germanium GeMPI detectors (GeMPI I, II, III) at the Gran Sasso Underground Laboratory. All numbers are within the radiopurity specifications.

Table 1. Radiopurity measurements of the Gd-BDK prepared for the neutrino targets of the Near and Far Detectors. Activity per detector in column three is calculated from column two.

Nuclide	Activity (mBq/kg)	Activity (mBq/detector)
Ra-226 (Rn-222)	<0.57	<26
U-235	<0.52	<24
Ra-228	<0.48	<22
Th-228 (2614 keV)	<0.54 (2 > detection limit)	<26
K-40	13.8 (7 > detection limit)	640

Eu-152	<1.0 (4 > detection limit)	<48

3.3 The TMHA carboxylic scintillator system

The TMHA complex of Gd has been studied over the course of five years (since 2003). Samples produced at 1 g/l for long term monitoring showed extremely good characteristics having good absorbance (>25 m, 440nm) and high light yield (64 % anthracene measured versus BC505). Samples up to 50 g/l Gd were readily produced; concentrated solutions at 10 g/l were made for for prototype tests. The procedure employs a pH controlled solvent extraction and remains pH stabilized. From these solutions, 100 liters of 1 g/l Gd 100 liters were produced for prototype testing in 2005. From various studies, it was found that initial degradation rates were approximately proportional to the Gd concentration at the start. Consequently a lifetime prediction of 15 meters to 5 meters absorbance was predicted over 5 years. A drawback of the TMHA system is the lability of the complexing anions suggesting its lower molecular stability relative to the BDK system, the absence of a purification mode from radiopurities via gaseous recrystallization and its dehydration characteristics. Consequently a long series of measurements (shown here for a sample observed for 4 years) indicates notable crystal formation during the 1000 and 1300 day interval.

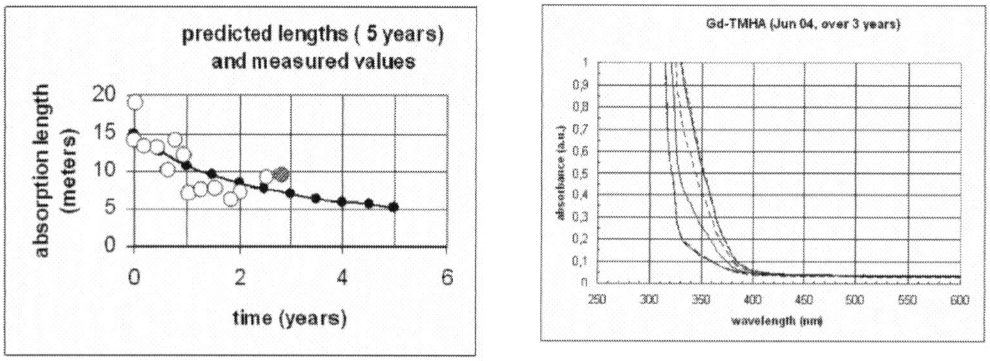

Figure 6. CBX system. After *ca.*1000 days (left) the system exhibits an instability attributed to changes in the molecular structure. Between 1000 and 1300 days the system was adjusted with a hydrogen bonding agent to see if it could be recovered. The system continues under study for 1500 days. Changes in absorbance (right) over the 1000 days are seen to occur (plotted at approx. 6 month intervals) and the absorbance slowly moves into the optical regime from "behind" the location of the UV absorption of the base fluor (here PXE). Nanogram to microgram crystals form after about 1000 days and are attributed to bridging. Afterwards, the UV spectrum is static.

Figure 7. Structural bridging. An explanation for the behavior of Gd-TMHA (CBX) degradation is the formation of bridging carboxylic acid anions which cross between Gd ions. Such bridging can occur (*ad infinitum*) due to the coordination sphere chemistry of Gd. Such bridging can lead to changes in the UV-vis properties due to electron delocalization. Perhaps (dotted line) aromatic effects can also enter.

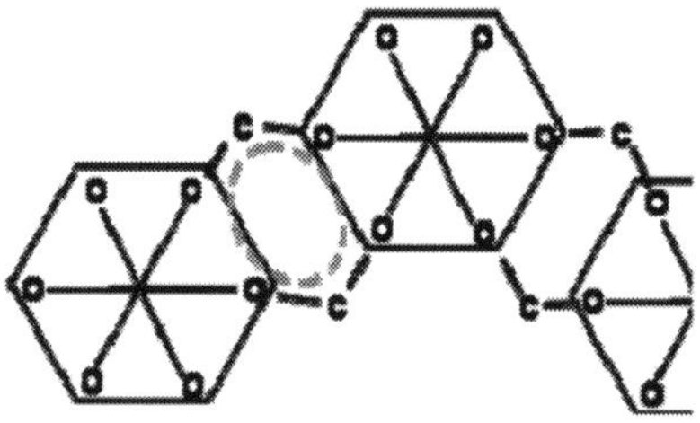

4.0 Summary

The Double Chooz Project is in the assembly phase and scheduled to come on-line in 2009. In 1.5 y, the Far Detector should reach a sensitivity level of $\sin^2 2\theta_{13} = 0.057$, $\theta_{13} = 7.0$ deg. The Near Detector then comes on-line; after an additional 3.5 y the sensitivity level is $\sin^2 2\theta_{13} = 0.030$,

$\theta_{13} = 5.0$ deg. A "3 sigma effect" discovery potential [20] is 5.8 deg in 3 y. A very stable (>5 y) scintillator is extremely important to the measurement of θ_{13}.

In this paper, we present the first detailed information about the target and gamma catcher scintillator for the Double Chooz project. This is the first such full detector system based on beta-diketone chemistry. For this reason we have sought to summarize a number of references related to details of the work and to record references to some of the background work as well.

The neutrino target and gamma catcher in Double Chooz have the light yield and densities matched. A light yield model has been developed for multi-component liquids that allows us to understand the energy transfer properties and as a result we are able to tune the pulse shapes. This feature is afforded by choice of the liquid components which allow us an extra degree of freedom.

Extensive work has been made since 2000 with the CBX system based on TMHA whence we developed the pH stabilized carboxylate system. Detailed studies now extend for 4 years. The systems have known degradations which are modeled by us. After about 1000 days in existing samples, we see gradual changes in the UV and the onset of nanogram-scale crystal formation in conformance with what is predicted from concentration studies. Continued and ongoing investigation search lead us to conclude, for the purposes of the TMHA system, that bridging is playing a role such that the system is not usable by 3.5 years.

We thus concluded that the BDK systems are best capable of putting the right nucleus in the right place at the right time for Double Chooz. Future investigations in neutrino physics, for example [21], are focused on the use of beta-diketones and proceed in this direction. Parallel work with Nd, of interest to double-beta decay, proceeds as well using both TMHA and BDK systems.

Acknowledgements
Acknowledgement is given to the Double Chooz Collaboration as regards relevant aspects. F X Hartmann acknowledges the support of the Max Planck Society in the conduct of this research. Research performed in the United States is under the auspices of Hartmann Scientific, a professional research entity liscensed and registered in the City of Virginia Beach, VA, USA for the purpose of the research and development of neutrino detector projects. Helpful conversations were had with T. Lasserre and D. Motta. The scintillator work described here is partly dependent on helpful interactions with S. Schoenert and the collaborative work of C. Buck. Table 1 and figure 5 (a),(c),(d) are provided by C. Buck. We express our thanks to Prof. Frank Avignone and the Physics Department of the University of South Carolina.

References

[1] Yao Y-M et. al. 2006 *Journal of Physics G* **33** 156-64.
[2] Grossman Y and Lipkin H 1997 *Phys. Rev. D* **55** 2760-7.
Nussimov S 1976 *Phys. Letts. B* **63** 201-3.
[3] Lasserre T and Sobel H 2005 "Reactor Neutrnios", *Comptes Rendus Physique, French Academy of Sciences* **6** 749-57.
[4] Apollonio et. al. (Chooz Collaboration) 1999 *Phys. Letts. B* **466** 415-30.
Apollonio et. al. (Chooz Collaboration) 2003 *Eur. Phys. C* **27** 331-74.
[5] Schreckenbach K, Golvin G, Gelletly W and von Feilitzsch F 1985 *Phys. Letts.B* **160** 325-30.
[6] Reines F and Cowan C L 1953 *Phys. Rev.* **90** 492-3.
Cowan C L, Reines R, Harrison F, Anderson E and Hayes F N 1953 *Phys. Rev.* **90** 493-4.
[7] Reines F and Cowan C L 1959 *Phys. Rev.* **113** 273-9.
[8] Piepke A G, Moser W S and Novikov V M 1999 *Nucl. Instrum. And Meth. A* **432** 392-8.
[9] The particular approach of pH stabilized production of the carboxylate scintillators was first presented in 2002: Hartmann F X 2002 "Organic liquid scintillators for low nu detection", LowNu 2002: Third International Workshop on Low Energy Solar Neutrinos, Heidelberg, Germany.
[10] The TMHA systems of Gd were first presented in 2003: Hartmann F X 2003 "Low Level Scintillators and Gadolinium", Workshop on Future Low Energy Neutrino Experiments, Technical University Munich, Munich 9-11 October, and
Hartmann F X, 2004 "Scintillator Development for a Reactor Experiment in Europe (Double

Journal of Physics: Conference Series **173** (2009) 012023 doi:10.1088/1742-6596/173/1/012023

Chooz), Third Workshop on Future Low Energy Neutrno Experiments, Niigata, Japan March 20-22.

[11] The BDK systems of Gd were first presented in 2003: Buck C, Hartmann F X, Motta D, Schoenert S and Schwan U, "Metal Beta-diketone Scintillators", Presentation at the Workshop on Future Low Energy Neutrino Experiments, Technical University Munich, Munich 9-11 October 2003.

[12] Springer P T, Bennett C L and P A Baisden 1987 *Phys. Rev. A* **35** 679-89.
Hartmann F X and Naumann R A 1985 *Phys. Rev. C (Rapid Comm.)* **31** 1594-6.

[13] Davis R and Rowley J K, private communications.

[14] Buck C, Hartmann F X, Schoenert S and Schwan U 2003 *J. Rad. and Nucl. Chem.* **258** 255-63
Buck C, Barabanov I, Besida O, Bezrukov L, Cattadori C, Danilov N, di Vacri A, Hartmann F X, Motta D, Schoenert S and Schwan U and Yanovich E *2005 Nucl. Phys. B (Proc. Suppl.)* **143** 487.
Danilov et. al. 2005 *Radiochemistry* **47** 487-93.
Motta D, Buck C, Hartmann F X, Lasserre T, Schoenert S and Schwan U 2005 *Nucl. Instrum. And Meth. A* **547** 368-88.

[15] Hartmann F X, 1985 Ph.D. Thesis, Princeton University, Princeton.

[16] The ball and stick chemical structure is courtesy of Jmol: an open-source Java viewer for chemical structures in 3D: //www.jmol.org.

[17] Buck C, Hartmann F X, Motta D and Schoenert 2007 *Chem. Phys. Letts.* **435** 252-6.
Buck C, Hartmann F X, Lasserre T, Motta D, Schoenert S and Schwan U 2004 *J. Lumin.* **106** 57-67.

[18] Aberle C, Optimierung der Fluoreszenzcharakteristik von Flussigszintillatoren des Double Chooz Reaktorneutrinoexperiments, 2008 Diplomarbeit at Max-Planck-Institut Kernphysik, Universitaet Heidelberg, Heidelberg.

[19] Buck C 2006 Journal of Physics: Conference Series **39** 291-3.

[20] Mention G 2005 *Nucl. Phys. B (Proc. Suppl.)* **145** 182-5.

[21] Buck C 2004 Development of metal loaded liquid scintillator for future detectors to investigate neutrino properties, Ph. D. Dissertation, University of Heidelberg, May 2004.

Carolina International Symposium on Neutrino Physics IOP Publishing
Journal of Physics: Conference Series **173** (2009) 012024 doi:10.1088/1742-6596/173/1/012024

First results of the ANTARES neutrino telescope

C Bigongiari on behalf of the ANTARES Collaboration

IFIC - Instituto de Física Corpuscolar, Edificio Institutos de Investigación de Paterna, CSIC - Universitat de València, Apartado de Correos 22085, E-46071, Valencia, Spain

E-mail: ciro.bigongiari@ific.uv.es

Abstract. The ANTARES Collaboration successfully completed in May 2008 the deployment of an underwater neutrino detector in the Mediterranean Sea, offshore the France coast, at 2475 m b.s.l. The main purpose of this experiment is the detection of high energy neutrinos produced in astrophysical sources. Neutrinos being neutral, stable and weakly interacting particles can travel undeflected trough the Universe reaching the Earth even from the inner core of very distant objects. They are therefore very powerful messengers which can provide us invaluable information about processes which are hardly accessible with other messengers like photons. The detection of astrophysical neutrinos is very challenging due to the very small neutrino interaction cross-section and the huge background produced by other cosmic rays. Very large instrumented volumes and a very efficient shielding are therefore needed to detect astrophysical neutrinos. Moreover a very good angular resolution is mandatory to trace detected neutrinos back to their origin. ANTARES with 0.04 km^2 muon effective area at $E_\nu > 10$ TeV shielded by more than 2000 m of water and an angular resolution of 0.3° at $E_\nu > 10$ TeV, perfectly fits these requirements. The ANTARES deployment started in 2006 and many data have been already collected with a partial detector. The detector in its final configuration is described and preliminary results of data analysis, especially about calibration issues, are shown. The completion of the ANTARES detector paved the way towards an even larger submarine neutrino telescope in the Mediterranean Sea like the one planned by the KM3NeT project.

1. Introduction

All astrophysical objects have been observed for centuries by naked-eye or by astronomical telescopes exploiting the visible light which reach the Earth. In the 20th century new detectors exploiting different ranges of the electromagnetic spectrum, from radio to gamma-rays, have been developed to study the Universe. Whenever a new detector came into play many new phenomena and completely new classes of astrophysical objects popped up. We can reasonably believe that detectors exploiting completely different astronomical messengers, like neutrinos, would provide us with at least the same number of discoveries.

Neutrinos are stable particles, like photons, so they don't decay while traveling to the Earth. They are also neutral particles, again like photons, so they are not deflected by the magnetic fields they go through along their path towards the Earth. Contrary to photons, neutrinos don't interact electromagnetically with the photons surrounding neutrino sources nor with the extragalactic background light. They can easily reach the Earth unabsorbed. Neutrinos are therefore ideal astronomical messengers which can provide us unaltered information from the very heart of their sources over cosmological distances.

We do know that astrophysical objects produce neutrinos because we can detect low energy neutrinos coming from the Sun [1, 2]. Moreover a burst of MeV neutrinos from SuperNova

SN 1987A was detected in 1987 by the KAMIOKANDE [3] and IMB [4] experiments. So far no astrophysical source of neutrinos above few GeV has ever been identified but physicists are confident that high energy neutrino sources exist because gamma-rays with such energies have already been detected from many astrophysical objects. A deep connection between high energy gamma-ray emission and neutrino production exists according to the "beam dump" model: protons or nuclei accelerated to relativistic energies interact with matter or photons which surround the source producing charged and neutral pions which in turn decay into neutrinos and gammas respectively [5].

$$\begin{array}{lllll} \mathrm{p} + \mathrm{X} \longrightarrow & \pi^{\circ} & + & \pi^{\pm} + \cdots \\ & \downarrow & & \downarrow \\ & \gamma\gamma & & \mu^{\pm} + \nu_{\mu}(\bar{\nu}_{\mu}) \\ & & & \downarrow \\ & & & e^{\pm} + \nu_e(\bar{\nu}_e) + \bar{\nu}_{\mu}(\nu_{\mu}) \end{array}$$

According to this scenario the flux of neutrinos from gamma-ray sources should be of the same order of the gamma-ray one. The neutrino flux could be even larger than the gamma-ray one in the case of very obscured sources, that is sources surrounded by a lot of matter or photons which absorb a large fraction of emitted gamma-rays while being nearly transparent for neutrinos [5]. There are also pure electromagnetic models for gamma-ray emission which do not foresee sizable neutrino flux but these models are disfavored for some sources. The combined radio, X-rays, and gamma-rays observation of the shell-type SuperNova Remnant RX J1713.7-3946 for example favors hadronic models [6] (but also pure leptonic process are possible [7]). The "orphan flares", that is a gamma-ray flares not associated with any X-ray flare, observed for the Blazar 1ES 1959+650 [8] are difficult to account for in purely leptonic models. Moreover it is well known that very high energy protons reach the Earth too, and recent results by AUGER [9] start to shed light over the sites of their production. It seems very unlikely that the high energy proton sources do not produce neutrinos as well.

The list of possible neutrino sources comprises both galactic and extragalactic objects. Concerning the galactic ones, the most promising are SuperNova Remnants, Pulsars and Micro-quasars, while the most probable extragalactic neutrino sources are Active Galactic Nuclei and Gamma-ray Bursts. The last ones are thought to emit neutrinos in short bursts. It would be helpful to spot their neutrino signal because a very short integration time $\sim 1 - 100\ s$ would eliminate nearly all background events.

Cosmic neutrinos can be produced also by self-annihilation of Dark Matter candidates like neutralinos. These particles lose their energy by elastic scattering and accumulate in massive objects, like the Sun or the Galactic Center, pulled by gravitational force. A neutrino telescope, therefore, can be used also for the indirect search of Dark Matter.

In spite of many efforts neutrino astronomy still lags behind gamma-ray astronomy due to the enormous difficulties encountered. Cosmic neutrinos are hard to detect and to distinguish from the background produced by the interaction of other cosmic rays with atmospheric nuclei. To detect a statistically significant amount of cosmic neutrinos in a reasonable time ($\sim$ 1 year), a very large instrumented volume is needed, due to the very low neutrino interaction cross section $\sigma_\nu \approx 5 \times 10^{-36}$ cm^2 at $E_\nu = 1$ TeV. Moreover, cosmic neutrinos have to be separated from the particles, especially muons and neutrinos, produced by the interaction of other cosmic-rays with atmospheric nuclei. The only known way to reduce this background is to bury the detector deep underground, under ice or underwater.

2. Detection principle

Cosmic neutrinos can interact via charged-current interaction in the medium which surrounds the detector producing a lepton which carries away a large fraction of neutrino energy. At high energy $E_\nu \gtrsim 10$ GeV the emitted lepton moves along a direction very close to the neutrino one. Muon neutrinos with energy $E_\nu > 30$ GeV produce muons whose range is $R_\mu > 100\ m$ and reach 10 km for PeV neutrinos. These muons can go through the detector even if produced far away. Since they travel at speed greater than the speed of light in sea-water they induce the emission of a Cherenkov light conical wave-front whose axis is the muon trajectory, see figure 1.

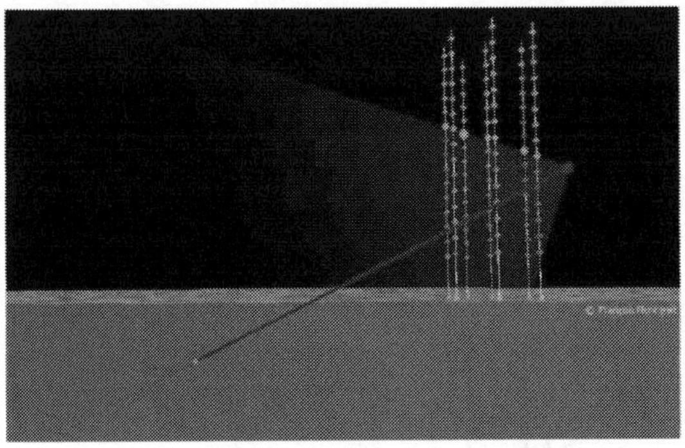

Figure 1. Detection principle of an underwater neutrino telescope. A muon neutrino (red line) interacting close to the detector produces an ultra relativistic (blue line) muon which induces the emission of Cherenkov light. This light can be detected by a 3D array of photon-detectors. Other kinds of events are detectable but harder to reconstruct.

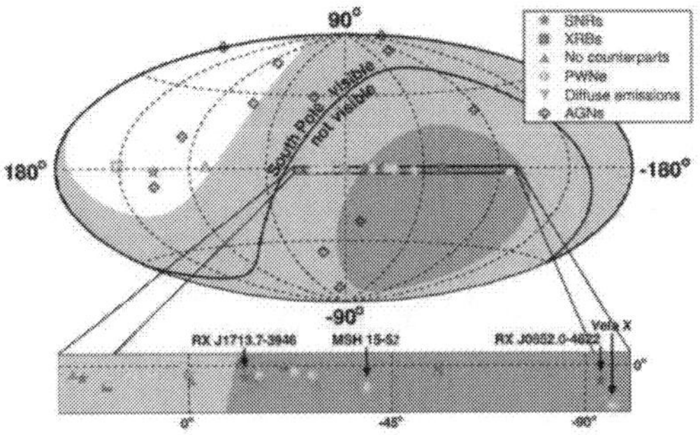

Figure 2. Field of view of the ANTARES neutrino telescope. Blue shaded area is observed for at least 75% of the time, while the light blue shaded area is observed for at least 25% of the time. The inset shows an enlarged view of the galactic plane region where many neutrino source candidates are present.

A 3-dimensional array of photon detectors can record the arrival times of Cherenkov photons at different points of the array. The muon track can be reconstructed knowing the position of the photon detectors and exploiting the geometry of Cherenkov light emission. Very good time and position resolution are needed to achieve a good angular resolution for muon tracks.

Electron neutrinos produce instead electrons that immediately interact again producing an electromagnetic shower. These showers induce Cherenkov emission too but in this case the determination of the primary electron track is much harder due to the smearing of the Cherenkov light front induced by the showering process. Moreover, these showers vanish very quickly and can be detected only if the primary interaction happens inside the detector itself or nearby.

Tau neutrinos produce τ particles which decay again very quickly due to the very short lifetime. The track length is too short for the track reconstruction unless the τ energy is $E_\tau \gtrsim 10^6$ GeV. In this case the event is indistinguishable from a muon one with lower energy. At lower energies possible signatures of τ events are the so-called "double-bang" and "lollipop" events. In the first case both particle showers, produced at ν_τ interaction point and at τ decay point, are inside the detector as well as the τ track. In the second case only one of the two showers and part of the τ track are inside the detector. These events will be very rare in ANTARES due to the limited extension of the detector.

3. The ANTARES detector

The ANTARES detector has been successfully deployed between March 2006 and May 2008 at 2475 m depth 40 km off La-Seine-Sur-Mer (French Mediterranean cost, close to Toulon) at 42°50′ N, 6°10′ E location. From this position ANTARES can observe the Northern Sky, see figure 2, and in particular a large fraction of the galactic plane and the galactic center itself.

This is a particularly interesting zone because many very high energy gamma-ray sources have been recently discovered in this region of the sky [10]. All these sources are promising neutrino source candidates. Moreover the Galactic Center can be an accumulation point for Dark Matter candidate particles, like the neutralinos, which could produce neutrinos by self-annihilation. A field of view covering the Galactic Center makes ANTARES suitable for the indirect Dark Matter search too.

The site had been long investigated before the beginning of the deployment. In particular, sea water properties have been extensively studied revealing a low light scattering [11] and an average noise trigger rate of 70 kHz per detection channel mainly due to bioluminescence and ^{40}K decay. The detector consists of 12 flexible mooring lines made of mechanically resistant electro-optical cables weighted to the sea bed and kept under tension by syntactic-foam buoys. A schematic view of the detector is shown in figure 3.

Figure 3. A schematic view of the ANTARES detector.The inset shows the basic element of the detector, the so-called storey.

The average distance between lines at sea-bed is about 70 m while their length is 450 m. Each line is equipped with 25 titanium frames, called storeys, which hold a triplet of optical modules (OM), pressure-resistant glass spheres housing a 10 inch hemispherical photomultiplier (Hamamtsu R7081-20) shielded by a μ-metal cage, the high voltage power supply, a LED system used to monitor the transit time of the photomultiplier and the associated electronics [12]. The storey holds the Local Control Module (LCM) too, a titanium cylinder housing the digitization electronics, the trigger discriminators and a compass/tilt-meter to measure the inclination and rotation of the storey. OMs are 120° spaced and oriented at 45° downward with respect to the vertical axis to ensure the maximum sensitivity to upward moving Cherenkov fronts. The inset of figure 3 shows a schematic view of a storey. The distance between adjacent storeys is 14.5 m and the lowest storey is placed at 100 m above the sea level to avoid the most turbid layer. One out of every five storeys is equipped with an optical beacon, a well controlled pulsed light, used for timing calibration purposes and five storeys of each line are equipped with a hydrophone which allows the measurement of these storey positions by acoustic triangulation.

An instrumented line is also present on site to monitor environmental parameters, like sea water temperature and salinity, sound velocity and the speed of sea current. A precise measurement of all these parameters is needed for an optimum track reconstruction.

All lines are connected through electro-optical cables to the junction-box which, in turn, is connected to the on-shore station by the 40 km long main electro-optical cable. The lines have been connected to the junction-box after their deployment by a remote operated submarine vehicle.

4. Trigger system

The ANTARES collaboration decided to maximize the flexibility of the data acquisition system choosing a large bandwidth connection with the on-shore station and applying to the events the less stringent trigger condition compatible with this bandwidth. In this way, only a very preliminary selection of the events, aimed at eliminating only a large fraction of events due to ^{40}K decay and the bioluminescence, is applied off-shore (L0 trigger) while a more stringent filtering is performed on-shore by a computer farm. The digitization of the OM signal is triggered when the pulse amplitude crosses a threshold set to a fraction (typically 0.3) of the single photoelectron average amplitude. Whenever this happens all data are sent to the on-shore station where data are firstly processed by the so-called DataFilter program which looks for correlated hits consistent with a specific physics signal. The DataFilter program incorporates different algorithms for different physics signals, like standard muon filter, magnetic monopole filter, optical beacon filter and also directional filters to look for events coming from specific positions in the sky (GRB filter, Center of Galaxy filter,etc.). The standard muon filter requires at least 2 L0 triggers in the same storey within 20 ns or one single L0 trigger with a pulse charge greater than 3 photoelectrons to consider one storey and at least 5 storeys whose hits are causally connected. The average L0 trigger rate is about 70 kHz for each OM while the overall event rate after the DataFilter selection is about 2 Hz.

5. Data Acquisition system

Each OM signal is processed by two Analogue Ring Samplers (ARS) in a token-ring configuration to minimize the dead time. Six ARSs measure arrival time and charge of the pulses for the three OMs present in a storey. Whenever the L0 trigger condition is fulfilled these data are sent through a 100 Mb/s link to a modified LCM (MLCM) which collects data from five storeys, merges them by an Ethernet switch and sends them through 1 Gb/s link to the string control module (SCM) at the bottom of the string, see figure 4. The SCM collects the data from all MLCMs in one line and multiplexes them onto one single optical fiber by Dense Wavelength Division Multiplexing technique (DWDM). The DWDM network enables the transmission of different streams of data along a single fiber using different wavelengths. In this way, data reach the junction-box and then the on-shore station where they are filtered and processed by a computer farm. A more detailed description of the ANTARES data acquisition system can be found in [13].

6. Detector Calibration

Good angular resolution is crucial to point back detected neutrinos to their sources. The pointing accuracy is closely related to the precision in the determination of the arrival time of Cherenkov photons at the OMs. Detector strings are not rigid but move slowly due to underwater currents, hence the second essential element to achieve a good angular resolution is precise monitoring of the OM positions. The ANTARES detector is equipped with redundant timing and positioning calibration systems which ensure a 0.5 ns relative time resolution and 0.2 m precision on OM position measurements. Actually, due to chromatic dispersion and scattering of Cherenkov light in sea water the attainable precision on the arrival time of photons at OMs depends on

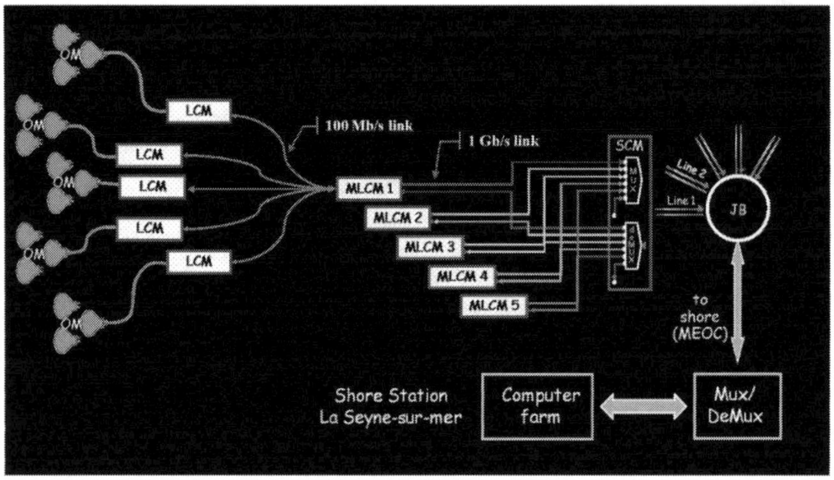

Figure 4. A schematic view of the data acquisition chain for one string.

the distance between the OM itself and the particle track. However, due to light absorption longer distances are highly disfavored and for distances usually involved in track reconstruction the time accuracy is about 1 ns. This value is well matched with the accuracy on the position measurement because the photon velocity is about 0.2 m/ns in sea-water: these two uncertainties contribute more or less equally to the uncertainty on the track direction.

6.1. Timing calibration

Relative time offsets between OMs of each line are measured before the deployment by illuminating them with a laser plus fibers system and measuring the arrival time of OM signals at read-out electronics (LCM). Once deployed, these offsets can be verified and monitored using the optical beacon system [14]. Optical beacons are controlled pulsed light sources which emit short intense light pulses at well-known times. Measuring the arrival time of OM signals at LCM and knowing the optical beacon emission time, the offsets between different OMs can be precisely determined whenever needed. In this way the RMS of the time offset distribution can be reduced to 0.7 ns and maintained stable over time, see figure 5. The time resolution of the ANTARES electronic chain can be measured by flashing OMs very close to an optical beacon. In this case, due to the large number of photons per flash which reach the OMs, all the contributions to the time resolution except the electronic one become negligible. Figure 6 shows the distribution of residuals $\Delta T = T_{Arrival} - T_{Emission} - T_{Direct}$ [1] for this setup, clearly demonstrating that a 0.4 ns time resolution has been achieved.

The ^{40}K present in the sea-water provides another cheap way for relative time calibration of close-by OMs. Electrons emitted in ^{40}K decay induce the emission of Cherenkov light which can reach two OMs of the same storey. The time distribution of coincidences between two OM signals shows a peak whose position is equal to the difference between the two OM time offsets. Same storey OMs can be inter calibrated in this way. The results obtained with ^{40}K method and the OB system show a remarkable agreement. A 20 MHz echo-based clock signal synchronized with a GPS is generated on-shore and sent to all LCMs. In this way the round-trip delay between the on-shore station and all LCMs can be measured and regularly monitored to avoid drifts. This clock provides the interline relative calibration and the absolute time calibration of the detector.

[1] $T_{Arrival}$ is the arrival time of photons at the OMs, $T_{Emission}$ is the emission time by the OB, and T_{Direct} is the time-of-flight of photons which travel strigthly from the OB to the OM.

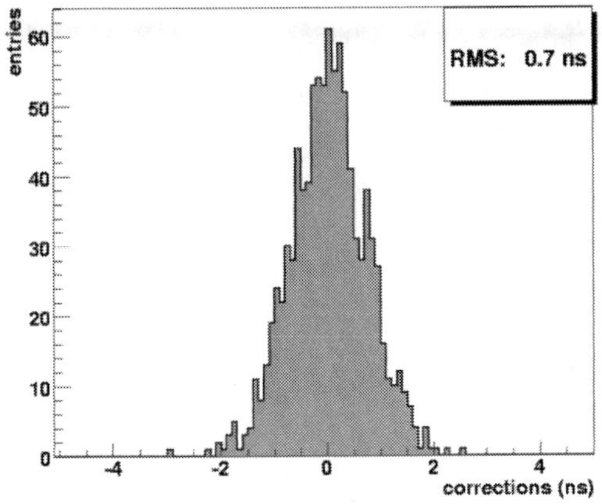

Figure 5. Distribution of time offsets between different detector channels as measured by the optical beacon system.

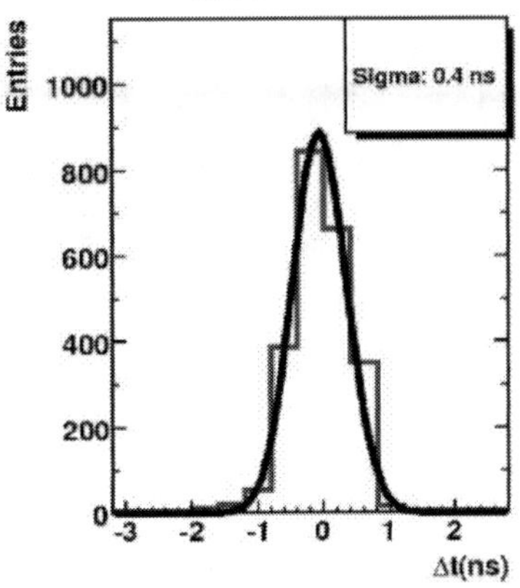

Figure 6. Distribution of time residuals for an OM very close to an OB. The standard deviation of 0.4 ns can be regarded as an estimation of the ANTARES electronic chain time resolution.

6.2. Positioning calibration
The storeys equipped with hydrophones receive sound pulses from three emitters fixed on the sea-bed. In this way, the positions of these storeys with respect to the emitters can be measured by triangulation. A precision on acoustic distance measurements of about 2 cm and an accuracy on the 3D hydrophones spatial positions better than 5 cm can be achieved [15]. Moreover, every storey is equipped with a tilt-meter/compass board which measures the local tilt angles with respect to the horizontal plane (pitch and roll), as well as the orientation with respect to the Earth Magnetic North (heading). Using all this information, a 3D mechanical model of the string which takes into account the weight and drag of all string elements can be fitted. In this way, the relative 3D position of all OMs can be measured with an accuracy better than 20 cm.

7. Comparison with MC

The data collected with a partial detector during the completion of the deployment allow a preliminary comparison between real data and Monte Carlo simulations. In particular the large amount of data taken in homogeneous conditions with 5 strings are being used for this cross-checks. Although ANTARES is shielded by more than 2000 m of sea-water many so-called atmospheric muons reach its active volume. These are muons produced in the decay of charged mesons produced by the interactions of primary cosmic rays with atmospheric nuclei. Atmospheric muons constitute the major background for ANTARES mainly because they can be wrongly reconstructed as up-going particles. On the other hand these numerous particles are very useful to check the Monte Carlo simulation of the detector response to the passage of charge particles. Figure 7 and figure 8 show respectively the azimuth and zenith angle distributions of reconstructed tracks (without any quality cut) for data (black points) and Monte Carlo simulation (red solid line). The red shaded area shows the systematic uncertainty on Monte Carlo prediction due to uncertainties on primary cosmic-ray flux and on the hadronic interaction model. Taking into account systematic uncertainties, data and MC agree pretty well.

8. Expected performances

Once validated trough its comparison with real data, Monte Carlo simulation chain has been used to estimate the full detector performances. The angular resolution for neutrinos with

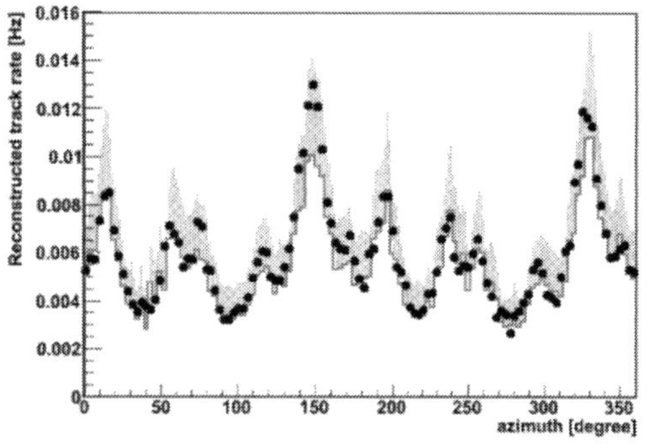

Figure 7. Azimuth angle distribution of reconstructed tracks (black points) compared to MC prediction (red solid line). The red shaded area shows a 40% systematic uncertainty on the MC prediction.

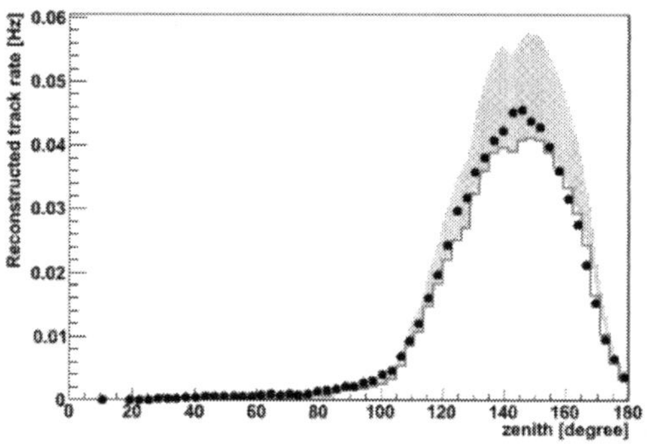

Figure 8. Zenith angle distribution of reconstructed tracks (black points) compared to MC prediction (red solid line). The red shaded area shows a 40% systematic uncertainty on the MC prediction.

$E_\nu > 10$ TeV results in 0.3°, see figure 9. Such a good angular resolution is extremely helpful for the search of point-like sources. Figure 10 shows the expected sensitivity for one year of data taking as a function of declination compared with flux upper limits already set by other experiments and the sensitivities of other neutrino telescopes.

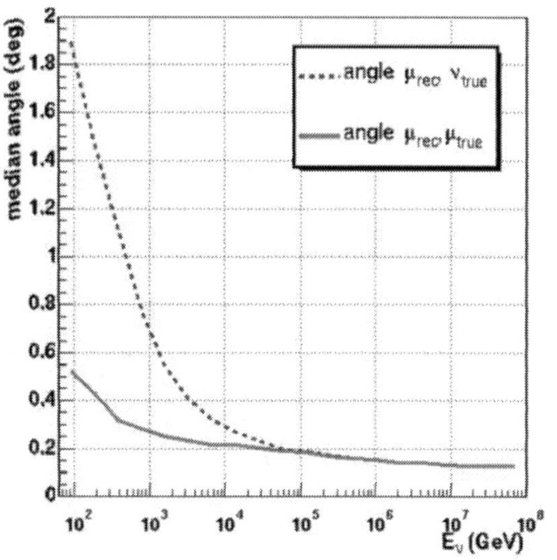

Figure 9. Angular resolution as a function of neutrino energy. Below 10 TeV the resolution is dominated by the angle between muon and neutrino trajectories. At higher energies the detector resolution dominates and the angular resolution improves as neutrino energy increases staying below 0.3°.

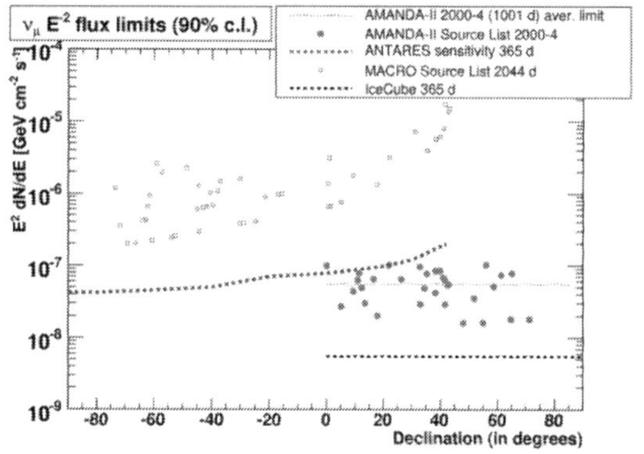

Figure 10. The ANTARES sensitivity for point-like sources for one year of data taking as a function of declination (blue dotted line). Magenta squares are 90% c.l. upper limits by MACRO [16]. Red circles are similar limits by AMANDA-II, while the red dotted line is the average limit [17]. The black dotted line shows the expected sensitivity for IceCube in one year [18].

9. Conclusions

The ANTARES collaboration successfully completed the deployment of a neutrino detector at 2475 m depth in the Mediterranean Sea. All the technological challenges involved in this project were met. An increasing fraction of the detector took data without major problems over the last two years. The analysis of the data acquired so far confirmed that the ANTARES detector can perform well within design specifications. In particular, the timing calibration system can achieve a 0.5 ns timing resolution while the acoustic positioning system can measure photon-detector positions with 20 cm accuracy. These values imply an angular resolution on neutrino tracks below 0.3° for $E_\nu > 10$ TeV and consequently very good sensitivity for the point-like source search. Thanks to its location ANTARES can observe the Northern Sky, complementing neutrino telescopes at South Pole. Notably a large fraction of the galactic plane, where many promising neutrino source candidates are present, can be observed for 75% of the observational time. The ANTARES detector is presently the neutrino telescope with the biggest instrumented volume and the largest effective area in the Northern Hemisphere representing a clear proof of concept for an even larger submarine neutrino telescope like the one pursued by the KM3NeT network [19].

Acknowledgments

The author acknowledges financial support from a Consejo Superior de Investigaciones Cientificas JAE-Doc research grant.

References

[1] Hirata K S *et al.* 1989 "Observation of ^8B solar neutrinos in the Kamiokande-II detector" *Phys. Rev. Lett.* **63** 16-19.
[2] Ahmad Q R *et al.* 2001 "Measurement of the Rate of $\nu_e + d \rightarrow p + p + e^-$ Interactions Produced by ^8B Solar Neutrinos at the Sudbury Neutrino Observatory" *Phys. Rev. Lett.* **87** 071301-06.
[3] Hirata K S *et al.* 1987 "Observation of a neutrino burst from the SuperNova SN 1987A" *Phys. Rev. Lett.* **58** 1490-93.
[4] Bionta R M *et al* 1987 "Observation of a neutrino burst in coincidence with SuperNova SN1987A in the Large Magellanic Cloud" *Phys. Rev. Lett.* **58** 1494-96.
[5] Halzen F and Hooper D 2002 "High-energy neutrino astronomy: the cosmic ray connection" *Rep. Prog. Phys.* **65** 1025-78.
[6] Berezhko E G and Völk H J 2006 "Theory of cosmic ray production in the SuperNova remnant RX J1713.7-3946" *Astron. Astrophys.* **451** 981-90.
[7] Plaga R 2008 "Arguments against a dominantly hadronic origin of the VHE radiation from the SuperNova remnant RX J1713-3946" *New Astron.* **13** 73-76.
[8] Krawczynski H *et al.* 2004 "Multiwavelength observations of strong flares from the TeV Blazar 1ES 1959+650" *Astrophys. J.* **601** 151-64.
[9] Abraham J *et al.* 2007 "Correlation of the highest-energy cosmic rays with nearby extragalactic objects" Science **318** 938-43.
[10] Aharonian F *et al.* 2006 "The H.E.S.S. survey of the Inner Galaxy in very high-energy gamma-rays" *Astrophys. J.* **636** 777-97.
[11] Aguilar J A *et al.* 2005 "Transmission of light in deep sea water at the site of the ANTARES Neutrino Telescope" *Astropart. Phys.* **23** 131-55.
[12] Amram P *et al.* 2003 "The ANTARES optical module" *Nucl. Instrum. Methods* A **484** 369-83.
[13] Aguilar J A *et al.* 2007 "The data acquisition system for the ANTARES Neutrino Telescope" *Nucl. Instrum. Methods* A **570** 107-16.
[14] Ageron M *et al.* 2007 "The ANTARES Optical Beacon System" *Nucl. Instrum. Methods* A **578** 498-509.
[15] Fehr F "Calibration systems and methods for the ANTARES neutrino telescope" *Preprint astro-ph/0711.2254*.
[16] Ambrosio M *et al.* 2001 "Neutrino astronomy with MACRO detector" *Astrophys. J.* **456** 1038-54.
[17] Ahrens J *et al.* 2004 "Search for extraterrestrial point sources of neutrinos with AMANDA-II" *Phys. Rev. Lett.* **92** 071102.
[18] Ahrens J *et al.* 2004 " Sensitivity of IceCube detector to astrophysical sources of high energy muon neutrinos" *Astropart. Phys.* **20** 507-32.

[19] de Wolf E 2008 "KM3NeT, a new generation neutrino telescope" *Nucl. Instrum. Methods* A **588** 86-91.

Journal of Physics: Conference Series **173** (2009) 012025 doi:10.1088/1742-6596/173/1/012025

Concluding Remarks: Neutrino Mixing, Majorana CP-violation, $(\beta\beta)_{0\nu}$-Decay and Beyond

S. T. Petcov [1]

Scuola Internazionale Superiore di Studi Avanzati and INFN, I-34014 Trieste, Italy

E-mail: `petcov@sissa.it`

Abstract. The problem of determination of the nature - Dirac or Majorana, of massive neutrinos is discussed. The physics potential of experiments, searching for $(\beta\beta)_{0\nu}$−decay, for providing information on the type of ν-mass spectrum, absolute scale of ν- masses and on the Majorana CP-violating phases in the PMNS neutrino mixing matrix U, is reviewed. The possibility that the CP-violation necessary for the generation of the baryon asymmetry of the Universe is due exclusively to the Majorana CP-violating phase(s) in U, is also briefly discussed.

1. Introduction

It is both an honor and a pleasure for me to speak at this Symposium organized in honor of Frank Avignone, Ettore Fiorini and the late Peter Rosen. I met Frank first in 1986, Ettore - in 1979 and Peter Rosen - in 1983. However, their names were familiar to me from some of the publications I was reading in the middle of the 1970's when I was a Ph. D. student. I was interested in neutrino physics and Frank, Ettore and Peter were actively working in this field. Their "passion" for neutrinos did not diminish with time. For Peter Rosen it continued until his last days. All three of them marked the field with their achievements: Frank and Ettore - in experimental neutrino physics, and Peter - in the theory of neutrinos.

Frank played an important role in the interpretation of the results of the $\bar{\nu}_e - e^-$ elastic scattering experiment of Gurr, Reines and Sobel at Savannah River Reactor. In 1978 he became interested in improving the sensitivity of Ge detectors for fundamental physics. In the early 1980s Frank started to work with R. Brodzinski and the Pacific Northwest Laboratory Group in lowering the background in Ge detectors by several orders of magnitude. In 1987 the group had published the first terrestrial sensitive search for cold dark matter (CDM), which eliminated massive Dirac neutrinos as a possible dominant component of the CDM in the halo of our galaxy. Later Frank with collaborators used these unique detectors to search for axions from the Sun. Starting in the late 80's, Frank has been actively involved in experiments searching for neutrinoless double beta ($(\beta\beta)_{0\nu}$-) decay: IGEX, CUORICINO, CUORE, MAJORANA. IGEX, in which Frank played a leading role, and CUORICINO, distinguished themselves with the limits obtained, while CUORE and MAJORANA belong to the next generation of detectors under preparation at present. Our hopes to determine the nature of neutrinos and thus resolve one of the most challenging among the fundamental problems in neutrino physics, are associated

[1] Also at: INRNE, Bulgarian Academy of Sciences, 1784 Sofia, Bulgaria.

with these new detectors. The MAJORANA project is to a large extent (if not entirely) due to Frank.

Ettore Fiorini's involvement in neutrino physics began in 1963, when he started the first $(\beta\beta)_{0\nu}$-decay experiment with ^{76}Ge. In 1970's Ettore participated in one of the most remarkable experiments in neutrino physics - "Gargamelle" at CERN, in which the neutral current weak interaction was discovered. Later in the second half of 1980's Ettore played a very active role in the GALLEX solar neutrino experiment, in which for the first time the lower energy component of the solar neutrino flux was observed. However, he never stopped thinking about how to improve the sensitivity of the experimental searches for $(\beta\beta)_{0\nu}$-decay. Intrigued by a suggestion in a preprint by G. Mitselmakher from JINR, Dubna, Russia, that the electrons from the $(\beta\beta)_{0\nu}$-decay might be detected thermally, Ettore together T. Niinikoski devised an approach that was applied in the first successful physics experiment using thermal spectrometers. In this experiment a new lower limit on the $(\beta\beta)_{0\nu}$-decay lifetime of ^{130}Te was obtained. The novel technique of thermal detection of beta-decay electrons was further developed and implemented in the successful CUORICINO experiment. It will be used in the CUORE detector which is under construction. Ettore is the leader of both these experiments.

Peter Rosen made pioneering contributions in the theory of $(\beta\beta)_{0\nu}$-decay. I was lucky to have met him in the early 1980's: the discussions with him were important for my scientific development. I also had the privilege to work with Peter on few scientific projects. As a curiosity let me mention that the results of one of these projects, done in collaboration with Boris Kayser, were widely known and quoted, but were never published. It is among the most quoted unpublished scientific works. Peter had also a strong indirect influence on my subsequent work with collaborators on the neutrino physics aspects of the $(\beta\beta)_{0\nu}$-decay.

The above sketchy description of the scientific activity of Frank, Ettore and Peter is far from being complete, of course. Frank and Ettore trained and continue to train young scientists, held important administrative positions, were (and still are) members of various scientific policy Committees, etc. Peter, in particular, held a high position in DOE, with responsibilities directly related to the funding of High Energy Physics.

Since many years Frank and Ettore are the driving force in the field of experimental searches for $(\beta\beta)_{0\nu}$-decay. Most of the scientific achievements of Peter are related to the development of the theory of $(\beta\beta)_{0\nu}$-decay: for many years Peter was one of the very few leading experts in this field. In view of this most of my talk will be devoted to the $(\beta\beta)_{0\nu}$-decay. I will review the current understanding of the neutrino physics aspects of $(\beta\beta)_{0\nu}$-decay and the physics potential of the $(\beta\beta)_{0\nu}$-decay experiments.

Establishing whether the neutrinos with definite mass ν_j are Dirac fermions possessing distinct antiparticles, or Majorana fermions, i.e., spin 1/2 particles that are identical with their antiparticles, is of fundamental importance for understanding the origin of ν-masses and mixing and the underlying symmetries of particle interactions (see, e.g. [1]). Let us recall that the neutrinos ν_j with definite mass m_j will be Dirac fermions if particle interactions conserve some additive lepton number, e.g., the total lepton charge $L = L_e + L_\mu + L_\tau$. If no lepton charge is conserved, the neutrinos ν_j will be Majorana fermions (see, e.g. [2]). The massive neutrinos are predicted to be of Majorana nature by the see-saw mechanism of neutrino mass generation [3], which also provides an attractive explanation of the smallness of neutrino masses and, through the leptogenesis theory [4], of the observed baryon asymmetry of the Universe. The observed patterns of neutrino mixing and of neutrino mass squared differences driving the solar and the dominant atmospheric neutrino oscillations can be related to Majorana massive neutrinos and the existence of an *approximate* symmetry in the lepton sector corresponding to the conservation of the *non-standard* lepton charge $L' = L_e - L_\mu - L_\tau$ [5]. Determining the nature (Dirac or Majorana) of massive neutrinos ν_j is one of the fundamental and most challenging problems in the future studies of neutrino mixing [1].

Extensive studies have shown that the only feasible experiments having the potential of establishing the Majorana nature of massive neutrinos at present are the $(\beta\beta)_{0\nu}$-decay experiments searching for the process $(A,Z) \rightarrow (A,Z+2)+e^- +e^-$ (for reviews see, e.g. [2,6–8]). The observation of $(\beta\beta)_{0\nu}$-decay and the measurement of the corresponding half-life with sufficient accuracy, would not only be a proof that the total lepton charge is not conserved, but might provide also unique information on the i) type of ν_j mass spectrum, [9, 10] (see also [11]), ii) absolute scale of ν_j masses (see, e.g. [12]), and iii) Majorana CP-violating (CPV) phases [9, 13–15] (see also, e.g. [16–18]).

In what follows we will consider the case of 3-ν mixing, for which we have compelling evidence from experiments with solar, atmospheric, reactor and accelerator neutrinos (see, e.g. [1,19] for extensive lists of references):

$$\nu_{l\mathrm{L}} = \sum_{j=1}^{3} U_{lj}\, \nu_{j\mathrm{L}}, \quad l = e, \mu, \tau, \tag{1}$$

where $\nu_{j\mathrm{L}}$ is the field of neutrino ν_j having a mass m_j and U is the Pontecorvo–Maki–Nakagawa–Sakata (PMNS) mixing matrix [20], $U \equiv U_{\mathrm{PMNS}}$. Assuming that the massive neutrinos ν_j are Majorana particles, we will use the standard parametrisation of the PMNS matrix:

$$U_{\mathrm{PMNS}} = \begin{pmatrix} c_{12}c_{13} & s_{12}c_{13} & s_{13}e^{-i\delta} \\ -s_{12}c_{23} - c_{12}s_{23}s_{13}e^{i\delta} & c_{12}c_{23} - s_{12}s_{23}s_{13}e^{i\delta} & s_{23}c_{13} \\ s_{12}s_{23} - c_{12}c_{23}s_{13}e^{i\delta} & -c_{12}s_{23} - s_{12}c_{23}s_{13}e^{i\delta} & c_{23}c_{13} \end{pmatrix} \mathrm{diag}(1, e^{i\frac{\alpha}{2}}, e^{i\frac{\beta}{2}}) \tag{2}$$

where $c_{ij} \equiv \cos\theta_{ij}$, $s_{ij} \equiv \sin\theta_{ij}$, $\theta_{ij} = [0, \pi/2]$, $\delta = [0, 2\pi]$ is the Dirac CP-violation (CPV) phase and α, β are two Majorana CPV phases [21]. The existing neutrino oscillation data allow us to determine the parameters which drive the solar neutrino and the dominant atmospheric neutrino oscillations, $\Delta m^2_{\odot} = \Delta m^2_{21}$, $\sin^2\theta_{12}$, and $|\Delta m^2_{\mathrm{A}}| = |\Delta m^2_{31}| \cong |\Delta m^2_{32}|$, $\sin^2 2\theta_{23}$, with relatively good precision, and to obtain rather stringent limits on the CHOOZ angle [22] θ_{13} (see, e.g. [23, 24]). The best fit values and the 95% C.L. allowed ranges of $|\Delta m^2_{\mathrm{A}}|$, $\sin^2 2\theta_{23}$, $\Delta m^2_{\odot}$ and $\sin^2\theta_{\odot}$ read:

$$(|\Delta m^2_{\mathrm{A}}|)_{\mathrm{BF}} = 2.5 \times 10^{-3}\ \mathrm{eV}^2, \quad 2.1 \times 10^{-3}\ \mathrm{eV}^2 \le |\Delta m^2_{\mathrm{A}}| \le 2.9 \times 10^{-3}\ \mathrm{eV}^2, \tag{3}$$

$$(\sin^2 2\theta_{23})_{\mathrm{BF}} = 1.0, \quad \sin^2 2\theta_{23} \ge 0.90, \tag{4}$$

$$(\Delta m^2_{\odot})_{\mathrm{BF}} = 8.0 \times 10^{-5}\ \mathrm{eV}^2, \quad 7.3 \times 10^{-5}\ \mathrm{eV}^2 \le \Delta m^2_{\odot} \le 8.5 \times 10^{-5}\ \mathrm{eV}^2, \tag{5}$$

$$(\sin^2\theta_{\odot})_{\mathrm{BF}} = 0.30, \quad 0.26 \le \sin^2\theta_{\odot} \le 0.36\,. \tag{6}$$

A combined 3-ν oscillation analysis of the global data gives [23, 24]

$$\sin^2\theta_{13} < 0.040\ (0.056) \quad \text{at} \quad 95\%\ (99.73\%)\ \text{C.L.} \tag{7}$$

The existing data allow a determination of, e.g. $\Delta m^2_{\odot}$, $\sin^2\theta_{\odot}$ and $|\Delta m^2_{\mathrm{A}}|$ at 3σ with an error of approximately 12%, 24%, and 28%, respectively. Future oscillation experiments will improve considerably the precision on these basic parameters: the indicated 3σ errors can be reduced to [25] 4%, 10% and 12%. There are several proposals for reactor experiments which could improve the current limit on $\sin^2\theta_{13}$ by a factor of (5-10) (see, e.g. [26]). Two of these experiments - Double CHOOZ and Daya Bay [27, 28], are under preparation at present. Oscillation experiments will provide data on the Dirac CPV phase δ as well - the latter is a source of CP-violation in ν-oscillations (see, e.g., [29, 30]). However, these experiments cannot give information on the absolute scale of neutrino masses. They are insensitive to the nature - Dirac or Majorana, of massive neutrinos ν_j and, correspondingly, to the Majorana CPV phases present in U [21, 31].

If ν_j are Majorana fermions, getting experimental information about the Majorana CPV phases in $U_{\rm PMNS}$ would be a remarkably difficult problem [9, 12, 14, 15, 32]. The phases α and β can affect significantly the predictions for the rates of (LFV) decays $\mu \to e+\gamma$, $\tau \to \mu+\gamma$, etc. in a large class of supersymmetric theories with the see-saw mechanism of ν-mass generation [33]. The Majorana CPV phase(s) in the PMNS matrix can play the role of the leptogenesis CPV parameter(s) at the origin of the baryon asymmetry of the Universe [34, 35] (see also [36]).

2. Majorana Neutrinos and $(\beta\beta)_{0\nu}$–Decay

Under the assumptions of 3-ν mixing, of massive neutrinos ν_j being Majorana particles, and of $(\beta\beta)_{0\nu}$-decay generated *only by the (V-A) charged current weak interaction via the exchange of the three Majorana neutrinos* ν_j having masses $m_j \lesssim$ few MeV, the $(\beta\beta)_{0\nu}$-decay amplitude has the form (see, e.g. [9]): $A(\beta\beta)_{0\nu} \cong <m> M$, where M is the corresponding nuclear matrix element (NME) which does not depend on the neutrino mixing parameters, and

$$|<m>| = \left| m_1 |U_{\rm e1}|^2 + m_2 |U_{\rm e2}|^2 e^{i\alpha} + m_3 |U_{\rm e3}|^2 e^{i\beta} \right| , \quad (8)$$

is the effective Majorana mass in $(\beta\beta)_{0\nu}$-decay, $|U_{\rm e1}|=c_{12}c_{13}$, $|U_{\rm e2}|=s_{12}c_{13}$, $|U_{\rm e3}|=s_{13}$. In the case of CP-invariance one has [37], $\eta_{21} \equiv e^{i\alpha}=\pm 1$, $\eta_{31} \equiv e^{i\beta}=\pm 1$, $\eta_{21(31)}$ being the relative CP-parity of Majorana neutrinos $\nu_{2(3)}$ and ν_1.

The predicted value of $|<m>|$ depends strongly on the type of neutrino mass spectrum [9, 10]. Depending on the sign of the neutrino mass squared difference which drives the atmospheric neutrino oscillations, $\Delta m^2_{\rm A} \equiv \Delta m^2_{31} \cong \Delta m^2_{32}$, the spectrum (in the standard convention) can be with *normal neutrino mass ordering*, $m_1 < m_2 < m_3$ ($\Delta m^2_{\rm A} \equiv \Delta m^2_{31} > 0$), or with *inverted neutrino mass ordering*, $m_3 < m_1 < m_2$ ($\Delta m^2_{\rm A} \equiv \Delta m^2_{32} < 0$). Depending on the sgn($\Delta m^2_{\rm A}$) and the value of the lightest neutrino mass, i.e., the absolute ν-mass scale, min(m_j) $\equiv m_{\rm min}$, the ν-mass spectrum can be

- *Normal Hierarchical (NH)*: $m_1 \ll m_2 \ll m_3$, $m_2 \cong (\Delta m^2_{\odot})^{\frac{1}{2}} \sim 0.009$ eV, $m_3 \cong |\Delta m^2_{\rm A}|^{\frac{1}{2}} \sim 0.05$ eV;
- *Inverted Hierarchical (IH)*: $m_3 \ll m_1 < m_2$, with $m_{1,2} \cong |\Delta m^2_{\rm A}|^{\frac{1}{2}} \sim 0.05$ eV;
- *Quasi-Degenerate (QD)*: $m_1 \cong m_2 \cong m_3 \cong m$, $m_j^2 \gg |\Delta m^2_{\rm A}|$, $m \gtrsim 0.10$ eV.

Information on the absolute scale of neutrino masses can be derived in ^{3}H β-decay experiments [38, 39] and from cosmological and astrophysical data. The most stringent upper bounds on the $\bar{\nu}_e$ mass were obtained in the Troitzk [38] and Mainz [39] experiments:

$$m_{\bar{\nu}_e} < 2.3 {\rm eV} \quad \text{at } 95\% \text{ C.L.} \quad (9)$$

We have $m_{\bar{\nu}_e} \cong m_{1,2,3}$ in the case of the QD ν-mass spectrum. The KATRIN experiment [39] is planned to reach a sensitivity of $m_{\bar{\nu}_e} \sim 0.20$ eV, i.e. it will probe the region of the QD spectrum. The CMB data of the WMAP experiment, combined with data from large scale structure surveys (2dFGRS, SDSS), lead to an upper limit on the sum of ν_j masses (see, e.g. [40]):

$$\sum_j m_j \equiv \Sigma < (0.4\text{–}1.7) \text{ eV} \quad \text{at } 95\% \text{ C.L.} \quad (10)$$

Data on weak lensing of galaxies, combined with data from the WMAP and PLANCK experiments, may allow Σ to be determined with an uncertainty of $\delta \sim 0.04$ eV [40].

It proves convenient to express [41] the three neutrino masses in terms of $\Delta m^2_{\odot}$ and $\Delta m^2_{\rm A}$, measured in neutrino oscillation experiments, and the absolute neutrino mass scale determined by min(m_j) [2]. In both cases of ν-mass spectrum with normal and inverted ordering one has (in the convention we use): $\Delta m^2_{\odot}=\Delta m^2_{21} > 0$, $m_2=(m_1^2 + \Delta m^2_{\odot})^{\frac{1}{2}}$. For normal ordering,

[2] For a detailed discussion of the relevant formalism see, e.g. [8, 9].

$\Delta m^2_{\rm A}{=}\Delta m^2_{31} > 0$ and $m_3{=}(m_1^2 + \Delta m^2_{\rm A})^{\frac{1}{2}}$, while if the spectrum is with inverted ordering, $m_{\rm MIN}{=}m_3$, $\Delta m^2_{\rm A}{=}\Delta m^2_{23} > 0$ and $m_1{=}(m_3^2 + \Delta m^2_{\rm A} - \Delta m^2_{\odot})^{\frac{1}{2}}$. Thus, given $\Delta m^2_{\rm A}$, $\Delta m^2_{\odot}$, $\theta_\odot$ and θ_{13}, $|{<}m{>}|$ depends on $min(m_j)$, Majorana phases α, β and the type of ν-mass spectrum.

The problem of obtaining the allowed values of $|{<}m{>}|$ given the constraints on the parameters following from ν-oscillation data, and more generally of the physics potential of $(\beta\beta)_{0\nu}$-decay experiments, was first studied in [41] and subsequently in a large number of papers[3]. Detailed analyses were performed more recently in [15, 18, 42]. The results are illustrated in Fig.

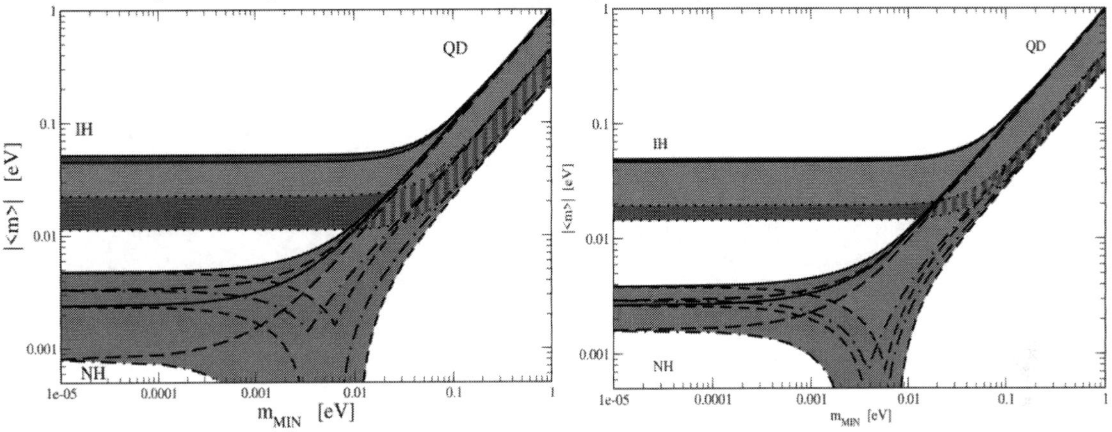

Figure 1. The value of $|{<}m{>}|$ as a function of $m_{\rm MIN} \equiv min(m_j)$, obtained using i) the 95% C.L. allowed ranges of $\Delta m^2_{\odot}$, $|\Delta m^2_{\rm A}|$, $\sin^2\theta_\odot$ and $\sin^2\theta_{13}$ (left panel), and ii) prospective 2σ uncertainty in $|{<}m{>}|$, corresponding to input 1-σ experimental errors in $\Delta m^2_\odot$ $\Delta m^2_{\rm A}$and $\sin^2\theta_\odot$ of 2%, 2% and 4% and $\sin^2\theta_{13} = 0.01 \pm 0.006$ (right panel). The regions shown in red/grey correspond to violation of CP-symmetry. (From [42].)

1. The main features of the predictions for $|{<}m{>}|$ are [9, 10, 12] (Fig. 1, left panel):
i) for NH spectrum, $|{<}m{>}| \cong |(\Delta m^2_\odot)^{\frac{1}{2}}\, s^2_{12} + (\Delta m^2_{\rm A})^{\frac{1}{2}}\, s^2_{13}\, e^{-i(\alpha-\beta)}| \lesssim 0.006$ eV;
ii) for IH spectrum, $|{<}m{>}| \cong (|\Delta m^2_{\rm A}|)^{\frac{1}{2}}\,(1 - \sin^2 2\theta_\odot\, \sin^2\frac{\alpha}{2})^{\frac{1}{2}}$,
thus $(|\Delta m^2_{\rm A}|)^{\frac{1}{2}}\cos 2\theta_\odot \lesssim |{<}m{>}| \lesssim (|\Delta m^2_{\rm A}|)^{\frac{1}{2}}$, or 0.013 eV $\lesssim |{<}m{>}| \lesssim$ 0.055 eV, the bounds corresponding to the CP-conserving values of $\alpha = 0;\ \pi$;
iii) for QD spectrum, $|{<}m{>}| \cong m_0\,(1 - \sin^2 2\theta_\odot\, \sin^2\frac{\alpha}{2})^{\frac{1}{2}}$, $m_0 \gtrsim |{<}m{>}| \gtrsim m_0\cos 2\theta_\odot \gtrsim 0.03$ eV, with $m_0 \gtrsim 0.1$ eV, $m_0 < 2.3$ eV [39] or $m_0 \lesssim 0.5$ eV [40].
For IH (QD) spectrum we have: $\sin^2(\alpha/2) \cong (1 - |{<}m{>}|^2/\tilde{m}^2)/\sin^2 2\theta_\odot$, $\tilde{m}^2 = |\Delta m^2_{\rm A}|\ (m_0^2)$. Thus, a measurement of $|{<}m{>}|$ (and m_0 for QD spectrum) allows us to determine α.

The experimental searches for $(\beta\beta)_{0\nu}$-decay have a long history [6]. The best sensitivity was achieved in the Heidelberg-Moscow ^{76}Ge experiment [43]: $|{<}m{>}| <$(0.35 - 1.05) eV (90% C.L.), where a factor of 3 uncertainty in the relevant NME (see, e.g. [44]) is taken into account. The IGEX collaboration has obtained [45]: $|{<}m{>}| < (0.33 - 1.35)$ eV (90% C.L.). A positive signal at $> 3\sigma$, corresponding to $|{<}m{>}| = (0.1{-}0.9)$ eV, is claimed to be observed in [46], while a more recent analysis reports evidence at 6σ for $(\beta\beta)_{0\nu}$-decay with $|{<}m{>}| = 0.32{\pm}0.03)$ eV [47]. Two experiments, NEMO3 (with ^{100}Mo and ^{82}Se) [48] and CUORICINO (with ^{130}Te) [49], designed to reach sensitivity to $|{<}m{>}| \sim (0.2 - 0.3)$ eV, obtained the following limits: $|{<}m{>}| < (0.7 - 1.2)$ eV [48] and $|{<}m{>}| < (0.16 - 0.68)$ eV [49] (90% C.L.), where estimated uncertainties in the NME are accounted for. Most importantly, a number of projects aim at sensitivity to $|{<}m{>}| \sim$(0.01–0.05) eV [50]: CUORE (^{130}Te), GERDA (^{76}Ge), SuperNEMO (^{100}Mo), EXO (^{136}Xe), MAJORANA (^{76}Ge), XMASS (^{136}Xe), etc. These experiments will probe the region

[3] An extensive list of references on the subject is given in [8].

corresponding to IH and QD spectra and test the positive result claimed in [46].

The existence of significant lower bounds on $|<m>|$ in the cases of IH and QD spectra [10], which lie either partially (IH spectrum) or completely (QD spectrum) within the range of sensitivity of the next generation of $(\beta\beta)_{0\nu}$-decay experiments, is one of the most important features of the predictions of $|<m>|$. These minimal values are given, up to small corrections, by $\Delta m^2_{\rm A} \cos 2\theta_\odot$ and $m_0 \cos 2\theta_\odot$. According to the combined analysis of the solar and reactor neutrino data [23,24], i) the possibility of $\cos 2\theta_\odot = 0$ is excluded at $\sim 6\sigma$, ii) the best fit value of $\cos 2\theta_\odot$ is $\cos 2\theta_\odot = 0.40$, and iii) at 95% C.L. one has $\sin^2\theta_{13} = 0$ (0.02), $\cos 2\theta_\odot \gtrsim 0.28$ (0.28). The quoted results on $\cos 2\theta_\odot$ together with the range of possible values of $|\Delta m^2_{\rm A}|$ and m_0 lead to the conclusion about the existence of significant and robust lower bounds on $|<m>|$ in the cases of IH and QD spectrum [10, 11]. At the same time one can *always* have $|<m>| \ll 10^{-3}$ eV in the case of spectrum with normal ordering [12]. As Fig. 1 indicates, $|<m>|$ cannot exceed $\sim$ 6 meV for NH ν-mass spectrum. This implies that $max(|<m>|)$ in the case of NH spectrum is considerably smaller than $min(|<m>|)$ for the IH and QD spectrum. This opens the possibility of obtaining information about the type of ν-mass spectrum from a measurement of $|<m>| \neq 0$ [10]. In particular, a positive result in the future $(\beta\beta)_{0\nu}$-decay experiments with $|<m>| > 0.01$ eV would imply that the NH spectrum is strongly disfavored (if not excluded). Prospective experimental errors in the values of the oscillation parameters (Fig. 1, right panel), in $|<m>|$ and the sum of neutrino masses, and the uncertainty in the relevant NME, can weaken but do not invalidate these results [4] [11, 14, 15].

As Fig. 1 indicates, a measurement of $|<m>| \gtrsim 0.01$ eV would either [12] i) determine a relatively narrow interval of possible values of the lightest ν-mass $m_{\rm MIN}$, or ii) would establish an upper limit on $m_{\rm MIN}$. If an upper limit on $|<m>|$ is experimentally obtained below 0.01 eV, this would lead to a significant upper limit on $m_{\rm MIN}$.

The possibility of establishing CP- violation in the lepton sector due to Majorana CPV phases has been studied in [12, 32] and in much greater detail in [14, 15]. It was found that it is very challenging: it requires quite accurate measurements of $|<m>|$ (and of m_0 for QD spectrum), and holds only for a limited range of values of the relevant parameters. More specifically [14,15], establishing at 2σ CP-violation associated with Majorana neutrinos in the case of QD spectrum requires for $\sin^2\theta_\odot$=0.31, in particular, a relative experimental error on the measured value of $|<m>|$ and m_0 smaller than 15%, a "theoretical uncertainty" $F \lesssim 1.5$ in the value of $|<m>|$ due to an imprecise knowledge of the corresponding NME, and value of the relevant Majorana CPV phase α typically within the ranges of $\sim (\pi/4 - 3\pi/4)$ and $\sim (5\pi/4 - 7\pi/4)$ (Fig. 2).

The knowledge of NME with sufficiently small uncertainty[5] is crucial for obtaining quantitative information on the ν-mixing parameters from a measurement of $(\beta\beta)_{0\nu}$-decay half-life.

3. Baryon Asymmetry from CP-Violating Majorana Phases in $U_{\rm PMNS}$

We will discuss next briefly the interesting possibility that the CP-violation necessary for the generation of the baryon asymmetry of the Universe, Y_B, in the leptogenesis scenario can be due exclusively to the Majorana (and/or Dirac) CPV phases in the PMNS matrix, and thus can be directly related to the low energy CP-violation in the lepton sector. Let us recall that leptogenesis [52] is a simple mechanism which allows us to explain the observed baryon asymmetry of the Universe. The simplest scheme in which this mechanism can be implemented is the "seesaw"(type I) model of neutrino mass generation [3]. In its minimal version it includes the Standard Model (SM) plus two or three right-handed (RH) heavy Majorana neutrinos, N_j. Thermal leptogenesis (see, e.g. [54]) can take place, e.g. in the case of a hierarchical spectrum

[4] Encouraging results, in what regards the problem of calculation of the NME, were reported recently in [44].
[5] A possible test of the NME calculations is discussed in [51].

of heavy Majorana neutrino masses, $M_1 \ll M_2 \ll M_3$. The lepton asymmetry is produced in out-of-equilibrium lepton number and CP nonconserving decays of the lightest RH Majorana neutrino, N_1, mediated by the neutrino Yukawa couplings, λ. The lepton asymmetry thus generated is converted into a baryon asymmetry by $(B-L)$-conserving but $(B+L)$-violating sphaleron interactions [53] which exist within the SM. In grand unified theories the masses of the heavy Majorana neutrinos are typically (by a few to several orders of magnitude) smaller than the scale of unification of the electroweak and strong interactions, $M_{\rm GUT} \cong 2 \times 10^{16}$ GeV. This range coincides with the range of values of M_j, required for a successful thermal leptogenesis [54].

At energies below the heavy Majorana neutrino mass M_1 and of the electroweak symmetry breaking scale, the neutrino Yukawa couplings generate a Majorana mass for the LH flavour neutrinos:

$$m^\nu = v^2 \lambda^T M^{-1} \lambda = U^* m U^\dagger, \tag{11}$$

where $M \equiv {\rm Diag}(M_1, M_2, M_3)$, $m \equiv {\rm Diag}(m_1, m_2, m_3)$, $M_j > 0$, $m_k \geq 0$, $U \equiv U_{\rm PMNS}$, and $v = 174$ GeV is the Higgs field vacuum expectation value. In our further discussion it is convenient to use the "orthogonal parametrisation" of the matrix of neutrino Yukawa couplings:

$$\lambda = v^{-1} \sqrt{M}\, R \sqrt{m}\, U^\dagger, \quad R\, R^T = R^T\, R = \mathbf{1}, \tag{12}$$

where R [55] is, in general, a complex matrix.

The possibility of existence of a connection between leptogenesis and the low energy CP-violation in the lepton (neutrino) sector has been widely discussed in recent years, the conclusion being generically negative (see, e.g. [54, 56] and the references quoted therein). Recent progress in the understanding of the importance of lepton flavour effects in leptogenesis [57, 58] (for earlier discussion see [59]) lead to the realization [34] that the CP-violation necessary for the generation of the baryon asymmetry of the Universe, Y_B, can be due exclusively to the Majorana and/or Dirac CPV phases in the PMNS matrix, and thus can be directly related to the low energy CP-violation in the lepton sector (e.g. in ν-oscillations, etc.). In the cases when the only source of CP-violation is respectively the Majorana or the Dirac phases in $U_{\rm PMNS}$, there exists a correlation between the baryon asymmetry $|Y_B|$ and i) the effective Majorana mass in $(\beta\beta)_{0\nu}$-decay, $|{<}m{>}|$, or ii) the rephasing invariant $J_{\rm CP}$ controlling the magnitude of CP-violation in neutrino oscillations [29].

As was shown in [57, 58], in the case of a hierarchical spectrum of the heavy Majorana neutrinos N_j, $M_1 \ll M_2 \ll M_3$, the flavour effects in leptogenesis can be significant for 10^8 GeV $\lesssim M_1 \lesssim (0.5 - 1.0) \times 10^{12}$ GeV. If the requisite lepton asymmetry is produced in this regime, the CP-violation necessary for successful leptogenesis can be provided entirely by the low energy Majorana CPV phases in $U_{\rm PMNS}$ [34]. Indeed, suppose that the mass of N_1 lies in the interval of interest, 10^9 GeV $\lesssim M_1 \lesssim 10^{12}$ GeV, and that the matrix R has real and/or purely imaginary elements: we are interested in the case when the CP-violation necessary for leptogenesis is due exclusively to the CPV phases in $U_{\rm PMNS}$. Under these assumptions Y_B generated via leptogenesis can be written as [57, 58] $|Y_B| \cong 3 \times 10^{-3}\, |\epsilon_\tau\, \eta|$, where ϵ_τ is the CPV asymmetry in the τ flavour (lepton charge) produced in N_1-decays [6],

$$\epsilon_\tau = -\frac{3M_1}{16\pi v^2} \frac{{\rm Im}(\sum_{jk} m_j^{1/2} m_k^{3/2} U^*_{\tau j} U_{\tau k} R_{1j} R_{1k})}{\sum_i m_i |R_{1i}|^2},$$

[6] We have given the expression for Y_B normalised to the entropy density, see, e.g. [34].

η is the efficiency factor [58], $|\eta| \cong |\eta(0.71\widetilde{m_2}) - \eta(0.66\widetilde{m_\tau})|$, $\widetilde{m_{2,\tau}}$ being the wash-out mass parameters, $\widetilde{m_2} = \widetilde{m_e} + \widetilde{m_\mu}$, $\widetilde{m_l} = |\sum_j m_j\ R_{1j}\ U^*_{lj}|^2$. An approximate analytic expression for $\eta(\widetilde{m})$ is given in [57, 58].

Consider the specific example of IH light neutrino mass spectrum, $m_3 \ll m_{1,2} \cong (\Delta m^2_{\rm A})^{\frac{1}{2}}$. Under the simplifying assumptions of $m_3 \cong 0$ and $R_{13} \cong 0$ (N_3 decoupling), leptogenesis can be successful for $M_1 \lesssim 10^{12}$ GeV only if $R_{11}R_{12}$ is not real [34, 60], so we consider the case of purely imaginary $R_{11}R_{12} = i\kappa|R_{11}R_{12}|$, $\kappa = \pm 1$. The requisite CP-violation can be due to the Majorana phase α in $U_{\rm PMNS}$. If we set $s_{13} = 0$, the maximum of $|Y_B|$ for, e.g. $\kappa = -1$, is reached for [34] $|R_{11}|^2 \cong 1.4$ ($|R_{12}|^2 = |R_{11}|^2 - 1 = 0.4$), and $\alpha \cong 2\pi/3; 4\pi/3$, and at the maximum $|Y_B| \cong 1.5\times 10^{-12}((\Delta m^2_{\rm A})^{\frac{1}{2}}/0.05\ {\rm eV})(M_1/10^9\ {\rm GeV})$ (Fig. 2, left panel). The observed

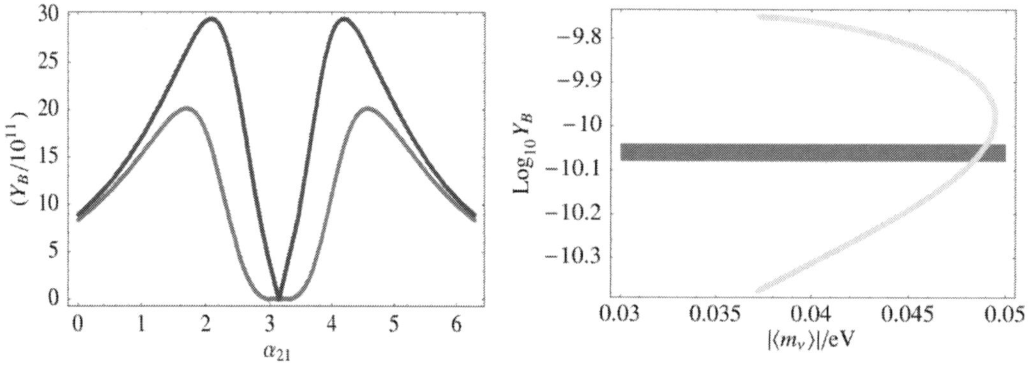

Figure 2. The asymmetry $|Y_B|$ versus i) the Majorana phase $\alpha_{21} \equiv \alpha$ (left panel), and ii) $|<m>|$ (right panel) when α is varied in the interval $[-\pi/2, \pi/2]$, for IH spectrum, purely imaginary $R_{11}R_{12} = i\kappa|R_{11}R_{12}|$, $|R_{11}|^2 - |R_{12}|^2 = 1$, and $M_1 = 2 \times 10^{11}$ GeV. The left (right) panel is obtained for $\delta = 0$, $\kappa = -1$ ($\kappa = +1$), $|R_{11}| = 1.20$ ($|R_{11}| = 1.05$), and $s_{13} = 0,\ 0.2$ - black and grey lines ($s_{13} = 0$). In the right panel the 2σ allowed range of $Y_{\rm B}$ is also shown. (From [34].)

baryon asymmetry, $|Y_B| \cong (8.0 - 9.2) \times 10^{-11}$, can be reproduced if $M_1 \gtrsim 5.3 \times 10^{10}$ GeV. Since both $|Y_B|$ and the effective Majorana mass $|<m>|$ depend on the Majorana phase α, there exists a correlation between the values of $|Y_B|$ and $|<m>|$ (Fig. 2).

As was shown in [34], we can have successful leptogenesis in the case of IH spectrum also if the source of CP-violation is the Dirac phase δ in $U_{\rm PMNS}$. In this case values of $|\sin\theta_{13}\sin\delta| \gtrsim 0.02$, or $|J_{\rm CP}| \gtrsim 4.6 \times 10^{-3}$, are required. Values of $|J_{\rm CP}|$ as small as 4.6×10^{-3}, can be probed in ν-oscillation experiments at neutrino factories [61]. Since both Y_B and $J_{\rm CP}$ depend on s_{13} and δ, there exists a correlation between the values of $|Y_B|$ and $J_{\rm CP}$ [34] if the other relevant parameters are fixed.

Similar results can be obtained [34] for the NH light neutrino mass spectrum, as well as in the case of quasi-degenerate in mass heavy Majorana neutrinos.

4. Conclusions

Determining the nature - Dirac or Majorana, of massive neutrinos is one of the most formidable and pressing problems in today's neutrino physics. The $(\beta\beta)_{0\nu}-$decay experiments have the potential of establishing the Majorana nature of neutrinos with definite mass. If the latter are Majorana particles, the $(\beta\beta)_{0\nu}-$decay experiments can provide information on the type of $\nu-$mass spectrum, on the absolute scale of neutrino masses and on the Majorana CPV phases present in the neutrino mixing matrix U. The CP-violation in leptogenesis, necessary for the generation of the baryon asymmetry of the Universe, can be due exclusively to the Majorana and/or Dirac CPV phase(s) in U. These results underline the importance of understanding the status of the CP-symmetry in the lepton sector and, correspondingly, of the experimental searches for leptonic CP-violation at low energies.

The developments described briefly in this talk would have pleased Peter and I am sure he would have contributed to at least some of them.

To Frank and Ettore: Happy Anniversary! Thank you for being the driving force in the field of experimental $(\beta\beta)_{0\nu}$-decay research and inspiration for the younger generation of researchers working in this field. My best wishes for personal happiness and many professional successes.

Acknowledgements.
This work was supported in part by the Italian MIUR and INFN programs on "Fisica Astroparticellare".

References

[1] S.T. Petcov, *Nucl. Phys. B (Proc. Suppl.)* **143** (2005) 159.
[2] S.M. Bilenky and S.T. Petcov, *Rev. Mod. Phys.* **59** (1987) 67.
[3] P. Minkowski, *Phys. Lett.* B **67** (1977) 421.
[4] M. Fukugita and T. Yanagida, *Phys. Lett.* B **174** (1986) 45.
[5] S.T. Petcov, *Phys. Lett.* B **110** (1982) 245;
P.H. Frampton *et al.*, *Nucl. Phys.* B **687** (2004) 31.
[6] A. Morales and J. Morales, *Nucl. Phys. Proc. Suppl.* **114** (2003) 141.
[7] C. Aalseth *et al.*, hep-ph/0412300.
[8] S.T. Petcov, *Physica Scripta* **T121** (2005) 94;
S. Pascoli and S.T. Petcov, hep-ph/0308034.
[9] S.M. Bilenky, S. Pascoli and S.T. Petcov, *Phys. Rev.* D**64** (2001) 053010 and 113003.
[10] S. Pascoli and S.T. Petcov, *Phys. Lett.* B **544** (2002) 239; *ibid.* B **580** (2004) 280.
[11] S. Pascoli, S.T. Petcov and W. Rodejohann, *Phys. Lett.* B **558** (2003) 141;
H. Murayama and Peña-Garay, *Phys. Rev.* D **69** (2004) 031301.
[12] Pascoli S, Petcov S T and Wolfenstein L 2002 *Phys. Lett.* B **524** 319.
[13] Bilenky S M *et al.* 1996 *Phys. Rev.* D **56** 4432.
[14] Pascoli S, Petcov S T and Rodejohann W 2002 *Phys. Lett.* B **549** 177.
[15] Pascoli S, Petcov S T and Schwetz T 2006 *Nucl. Phys.* B **734** 24.
[16] Bilenky S M *et al.* 1999 *Phys. Lett.* B **465** 193;
Barger V and Whisnant K 1999 *Phys. Lett.* B **456** 194;
Vissani F 1999 *JHEP* **06** 022;
Czakon M *et al.* 2000 Preprint hep-ph/0003161.
[17] Rodejohann W 2001 *Nucl. Phys.* B **597** 110;
Matsuda K *et al.* 2000 Preprint hep-ph/0007237;
Fukuyama T *et al.* 2001 *Phys. Rev.* D **64** 013001; Joaquim F R 2003 Preprint hep-ph/0304276.
[18] Choubey S and Rodejohann W 2005 *Phys. Rev.* D **72** 033016;
Lindner M, Merle A and Rodejohann W 2006 *Phys. Rev.* D **73** 053005.
[19] Fogli G L *et al.* 2006 *Prog. Part. Nucl. Phys.* **57** 71.
[20] Pontecorvo B 1957 *Zh. Eksp. Teor. Fiz. (JETP)* **33** (1957) 549;
Pontecorvo B 1958 *Zh. Eksp. Teor. Fiz. (JETP)* **34** (1958) 247;
Pontecorvo B 1967 *Zh. Eksp. Teor. Fiz. (JETP)* **53** (1967) 1717;
Maki Z, Nakagawa M and Sakata S 1962 *Prog. Theor. Phys.* **28** 870.
[21] Bilenky S M, Hosek J and S.T. Petcov S T 1980 *Phys. Lett.* B **94** 495.
[22] Apollonio M et al. 1999 *Phys. Lett.* B **466** 415.
[23] Bandyopadhyay A et al. 2005 *Phys. Lett.* B **608** 115;
Bandyopadhyay A et al. 2008 Preprint arXiv:0804.4857.
[24] Schwetz T, Tortola M and Valle J W F 2008 Preprint arXiv:0808.2016.
[25] Choubey S and Petcov S T 2004 *Phys. Lett* B **594** 333;
Bandyopadhyay A *et al.* 2005 *Phys. Rev.* D **72** 072002;
Huber P *et al.* 2004 *Phys. Rev.* D **70** 073014.
[26] Anderson K *et al.* 2004 Preprint hep-ex/0402041.
[27] Ardellier F *et al.* [Double Chooz Collaboration] 2006 Preprint hep-ex/0606025.
[28] See, e.g., Heeger K M 2006, talk given at Neutrino'06 International Conference, June 13 - 19, 2006, Sant Fe, U.S.A.
[29] Krastev P I and S.T. Petcov S T 1988 *Phys. Lett.* B **205** 84;

Freund M *et al.* 2000 *Nucl. Phys.* B **578** 27.
[30] Bandyopadhyay A *et al.* 2007 Preprint arXiv:0710.4947;
Albright C *et al.* 2004 Preprint physics/0411123;
Itow Y *et al.* 2001 Preprint hep-ex/0106019.
[31] Langacker P *et al.* 1987 *Nucl. Phys.* B **282** 589.
[32] Barger V *et al.* 2002 *Phys. Lett.* B **540** 247;
De Gouvea A, Kayser B and Mohapatra R 2003 *Phys. Rev.* D **67** 053004.
[33] Pascoli S, Petcov S T and Yaguna C E 2003 *Phys. Lett.* B **564** 241;
Petcov S T, Shindou T and Takanishi Y 2006 *Nucl. Phys.* B **738** 219; Petcov S T and Shindou T 2006 *Phys. Rev.* D **74** 073006.
[34] Pascoli S, Petcov S T and Riotto A 2007 *Phys. Rev.* D **75** 083511 and *Nucl. Phys.* B **774** 1.
[35] Molinaro E *et al.* 2008 *Nucl. Phys.* B **797** 94.
[36] Petcov S T and Shindou T 2006 Preprint hep-ph/0605204.
[37] Wolfenstein L 1981 *Phys. Lett.* B **107** 77;
Bilenky S M *et al.* 1984 *Nucl. Phys.* B **247** 61;
Kayser B 1984 *Phys. Rev.* D **30** 1023.
[38] Lobashev V *et al.* 2003 *Nucl. Phys.* A **719** 153c.
[39] Eitel K *et al.* 2005 *Nucl. Phys. Proc. Suppl.* **143** 197.
[40] Tegmark M 2005 Preprint hep-ph/0503257;
Hannestad S *et al.* 2006 Preprint astro-ph/0603019.
[41] Petcov S T and Smirnov A Yu 1994 *Phys. Lett.* B **322** 109.
[42] Pascoli S and Petcov S T 2008 *Phys. Rev.* D **77** 113003.
[43] Klapdor-Kleingrothaus H V 2001 *et al.*, *Nucl. Phys. Proc. Suppl.* **100** 309.
[44] RodinV A *et al.* 2003 *Phys. Rev.* C **68** 044302 and Preprint nucl-th/0503063;
Caurier E *et al.* 2007 Preprints arXiv:0709.2137 and arXiv:0709.0277 (nucl-th).
[45] Aalseth C E *et al.* 2000 *Phys. Atomic Nuclei* **63** 1225.
[46] Klapdor-Kleingrothaus H V *et al.* 2004 *Phys. Lett.* B **586** 198.
[47] Klapdor-Kleingrothaus H V and Krivosheina I V 2006 *Mod. Phys. Lett.* A **21** 1547.
[48] Barabash A *et al.* 2004 *JETP Lett.* **80** 377.
[49] Amaboldi C *et al.* 2008 Preprint arXiv:0802.3439.
[50] Avignone A 2005 *Nucl. Phys. Proc. Suppl.* **143** 233.
[51] Bilenky S M and Petcov S T 2004, Preprint hep-ph/0405237.
[52] Fukugita M and Yanagida Y 1986 *Phys. Lett.* B **174** 45.
[53] Kuzmin V A, Rubakov V A and Shaposhnikov M E 1985 *Phys. Lett.* B **155** 36.
[54] Buchmuller W *et al.* 2002 *Nucl. Phys.* B **643** 367; Buchmuller W *et al.* 2005 *Annals Phys.* **315** 305.
[55] JCasas J A and Ibarra A 2001 *Nucl. Phys.* B **618** 171.
[56] Pascoli S, Petcov S T and Rodejohann W 2003 *Phys. Rev.* D **68** 093007.
[57] Abada A. *et al.* 2006 *JCAP* **0604** (2006) 004;
Nardi E *et al.* 2006 *JHEP* **0601** (2006) 164.
[58] Abada A. *et al.* 2006 JHEP **0609** (2006) 010.
[59] Barbieri R *et al.* 2000 *Nucl. Phys.* B **575** 61.
[60] Petcov S T *et al.* 2006 *Nucl. Phys.* B **739** 208.
[61] See, for instance, De Lellis G *et al.* 2005 "Neutrino factories and superbeams", Proceedings, 7th International Workshop, NuFact05, Frascati, Italy, June 21-26, 2005.

AUTHOR INDEX

AUTHOR INDEX

Institute of Physics
Dirac House, Temple Back
Bristol BS1 6BE UK